The Perception of Visual Information

Second Edition

Springer
New York
Berlin
Heidelberg
Barcelona
Budapest
Hong Kong
London
Milan
Paris
Santa Clara
Singapore
Tokyo

William R. Hendee
Peter N.T. Wells
Editors

The Perception of Visual Information

Second Edition

With 170 Illustrations

 Springer

William R. Hendee
Medical College of Wisconsin
8701 Watertown Plank Road
Milwaukee, WI 53226
USA

Peter N.T. Wells
Bristol General Hospital
Guinea Street
Bristol B51 65Y
United Kingdom

Library of Congress Cataloging-in-Publication Data
Perception of visual information / [edited by] William R. Hendee,
    Peter N.T. Wells – – 2nd ed.
        p.cm.
    Includes bibliographical references and index.
    ISBN 0-387-94910-0 (hardcover : alk. paper)
    1. Vision.   2. Visual perception.   3. Computer vision.
  I. Hendee, William R., II. Wells, P.N.T. (Peter Neil Temple)
    [DNLM: 1. Visual Perception, 2. Vision– –physiology. 3. Image
Processing, Computer-Assisted. WW 105 P4285 1997]
QP475.P36 1997
152.16– –dc21
DNLM/DLC
for Library of Congress   97-10231
                                                               CIP

Printed on acid-free paper.

Production coordinated by Black Hole Publishing and managed by Terry Kornak; manu-
    facturing supervised by Jacqui Ashri.
Typeset by J. Kaiping/Black Hole Publishing, Berkely, CA.
Printed and bound by Edwards Brothers, Inc. Ann Arbor, MI.
Printed in the United States of America.

9 8 7 6 5 4 3 2 1

ISBN 0-387-94910-0 Springer-Verlag, New York Berlin Heidelberg   SPIN 10559742

Preface to the Second Edition

The first edition of *The Perception of Visual Information* was published in 1993. Interest in the book exceeded our expectations and those of the publisher, and the first printing was quickly sold out. As we contemplated a second printing, we realized that we had the opportunity to prepare a second edition with revisions to update each of the chapters. We were also eager to add a chapter on virtual and augmented reality and the present and future contributions of these concepts to clinical medicine. Although it is rare to have editions of an advanced text separated by only 4 years, we decided in favor of a second edition because of the rapidity with which the understanding and modeling of human vision is proceeding. We hope only that the reception accorded this edition is somewhere close to the response to the first edition.

In this new book, all of the authors of the earlier edition reviewed their chapters. A few added only occasional paragraphs scattered throughout their chapters, while others made many revisions and added major sections. The chapter on virtual and augmented reality is a valuable addition to the book, and provides insight into how this technology promises to profoundly impact both diagnostic and therapeutic medicine, but in all disciplines strongly dependent on the use of visual information for orientation and guidance.

Preparing this second edition has been a pleasurable experience for the editors. The result is a book that we hope will help the reader understand and appreciate the current state of knowledge about the perception, interpretation and understanding of visual information.

William R. Hendee, Ph.D.
Milwaukee, Wisconsin, USA
March 1, 1997

Peter N.T. Wells, Ph.D., D.Sc., F.Eng.
Bristol, U.K.

Preface to the First Edition

Human knowledge is primarily the product of experiences acquired through interactions of our senses with our surroundings. Of all the senses, vision is the one relied on most heavily by most people for sensory input about the environment. Visual interactions can be divided into three processes: (1) detection of visual information; (2) recognition of the "external source" of the information; and (3) interpretation of the significance of the information. These processes usually occur sequentially, although there is considerable interdependence among them.

With our strong dependence on the processes of visual interactions, we might assume that they are well characterized and understood. Nothing could be further from the truth. Human vision remains an enigma, in spite of speculations by philosophers for centuries, and, more recently, of attention from physicists and cognitive and experimental psychologists. How we see, and how we know what we see, remains an unsolved mystery that challenges some of the most creative scientists and cognitive specialists.

The presentation of information for visual interpretation is critical to almost every endeavor of modern technology, from space exploration and military surveillance to process engineering and diagnostic and therapeutic medicine. Today visual images can be presented in almost any format desirable, with enhanced spatial detail, altered contrast and color scales, suppressed statistical noise, and sharpened edges. The challenge is not in discovering new ways to present images; it is instead in understanding how visual information can be displayed so that its use is optimized for the observer.

This text examines what is known or thought about the process of human vision, and how visual images can be presented to facilitate their use by observers. It begins by focusing on the anatomical and physiological properties of the eye as "the window through which the mind perceives the world around it," and examines the current state of knowledge concerning how visual information is detected and recognized. Several models of human vision are described, and approaches to quantifying the visual responses of individuals are considered. Theories are then explored of how visual information is interpreted and the cognitive responses of observers to images are discussed.

As explained in the second half of the book, images can be presented in a multitude of formats, with virtually every characteristic of the images susceptible to modification according to the wishes of the observer. Image manipulation has, in fact, become almost a scientific discipline in itself, and some researchers devote their time and energy full-time to this area. This research effort is important, because decisions in a variety of areas, including air travel, product manufacturing and medical diagnosis, depend upon the presentation and interpretation of visual information, often with the assistance of computers and without much, if any, direct human intervention. The flexibility of image presentation gives rise to the need for optimized environments for the display and interpretation of visual information. The challenge of workstation environments for interactions with visual information is examined in the penultimate chapter of the book.

This text is written for anyone with an interest in the visual sciences who has a background in science or engineering at the undergraduate level. Authors have been encouraged to write beyond their specialized areas of research in an effort to develop principles and themes that not only cross interdisciplinary boundaries, but also penetrate barriers between different applications such as space exploration, manufacturing, surveillance, air science and medicine. Although the authors are recognized experts in their own disciplines, they have been chosen for this text because of their knowledge about human vision and their ability to write clearly and concisely about a complex and technical topic.

In preparing this book, we have been ably assisted by many persons, including Ms. Diane Reuter and Ms. Terri Komar. Special thanks go to Ms. Vickie Grosso, our editional assistant; without her diligent nurturing of the book, its editors and its authors, the project would have taken longer and been much less enjoyable. Ms. Jan Fish was responsible for editorial work in the U.K.

The editors have gained considerable pleasure and knowledge from the experience of putting this text together. We hope that readers will share our pleasure in part, and that the text will help them appreciate the challenges and rewards of exploring the mysteries of human vision.

William R. Hendee, Ph.D.   Peter N.T. Wells, Ph.D., D.Sc., F.Eng.
Milwaukee, Wisconsin, USA   Bristol, U.K.

August 5, 1991

Acknowledgments

Chapter 4 Acknowledgments. Most of the ideas are based on the "school" of Koenderink and van Doorn and on their pioneering and profound work. We thank Blom, Bel, van Damme, Frens, Bakker, Bookelman, Salden, and Viergever for many contributions, implementations, and discussions. This work was supported by the Dutch Ministry of Economic Affairs Grant No. [V3-3DM]-50249/89-01.

Chapter 9 Acknowledgments. Supported in part by National Institutes of Health grants R01 CA 42453 and R01 CA 62362.

Contents

12 Virtual Reality and Augmented Reality in Medicine   361
David Hawkes

13 Problems and Prospects in the Perception of Visual Information   391
Peter N. T. Wells

Index   401

Contributors

Kevin S. Berbaum   Department of Radiology, University of Iowa, Iowa City, IA 52242, USA.

Paul S. Cho   Department of Radiation Oncology, University of Washington Medical Center, Seattle, WA 98195, USA.

Kunio Doi   Kurt Rossmann Laboratories for Radiologic Image Resonance, Department of Radiology, University of Chicago, Chicago, IL 60637, USA.

Donald D. Dorfman   Department of Radiology, University of Iowa, Iowa City, IA 52242, USA.

Luc Florack   Computer Vision Research Group, Department of Radiology, Utrecht University Hospital, 3564 CX Utrecht, THE NETHERLANDS.

Alastair G. Gale   Academic Radiology, University of Nottingham, Queen's Medical Centre, Nottingham NG7 2U11, UNITED KINGDOM.

Maryellen L. Giger   Kurt Rossmann Laboratories for Radiologic Image Research, Department of Radiology, University of Chicago, Chicago, IL 60637, USA.

Arthur P. Ginsburg   Director, Research and Development, Vision Sciences Research Corporation, 130 Ryan Industrial Court, Suite 105, San Ramon, CA 94583, USA.

David Hawkes   Division of Radiological Science, UMDS, Guy's Hospital, London Bridge, London SE1 9RT, UNITED KINGDOM.

William R. Hendee   Senior Associate Dean and Vice President, Medical College of Wisconsin, Milwaukee, WI 53226, USA.

H.K. Huang   Department of Radiology, University of California, San Francisco, CA 94143, USA.

Charles A. Kelsey   Department of Radiology, School of Medicine, The University of New Mexico, Albuquerque, NM 87131, USA.

Mark Madsen   Department of Radiology, University of Iowa, Iowa City, IA 52242, USA.

Russell Philips   Department of Ophthalmology, University of Aberdeen, Aberdeen AB9 2ZD, UNITED KINGDOM.

Ronald R. Price   Radiology Department, Vanderbilt University, Nashville, TN 37232, USA.

Ulrich Raff VCHSC, Cardiac Cath. Laboratory, University Hospital, 4200 E. 9th Ave., Box B-132, Denver, CO 80262, USA.

Bart M. Ter Haar Romeny Computer Vision Research Group, Department of Radiology, Utrecht University Hospital, 3564 CX Utrecht, THE NETHER-LANDS.

Peter F. Sharp Department of Biomedical Physics and Bioengineering, University of Aberdeen, Aberdeen AB9 2ZD, UNITED KINGDOM.

Henry A. Swett Department of Diagnostic Radiology, Yale University School of Medicine, New Haven, CT 06510, USA

Peter N.T. Wells Department of Medical Physics and Bioengineering, Bristol General Hospital, Guinea Street, Bristol BS1 6SY, UNITED KINGDOM.

1

Physiological Optics

Peter F. Sharp and Russell Philips

1.1 Introduction

The earliest theories about the visual system regarded the eye as emitting "psychic stuff"; vision was a process of feeling the scene. This led the early anatomists to regard the optic nerve as a hollow tube through which this "stuff" was transmitted.

The first firm outline of optics as we know it was given by Kepler in 1604, in "Physiological Optics." It was Kepler who first compared the eye to a "camera," a darkened chamber with the image focused on its rear surface. Descartes in his book "La Dioptrique," published in 1637, described an experiment in which part of the rear of the eye of an ox was scraped to make it translucent; an inverted image of the scene was then observed on this area.

In fact this observation created its own problems. Because we do not see images as inverted, the active region for seeing was postulated to be the lens rather than the retina. This confusion arose through a lack of realization that we do not see the retinal image but rather see with the aid of it. The eye is not simply an instrument for recording an image; rather it acts as the optical interface between the environment and the neural elements of the visual system. It provides the basic attributes of vision, namely form, field, color, motion, and depth.

The refractive power of the eye is a function of its specialized anatomy, which focuses an image on the retina, where the photoreceptors convert light energy into electrical impulses. These electrical signals are preprocessed by a complex retinal neural network and are passed by highly anatomically and functionally organized pathways to the visual cortex of the brain, where they are further processed prior to interpretation. Efferent pathways from the visual cortex influence the neural centers that control eye movements, thus allowing the eye to fixate and follow an object.

In this chapter, we start by describing the optics, then the anatomy and physiology of the retina and the visual pathways. Finally, the performance of the visual system is considered.

1.2 Optical Anatomy of the Eye

The anatomy of the eye is illustrated in Fig. 1.1.[1.1] The transparent elements, namely the cornea, aqueous, lens, and vitreous are known collectively as the ocular refractive media, and the clarity of the media is essential for the maintenance of

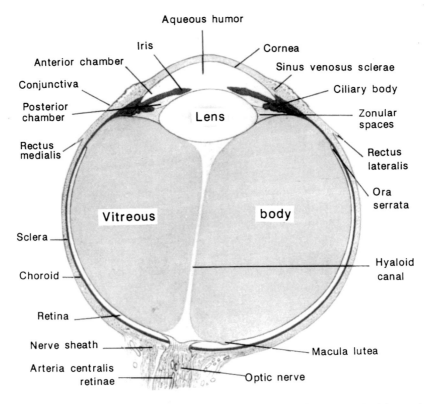

FIGURE 1.1. Horizontal section through right human eyeball. Reproduced, with permission, from reference 1.1.

optimal visual performance. The effective power of the eye is approximately 60 diopters (i.e., a focal length of approximately 1.6 cm). A diagram of the optical components of the eye is shown in Fig. 1.2.[1.2]

Light entering the eye is first refracted at the anterior surface of the cornea, passes through the pupil, which acts as an aperture stop, and is refracted by the lens to form a focused image on the retina. The amount that light is refracted depends on the difference between the refractive indices of the media to either side of the surface it is traversing, and the angle at which the light meets the surface. The refractive index of the cornea is 1.376 (air has an index of 1), and the radius of curvature of its anterior surface centrally is approximately 7.7 mm. Therefore the cornea is the strongest refractive element of the eye, with an effective power of approximately 45 diopters. The posterior surface of the cornea normally has a negligible effect on refraction because of the small difference between the refractive indices of the cornea (1.376) and the aqueous (1.336).

The structure of the lens is anatomically and optically more complex than it would at first appear. The anterior surface of the lens has a radius of curvature of approximately 10 mm, and the posterior surface a radius of curvature of approxi-

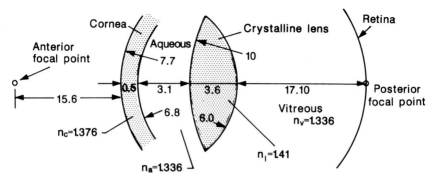

FIGURE 1.2. Diagram of the optical components of the eye showing relative distances and radii of curvature in millimeters, and the refractive indices of the ocular media. Reprinted with permission from reference 1.2.

mately 6 mm. The protein fibrils of the lens are arranged in such a way that it has a central "nucleus" which is almost spherical in shape and of higher refractive index (1.41) than the more peripheral lens cortex (1.381). The structural arrangement of the lens fibrils gives the lens as a whole an effective refractive index of 1.41, and a power in the unaccommodated state of approximately 15 diopters.[1.3] The process of accommodation by which near objects are brought into focus is achieved by contraction of the ciliary muscles, which reduces the tension in the zonules. The elasticity of the lens capsule causes the anterior surface of the lens to bow forward, reducing its radius of curvature, shallowing the anterior chamber, and increasing the thickness of the lens.[1.4] These changes in the lens are accompanied by constriction of the pupil, which increases the depth of field of the eye.

1.3   Aberrations of the Eye

The eye is subject to all the aberrations[1.5] inherent in any optical system, but has a variety of mechanisms by which degradive effects are reduced.

1.3.1   Spherical Aberration

Spherical aberration results from the prismatic effect of the peripheral parts of a lens, which cause rays passing through the periphery of the lens to deviate more than rays passing through the axial and paraxial parts of the lens. In the eye, spherical aberration is reduced in a number of ways. First, the curvature of the cornea is steepest centrally, and becomes flatter more peripherally, i.e., it is aplanatic. Second, the axial part of the lens has a greater refractive power than the periphery of the lens. Third, the iris acts as a stop, limiting the entry of peripheral rays into the eye. The optimum pupil size is between 2 and 2.5 mm; above this value, image quality is degraded by spherical aberration, while below it, quality is degraded by diffraction. Finally, retinal cones are more sensitive to

light which strikes them axially rather than obliquely, a phenomenon known as the Stiles-Crawford effect.[1.6] This also reduces the deleterious effect caused by the scattering of light as it hits the retina.

1.3.2   Chromatic Aberration

Chromatic aberration is caused by the dispersion of light by a lens. Shorter wavelengths (i.e., blue) are refracted more by a lens than are longer wavelengths. The eye is normally focused so that the intermediate wavelengths form the most sharply focused image, thus taking advantage of the wavelength sensitivity peaks of the cones, as described in Sec. 1.4.1.

1.3.3   Oblique Astigmatism and Coma

Oblique astigmatism is an aberration which results from the passage of light rays obliquely through a spherical lens. This aberration is minimized in the human eye by several factors. First, the aplanatic surface of the cornea reduces oblique astigmatism. Second, the curvature of the retina is such that in an emmetropic eye (an eye with no refractive error) the circle of least confusion of the astigmatic image falls upon the retina. Third, astigmatic images tend to fall upon the peripheral retina, which has relatively poor resolution, thus reducing appreciation of the image.

Coma is a type of spherical aberration caused by rays of light coming from points not lying on the principal axis of the lens. The result is unequal magnification of the image formed by different zones of the lens, producing not a circular, but comet-shaped composite image. The factors which reduce oblique astigmatism in the eye also minimize the importance of coma.

1.3.4   Curvature of Field

Curvature of field occurs when a plane object gives rise to a curved image. It is dependent on the refractive index and curvature of the lens surfaces, but is independent of spherical aberration, oblique astigmatism, and coma. The curvature of the retina compensates for the curvature of field caused by the optical system of the eye.

1.4   The Visual Pathways

1.4.1   Photoreceptors

The initial event in the visual process is absorption of a photon of light by visual pigment contained in the photoreceptors of the retina. There are two types of photoreceptors, the rods and the cones, so named because of their shapes. There

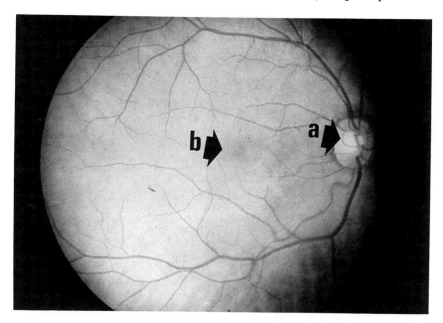

FIGURE 1.3. Photograph of normal right retina showing the optic nerve (a) and fovea (b). The macula is the area, of about four times the diameter of the optic nerve, surrounding the fovea.

are approximately 120 million rods in the retina, all situated outside the fovea, and 6 million cones, mostly in the foveal and parafoveal regions.[1.7] See Figs. 1.3–1.4. The visual pigment associated with rods is rhodopsin. This consists of a lipoprotein called opsin and a chromophore (light absorbing chemical) called 11-cis-retinal, and has a spectral sensitivity peak at 500 nm. The cones contain three visual pigments with spectral sensitivity peaks at 450 (blue cones), 525 (green cones), and 555 nm (red cones).[1.8] See Fig. 1.5(b). The three cone pigments share the same chromophore as the rods, and their different spectral sensitivities result from differences in the opsin.

In the dark-adapted eye, the chromophore is in the 11-cis-retinal form. Light causes the isomerization of the cis-retinal to the trans form; this is accompanied by hyperpolarization of the photoreceptor membrane, thus generating a transmissible neural impulse. The chromophore is then enzymically reisomerized to its 11-cis form, ready for repetition of the cycle.

1.4.2   Bipolar, Ganglion, Horizontal, and Amacrine Cells

Apart from the photoreceptors, the retina contains four other groups of cells involved in the visual process. These are bipolar cells, ganglion cells, horizontal cells, and amacrine cells, as shown in Fig. 1.6. There are subtypes of each of these cells, and, in fact, the retina contains at least 50 morphologically and functionally

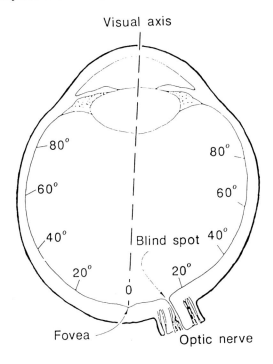

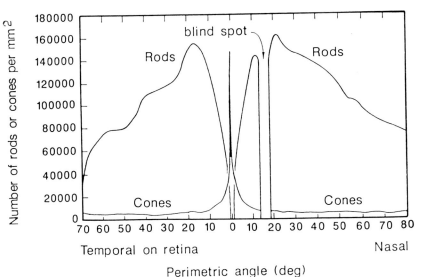

FIGURE 1.4. Distribution of rods and cones across the retina. Note the absence of photoreceptors at the blind spot, where the nerve connections from the receptors leave the eye forming the optic nerve. The highest visual acuity is achieved at the fovea which consists solely of cones and is situated on the visual axis. The fovea and blind spot are demonstrated in Fig. 1.3. Reprinted, with permission, from reference 1.7.

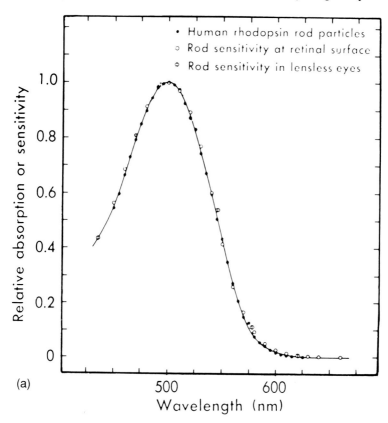

FIGURE 1.5. Spectral sensitivity curve for (a) rods and (b) the three different types of cones. After reference 1.8.

distinct types of neural cells. All visual signals pass through the bipolar cells that link the photoreceptors on the outer nuclear layer of the retina to the ganglion cells within the inner plexiform layer. The impulse generated by the photoreceptors passes ultimately to the ganglion cells, which are the final common pathway for the neural output from the eye. The axons of the ganglion cells form the optic nerve. The fact that there are over 120 million photoreceptors and only about 1 million ganglion cells implies a considerable convergence of photoreceptor output. This convergence plays a part in the preprocessing of raw visual input that is carried out within the retina.

Experimental work recording the output from single ganglion cells shows them to have specific receptive fields in which a stimulus evokes a train of neural impulses. In some, the receptive fields are of a concentric nature; thus, in one type, characterized as on-center off-surround, illumination of the center of the receptive field causes an increase in neural pulse rate above the resting level, as illustrated in Fig. 1.7, top left. In contrast, when the concentric ring "surround" part of the receptive field is illuminated, there is a response only when the illumination is

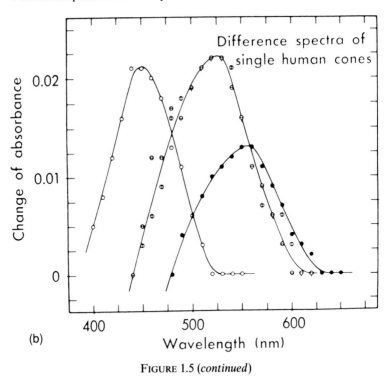

(b)

FIGURE 1.5 *(continued)*

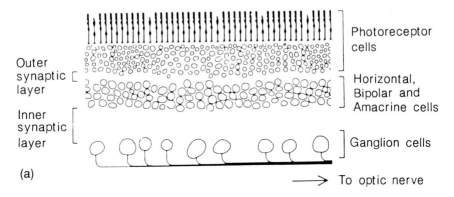

(a)

FIGURE 1.6. Diagrammatic representation of the cellular architecture of the retina. (a) The photoreceptors are the deepest layer; light has to traverse the other layers of the retina to reach them. (b) and (c) The neural pathways from the photoreceptors to the ganglion cells. Reprinted, with permission, from reference 1.9.

switched off, as shown in Fig. 1.7, top right. There are also off-center, on-surround receptive fields in which the roles of the center and surround are reversed.[1.9] In either case, illumination of the entire receptive field provokes little or no response. Another type of ganglion cell has been found that responds to a stimulus moving

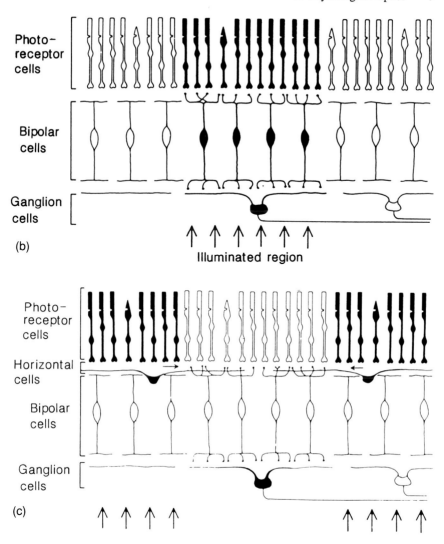

FIGURE 1.6. (*continued*)

within its receptive field, with the response being directionally selective. The receptive fields in the peripheral retina are large compared with those at the foveal and parafoveal regions.

The refinement of retinal signal processing is brought about by the lateral neurones, namely the horizontal and amacrine cells. The horizontal cells laterally inhibit the output of the bipolar cells, thus producing the concentric receptive field pattern mentioned above. The amacrine cells (of which there are at least 30 types!) are associated with temporal responses shown by some ganglion cells. For example, some ganglion cells show a burst of activity when their receptive field is first illuminated; but if the stimulus remains unchanged, the ganglion cell activity

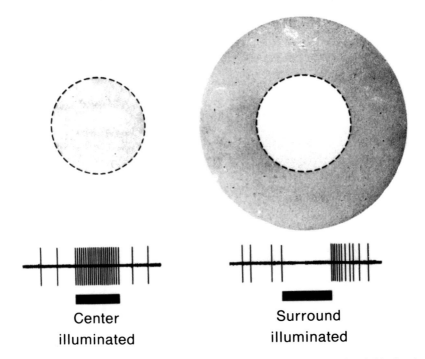

FIGURE 1.7. Diagram of a typical center-surround antagonistic receptive field of a single ganglion cell. The neural pulse sequence which results from illumination within the receptive field. Reprinted, with permission, from reference 1.9.

dies away. The amacrine cells also mediate in the rod pathways, with many rod bipolar cells synapsing not directly with ganglion cells, but via amacrine cells. Along with convergence of output from a number of rods to a single ganglion cell, this plays a role in summating the input from rods under conditions of low illumination and thus contributes to the very wide dynamic range of the eye. Within the fovea, the ratio of cones to ganglion cells is 1 : 1 and it is this low convergence ratio which is responsible for the visual system's adeptness at spatial resolution.

There are two main types of ganglion cells, called P_α and P_β. They each carry information of a different nature. The axons of the ganglion cells leave the eye as the optic nerve, which contains approximately 1 million fibers. At the optic chiasm, the optic nerves from the two eyes meet, and the axons subserving the nasal half of the retina cross into the contralateral optic tract.[1.10] It is this crossover, illustrated in Fig. 1.8, which produces the visual field topography which is of great clinical importance.

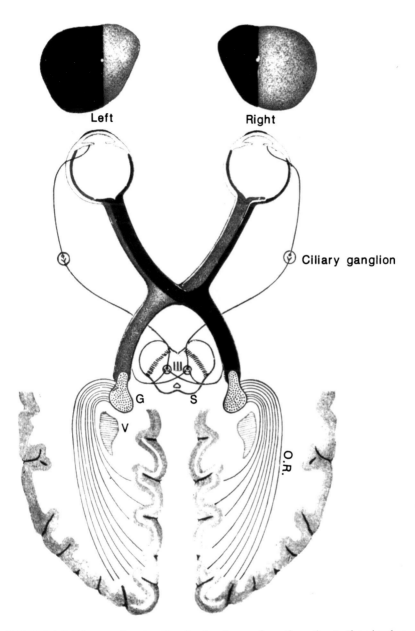

FIGURE 1.8. Diagram showing the visual pathways from the retina to the visual cortex illustrating the decussation of the fibers from the nasal half of the retina at the optic chiasm. The reference to the ciliary ganglion and the oculomotor nucleus in the midbrain relate to the pupillary light reflex which is not discussed in this chapter. G: lateral geniculate body; S: superior colliculus; III: oculomotor nucleus; V: posterior horn of lateral ventricle; OR: optic radiations. Reprinted, with permission, from reference 1.10.

FIGURE 1.9. Histological section of monkey lateral geniculate body showing the six distinct cytoarchitectural layers. Reprinted, with permission, from reference 1.11.

1.4.3   Lateral Geniculate Nucleus

The optic nerve axons pass via the optic tracts to synapse in the lateral geniculate nucleus (LGN), a subnucleus of the thalamus. The LGN contains six distinct cytoarchitectural layers, as shown in Fig. 1.9.[1.11] Layers 1, 4, and 6 are derived from the contralateral eye, and layers 2, 3, and 5 from the ipsilateral eye. The P_α ganglion cells project to large (magnocellular or M) cells in layers 1 and 2 in the LGN. The P_β ganglion cells project to small (parvocellular or P) cells in layers 3, 4, 5, and 6 of the LGN. The LGN cells exhibit similar receptive field responses to the ganglion cells from which they receive their input. Many of the LGN cells with center-surround antagonistic receptive fields exhibit color opponency within these

TABLE 1.1. Functional Distinctions Between the Large (M) and Small (P) Cells in the Lateral Geniculate Nucleus.

M — not wavelength selective
— faster, transient responses
— greater sensitivity to contrast (respond to 1%–2% contrast between center and surround, but response levels off at 10%–15% contrast)
— larger receptive fields
P — wavelength selective
— slower, tonic response
— lower contrast sensitivity (respond to 1%–2% between center and surround, but saturates at much higher levels)
— smaller receptive fields

fields.[1.12] For example, a maximal response might be obtained by a red center and green surround. Functional distinctions between the P and M cells of the LGN are demonstrable (see Table 1.1).

1.4.4 Primary Visual Cortex

From the LGN, the visual pathway consists of the optic radiations through the parietal and temporal lobes, as illustrated in Fig. 1.10.[1.13] The optic radiations terminate in the primary visual cortex (Vl) in the occipital lobe (Brodmann's area 17); see Fig. 1.11.[1.14]

The primary visual cortex is known as the striate cortex because of its distinctive cytoarchitecture. Six layers are visible, layer 1 being nearest the surface of the brain, and layer 6 nearest the white matter. As shown in Fig. 1.12, layer 4 is the most prominent as a dark stripe.[1.15] Output from the striate cortex is to the prestriate cortex (Brodmann's area 18), which itself is subdivided into several zones known as $V2$, $V3$, $V4$, and $V5$; see Fig. 1.12. The striate and prestriate cortex have a modular anatomical structure which is involved in the functional segregation of the M and P systems summarized in Fig. 1.13.

Layer 4 of $V1$ is subdivided into $4A$, $4B$, $4C_\alpha$, and $4C_\beta$. Inputs from the two eyes occupy separate territories within $V1$ layer $4C$, but converge thereafter to binocularly driven cells. Within $V1$, the M and P pathways are distinguishable anatomically and functionally. The M input to $V1$ projects to $4C_\alpha$ and from there to cells in layer $4B$, which exhibit orientation selectivity and usually selectivity to the direction of movement of a stimulus. Few of these cells exhibit selectivity to the length of a stimulus ("end stopping"). They receive a binocular input and respond strongly to binocular retinal disparity, and thus are involved in stereoscopic vision.[1.12,1.15]

The P input to $V1$ projects to $4C_\beta$ and from there to layers 2 and 3. Histological staining of these areas with markers for metabolic enzymes reveals a distinct

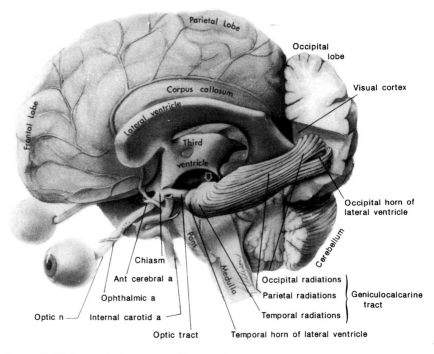

FIGURE 1.10. Anatomical diagram to illustrate the relationship of the visual pathways and cortex to the remainder of the brain. The small open arrow points to the lateral geniculate body; the midbrain is the most rostral part of the brainstem and is situated in the area between the pons and the optic tract shown on the diagram. Reprinted, with permission, from reference 1.13.

morphology. Darkly staining columns are visible which form "blobs" separated from each other by lighter staining areas named "interblobs." The cells within the blobs show no orientation selectivity, but are sensitive to wavelength or luminance. Many of them have double opponent receptive fields, which may differ in their spectral wavelength and luminance characteristics from the receptive fields of the LNG.[1.15] The interblob cells are orientation-selective and respond to contours generated by differences in wavelength or luminance.[1.12, 1.15] Between 10% and 20% of them exhibit end stopping. These cells are not explicitly color coded and do not exhibit color opponency. Unlike cells within the M system, the interblob cells respond to a border of two colors even when the colors are of equal luminance. The blob cells are thus more sensitive to wavelength information at low spatial frequencies and the interblob cells are more sensitive to luminance information at higher spatial frequencies.

Both M and P systems project to $V2$ which, when stained for cytochrome oxidase, reveals its own distinct pattern of thick and thin dark staining stripes separated by paler stripes (interstripes). The thick stripes receive their input from the M cells of $V1$ layer $4B$. The cells within the thick stripes are orientation and directionally selective, but are seldom end stopped. They show a strong response

(a)

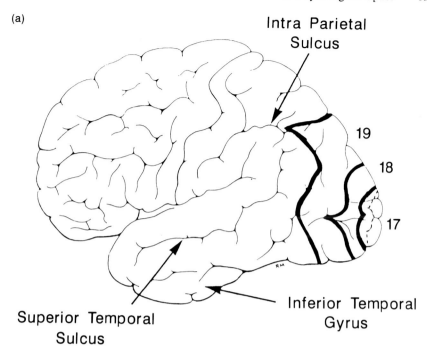

Intra Parietal
Sulcus

19

18

17

Superior Temporal
Sulcus

Inferior Temporal
Gyrus

(b)

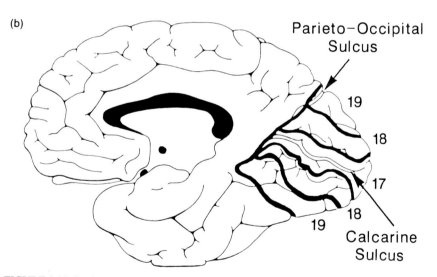

Parieto-Occipital
Sulcus

19

18

17

18

19

Calcarine
Sulcus

FIGURE 1.11. Surfaces of the cerebral hemispheres showing the visual areas in the occipital lobe. The striate, parastriate, and peristriate areas correspond to Brodmann's areas 17, 18, and 19, respectively. a) Superlateral surface; b) medial surfaces. Reprinted, with permission, from reference 1.14.

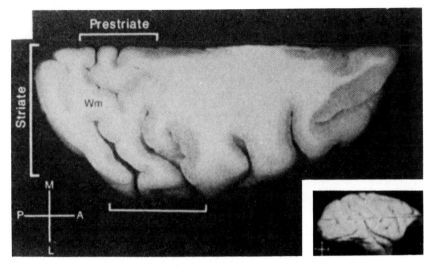

(a)

FIGURE 1.12. The visual cortex. In both man and macaque monkey (shown here), the cerebral cortex is a highly convoluted sheet. a) Much of the cerebral cortex is buried inside folds or sulci; the level of this section is indicated in the inset photograph. (Wm is white matter.) b) The visual cortex occupies the posterior third of the cortex and can be divided into many distinct visual areas; this diagram corresponds to the same level as that shown in a). c) One of the oldest methods used to identify areas anatomically is the study of cytoarchitecture, the grouping of cells into different layers. This is a section through the posterior third of the brain, treated to stain the cell bodies. It shows two distinctive regions, the primary visual striate cortex area $V1$ (left) and the prestriate cortex lying in front of it. The border is indicated by an arrowhead. The striate cortex, so named because of its richly laminated appearance, receives all the fibers from the retina through the lateral ginculate nucleus, a relay nucleus below the cortex. d) The prestriate cortex has a less elaborate cytoarchitecture. Staining for the activity of the metabolic enzyme cytochrome oxidase reveals complex subdivisions in the striate and prestriate areas and functional studies reveal several distinct areas in the prestriate cortex ($V2 - V5$), each specialized to process different attributes of the visual scene. Reprinted, with permission, from reference 1.15.

to retinal disparity. The thin stripes receive their input from the blob cells of the P system in layers 2 and 3 of $V1$. The thin stripe cells show no orientation selectivity, but a high proportion of them are color coded, with double opponent center-surround type receptive fields. The interstripes of $V2$ receive their input from the P cells in the interblobs of layers 2 and 3 of $V1$. The interstripe cells are orientation but not directionally selective and approximately 50% are end stopped. They respond to differences in color or luminance but are not specifically color coded.

Cortical areas $V3$, $V4$, and $V5$ show no modular cytoarchitectural features. The cells of $V3$ receive input from the M pathway from the thick stripes of $V2$ and directly from layer $4B$ of $V1$. These cells are concerned primarily with orientation selectivity. The $V4$ cells receive their input from the thin stripes and interstripes of $V2$ and are thus involved in both color and form vision. $V4$ is the first cortical

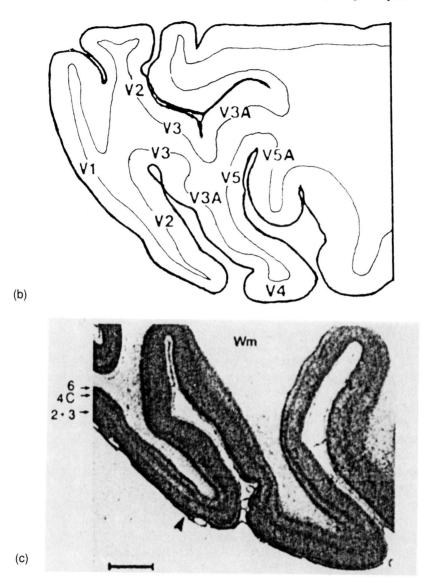

(b)

(c)

FIGURE 1.12. (*continued*)

region in which wavelength discrimination is perceived as color. It is not known whether the segregation of color and orientation selectivity seen in the P pathway of $V2$ is maintained within $V4$. $V5$ was previously known as the "motion area" because its cells show a high degree of directional selectivity. They also show a strong response to retinal disparity and thus are involved in stereopsis. The $V5$ cells receive their input from the thick stripes of $V2$ and directly from layer $4B$ of $V1$. These pathways are summarized in Figs. 1.13[1.14] and 1.14.[1.16]

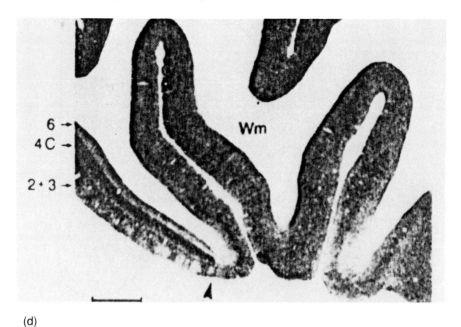

(d)

FIGURE 1.12. (*continued*)

The specialized prestriate areas project to the visual association cortex in the parieto-occipital and temporo-occipital areas. V4 and V5 have projections to the intraparietal sulcus of the parietal lobe where the position and motion of objects are recognized. Recent experiments using Positron Emission Tomography (PET) have shown activation of this and adjacent areas in the perception of optical flow (forward motion in depth) and three-dimensional structure from motion. It thus appears that the parietal areas of the visual association cortex provide information for spatial awareness and visuomotor coordination.[1.17] $V5$ projects to the superior temporal sulcus, where object recognition occurs. $V4$ projects to the inferior temporal gyrus of the temporal lobe, an area concerned with learning to identify objects such as faces and letter strings by their appearance.[1.15, 1.18] See Fig. 1.11. Color information from area $V4$ projects to the posterior fusiform gyrus located postero-medial to the inferior temporal gyrus. See Fig. 1.11.

This description of the functional and anatomical segregation of the P and M systems is a simplified one. In fact, there are many forward, backward, and lateral interconnections between the specialized prestriate areas. These interconnections provide feedback information and are also responsible for the functional convergence which tends to enlarge receptive fields, so that more information can be extracted from a single field. This is illustrated by the observation that the output from the 2500 foveal cones influences approximately 2.5 million cells within $V1$.[1.19]

To summarize, in man the M system is the more evolutionary primitive part of the visual system and may well be homologous with the entire visual system of

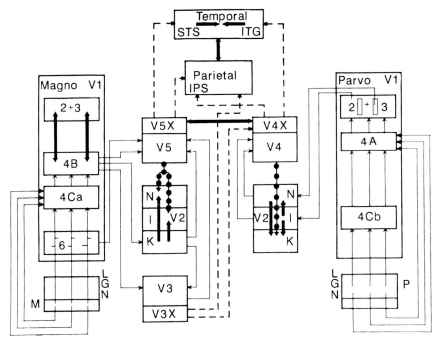

FIGURE 1.13. Diagrammatic summary of the neural input of the M and P systems to the visual cortex (thin black arrows). Forward connections are shown in dashed lines; like all the thin black lines they have reciprocating backward connections (not shown). Backward connections are show with dotted lines and lateral connections are shown with thick black lines. The diagram shows the lateral geniculate nucleus (LGN) subdivided into magnocellular (M) and parvocellular (P) layers. $V1$ is subdivided into layers 6, $4Cb$, $4Ca$, $4B$, $4A$, 3, 2, with cytochrome oxidase blobs (small oblongs) in layers 2 and 3 in the P system on the right; $V2$ is subdivided into cytochrome oxidase thick (K), thin (N), and inter (I) stripes; areas $V3$, $V4$, and $V5$ as part of the third, fourth, and fifth visual complexes are denoted by $V3X$, $V4X$, and $V5X$ (connections shown to arise from $V3X$, $V4X$, and $V5X$ include the whole complex and do not exclude the areas $V3$, $V4$, and $V5$ proper); the highest visual areas are in the parietal and temporal lobes, the former including the intraparietal sulcus (IPS), and the latter the superior temporal sulcus (STS) and the inferior temporal gyrus (ITG), all of which probably consist of several different areas. Reprinted, with permission, from reference 1.15.

nonprimates. It has large receptive fields and is specialized to perceive the form of moving objects. The P system is well developed only in primates and in man is approximately ten times the size of the M system. It uses orientation, wavelength, and luminance information to produce the fine-grain construction of a static image.

1.5   Mechanisms of Viewing

When the eye fixates a stationary object, three types of small-amplitude involuntary eye movements occur, resulting in a constant movement of the image over the

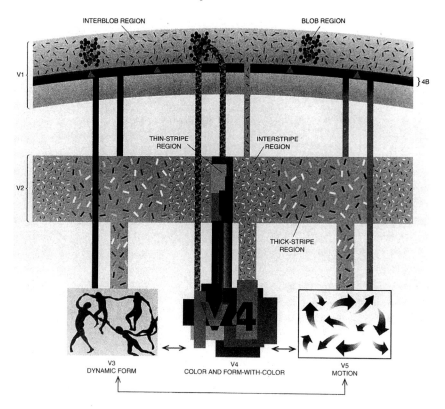

FIGURE 1.14. Four perceptual pathways have been identified within the visual cortex. Color is seen when the wavelength-selective cells in the blob regions of $V1$ send signals to the specialized area $V4$ and also to the thin stripes of $V2$, which connect with $V4$. Form in association with color depends on connections between the interblobs of $V1$, the interstripes of $V2$ and area $V4$. Cells in layer $4B$ of $V1$ send signals to the specialized areas $V3$ and $V5$ directly and also through the thick stripes of $V2$; these connections give rise to the perception of motion and form. Reprinted, with permission, from reference 1.16.

retina. These are slow drift, fast click (or flicker),and a superimposed high-frequency tremor, as illustrated in Fig. 1.15.[1.20] The constant minute movement of the image on the fovea prevents the image from fading, as discussed in Sec. 1.8.

The human visual system can also move the eyes in a coordinated fashion to fixate an eccentrically located object of interest on the fovea, and to follow a moving object, matching its velocity so that foveal fixation is maintained. These motions, collectively known as conjugate eye movements, are, in the first case, a rapid movement known as a saccade, and, in the second case, a slower movement known as pursuit.

Before discussing the neural pathways involved in the control of eye movement, an understanding of the basic anatomy of the muscles which move the eyes and

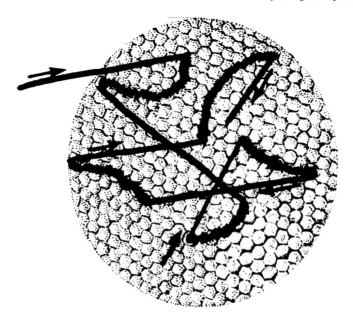

FIGURE 1.15. Diagram of an area of the fovea approximately 50 μm in diameter showing the drift, flicker, and high frequency tremor which allow stable viewing of a stationary object without fading of the scene. Reprinted, with permission, from reference 1.20.

their innervation is essential. Figure 1.16 illustrates the names and positions of the extraocular muscles and details their innervation.[1.21] Each of the three nerves which innervate the extraocular muscles has its cell bodies within the brainstem. The sixth nerve nucleus is located in the pons and the third and fourth nerve nuclei are in the midbrain. The third nerve nucleus is subdivided into subnuclei, one for each muscle it innervates.

For both saccadic and pursuit horizontal conjugate eye movements, the common "control center" lies within the paramedian pontine reticular formation (PPRF) within the pons. Cortical control for saccadic (rapid) eye movements is centered in the contralateral frontal lobe, and decussates (crosses over) in the midbrain to reach the PPRF. The PPRF then sends stimuli to its ipsilateral sixth nerve nucleus and its contralateral medial rectus subnucleus of the third nerve. Pursuit movements originate in the ipsilateral parieto-occipital visual association cortex.[1.22] See Fig. 1.17 for summary. Vertical conjugate gaze movements are under bilateral cortical and midbrain control, but the exact anatomical pathways involved are poorly understood.

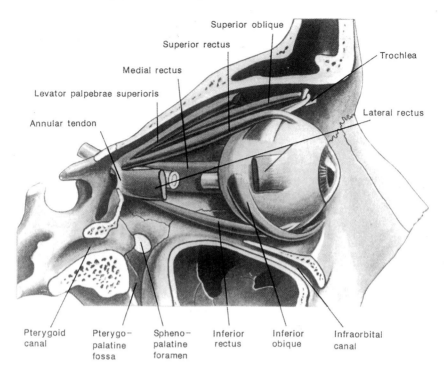

FIGURE 1.16. Illustration of right globe and orbit showing the positions of the extraocular muscles. The third cranial nerve (oculomotor nerve) innervates the superior, medial, and inferior recti, and the inferior oblique. The fourth cranial nerve (trochlear nerve) innervates the superior oblique and the sixth cranial nerve (abducent nerve) innervates the lateral rectus. The third and fourth nuclei are located in the midbrain, and the sixth nucleus is located in the pons (see Fig. 1.10). Reprinted, with permission, from reference 1.21.

1.6   Color Vision

The presence of three types of cones with different spectral responses was mentioned in Section 1.4.1. These are often referred to as the red, green, and blue receptors but, as is obvious from Fig. 1.5(b), they have broad band responses. The so-called trichromatic theory of color vision, the foundations of which were laid down by Thomas Young in the early 19th century, is based on the observation that any color can be matched by a mixture of three other primary colors: red, green, and blue.[1.23] To achieve this match, however, it may be necessary to subtract one color from the others, not simply to add them. This theory thus assumes that the ratio of the outputs from the red, green, and blue receptors provides the information on hue, while the sum of the outputs determines saturation and brightness.

Although the trichromatic theory explains color matching, it does not address the problem of how a color looks to an observer. For example, a color never appears both reddish and greenish, although it may look both greenish and bluish. In 1870 Hering proposed the opponent theory of color vision.[1.24] This model is

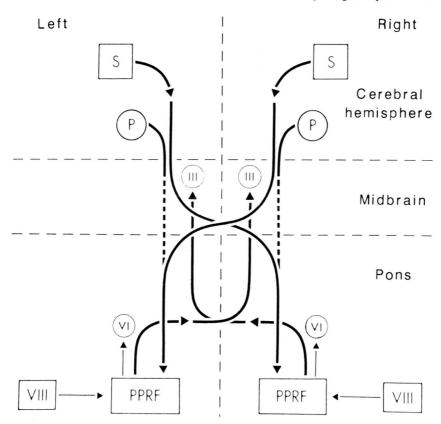

Left

Right

FIGURE 1.17. Diagram summarizing the supranuclear control of horizontal conjugate eye movements. *S*: saccadic center in frontal lobe; *P*: pursuit center in parieto-occipital region; III: third nerve nucleus; VI: sixth nerve nucleus; PPRF: paramedian pontine reticular formation; VIII: vestibular input, which also influences the coordination of all eye movements, especially in relation to head and body posture and motion. Reprinted, with permission, from reference 1.22.

based on the presence of three pairs of detectors acting in opposition; two of these deal with color, one corresponding to red and green and the other to yellow (red plus green) and blue. The third is a black-white opponent system, primarily for brightness and saturation. The red- green detector signals the presence of either red or green wavelengths, but gives no indication of the yellow or blue ones. If the red input predominates, then the output of the red-green system is positive, i.e., above base-line levels; if the green is greater then the output is negative. At the time this theory was proposed, Hering had no physiological evidence for the existence of such an opponent system. As explained in Sec. 1.4.3, there is now support for the presence of a mechanism at the level of the lateral geniculate nucleus.

At present, the most likely model for color vision seems to be one incorporating elements of both the component and opponent theories. The red, green, and blue receptor signals are handled separately up to the level of the lateral geniculate nucleus, where they are then combined via the lateral inhibitory networks.[1.25] Opponent color coding is then assumed to continue to the cortex.

A third channel is required to signal luminance. The red and green, but probably not the blue, signals are added. This channel is also opponent coded and signals not the absolute luminance but how the local value compares with the luminance across the rest of the image. Comparisons are made across space rather than color. This explains color constancy, where our perception of color depends largely on the object's spectral reflectance characteristics rather than on the spectral composition of the illuminating light. Land's retinex theory of color vision is also relevant to this.[1.26]

1.7   Physical Performance of the Visual System

1.7.1   Spectral Sensitivity

The visual system is sensitive to electromagnetic radiation of wavelengths from 400 to 700 nm, as illustrated in Fig.1.5. High-frequency radiation is filtered by the cornea and lens and this is obviously important to avoid damage to molecules by ultraviolet light. Low-frequency radiation is transmitted by the optics of the eye, but causes no effect in the photoreceptor.[1.27] Wavelength shifts of between 1 and 2 nm are detectable in the middle of the visible range, with the value increasing to 6 nm toward the ends of the range.[1.28]

1.7.2   Incremental Brightness Sensitivity

The smallest change in intensity that can be seen depends on a number of factors. For example, we know that the retina contains two classes of photoreceptors with different sensitivities (Sec. 1.4.1) and whose distribution varies across the retina (Fig. 1.4). Also, the eye has automatic gain mechanisms whereby the sensitivity can alter with time. See Sec. 1.7.5. Thus, the precise results obtained depend upon where on the retina measurements are made and what measurement procedure is used. Such an experiment might consist of asking the subject to report the presence of a disc of light on an illuminated background[1.29] or of presenting two discs of slightly different luminance and asking the subject to report which is the brighter.[1.30] It is found that at very low levels of background luminance, I, the smallest perceptible change in the luminance of the disc, ΔI, varies little with the value of I. The absolute value of ΔI is highest in the area of retina close to the fovea which is dominated by the less sensitive cones. As I is increased, however, it is found that ΔI increases linearly with I, a behavior known as the Weber-Fechner law.

1.7.3   Spatial Response

The ability of an imaging system to record spatial information is expressed in terms of its optical transfer function. This assumes that linear systems theory can be applied, so that the system output is linearly related to the input and the imaging system gives the same response at all points in its field of view (i.e., it is stationary). Unfortunately, these conditions do not hold for the visual system. While its response is reasonably homogeneous near the optic axis, in other parts of the visual field it is anisoptropic, that is, the response to a test pattern depends upon its orientation. The response is changed little if the pattern is rotated through a right angle, but altered significantly at intermediate degrees of rotation. Such behavior is found even after correction has been made for astigmatism of the optics.[1.31] The eye's response to spatial detail also depends on the contrast of the input pattern. Thus, in the absence of both stationarity and linearity, the response of the visual system to spatial information cannot be described precisely by a single curve.

If measurements are confined to one area of the retina and made at low contrast, then a curve such as that shown in Fig. 1.18 is produced.[1.32] This shows not the optical transfer function but the relative sensitivity of response of the visual system to a sine-wave pattern. The precise shape of the curve and its cut-off frequency depends upon how measurements are made and the absolute brightness levels used.[1.33-1.35] A similar shape of curve is produced for chromatic response but shifted towards the lower frequencies.[1.36]

The visual system is most sensitive to spatial frequencies of about 15 cycles per degree and falls off rapidly at both higher and lower frequencies, the maximum frequency being about 60 cycles per degree. This decrease is not dependent on the properties of the cornea and lens as a similar response is found when the test pattern is formed directly on the retina.[1.37] The decrease is probably a function of the size of the receptors. At high spatial frequencies, the summation of outputs from rods (see Sec. 1.4.2) degrades resolution. Light scatter has the same effect but it is reduced by the Stiles-Crawford effect (see Sec. 1.3.1). The low-frequency response is a result of the process of lateral inhibition (see Sec. 1.4.2), which effectively filters out low spatial frequencies.

Although an observer may not be aware of this limitation of the visual system in reproducing spatial information in a scene, it can lead to some unusual effects in images. The most notable of these is the Mach band effect,[1.38] illustrated in Fig. 1.19.

1.7.4   Temporal Response

Whether the eye sees a single flash of light depends not only on the number of quanta absorbed but also the length of time for which the flash lasts. The maximum time over which the effect of quanta can be integrated is known as the critical duration of vision, and is of the order of 0.1 s. The actual value of the critical duration (t) depends on the luminance (B) of the light; the relationship

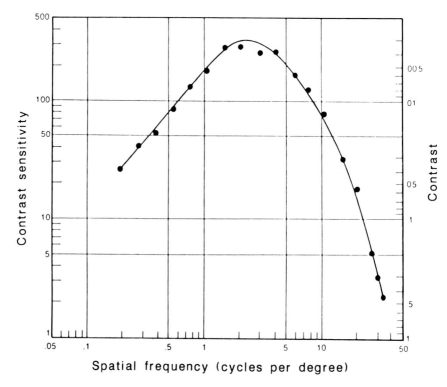

FIGURE 1.18. Contrast sensitivity curve for the human eye. The eye shows an optimal response, i.e., high contrast sensitivity, at frequencies of a few cycles per degree, but to see frequencies of a fraction of a cycle or several tens of cycles the contrast needs to be a factor of 1 or 2 orders of magnitude greater. Reprinted, with permission, from reference 1.32.

between them is given by Bloch's law, which states that

$$B \times t = \text{constant} \tag{1.1}$$

If the stimulus is flickering, rather than a single flash, then at a certain frequency the eye sees it as a steady light. This value is known as the critical flicker frequency or critical fusion frequency.[1.39] The temporal equivalent to Fig. 1.18 can be measured, spatial frequency being replaced by temporal frequency, as shown in Fig. 1.20. As can be seen, the maximum flicker rate detectable by the eye is about 60 cycles per second. This effect is, of course, used in ciné presentation of images.

1.7.5   Dynamic Range

The eye is capable of performing over a dynamic range of about 10^{13} in brightness. For a dynamic range of about 10^4 its performance remains constant, while over the remainder it gradually deteriorates.

This performance is achieved by a combination of four mechanisms. First, the rods and cones have different sensitivities to light,[1.40] as shown in Fig. 1.21. Vision

FIGURE 1.19. Mach band effect. In the left-hand pattern, a bright vertical line is visible in the middle of the image, while in the right-hand one, a dark vertical line is seen. This is an illusion created by the nonlinear response of the visual system to spatial frequency, as shown in Fig. 1.18. Reprinted, with permission, from reference 1.38.

at low light levels primarily uses the rods, with consequent loss of acuity. Second, there is summation of output from many rods to a single ganglion cell, which improves performance at low light levels. Third, there is a chemical gain control (dark adaptation, where unbleached 11-cis is resynthesized[1.41]), but this changes relatively slowly. Finally, the lens aperture (the iris) can vary rapidly, giving a gain control of a factor of about 64.

1.8   Information Transfer Rates

From the preceding description, it is obvious that the visual system must do a great deal of image processing in order to minimize the amount of information that requires processing by the visual system. For example, as the eye responds to spatial frequencies up to about 60 cycles per degree and at a maximum rate of about 60 cycles per second, a maximum of 5.5×10^{10} samples per second could be available to be sent to the cortex.[1.19] This exceeds the bandwidth available in the optic nerve.

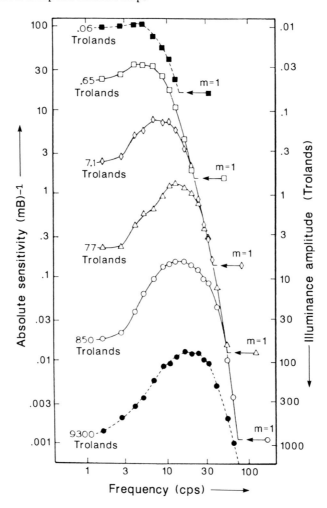

FIGURE 1.20. Temporal response curves. The brightness of a uniform disk was varied sinusoidally at a particular frequency (1 cps = 1 Hz). The amplitude of the sine wave (illumination amplitude) was varied until the changes in intensity could just be seen (The *Troland* is the retinal illumination produced by a surface luminance of 1 Cd m^{-2} when the aperture of the eye is 1 mm^2). The experiment was repeated for different frequencies and various levels of average disk brightness. The shape of the curves at all but the lowest average brightness levels is similar to that of Fig. 1.18. Note that at high frequencies the ability to perceive flicker is independent of average brightness, while at low frequencies it is highly dependent. Reprinted, with permission, from reference 1.39.

A number of strategies are used to compress the data. Rather than simply transmitting all the information about the visual scene, it is sufficient to signal only changes in the information. The on-center off-surround type of receptors effectively act as a bandpass filter, removing the dc and low spatial frequency

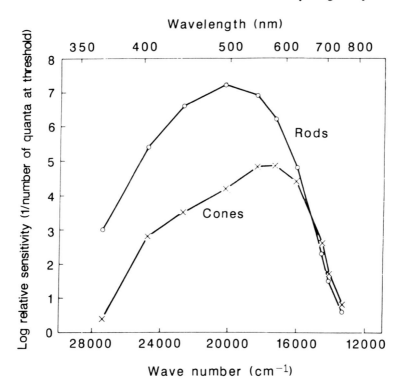

FIGURE 1.21. Photopic (cone) and scotopic (rod) spectral sensitivity curves. Note that the rods are more sensitive than the cones at all but the highest wavelengths. The cone response curve is a mixture of three different spectral responses [see Fig. 1.5(b)]. Reprinted, with permission, from reference 1.40.

information from the scene and transmitting information about the change in the spatial data in the retinal image. One unfortunate result of this would be that, if the scene did not change, then the image would fade away. Fortunately, this does not happen in practice because, even when steadily fixating a stationary object, the eye is not truly stationary. During fixation there are continuous small-amplitude movements (drift and flicker) and a superimposed tremor of up to 150 cycles per second with an amplitude of about half a cone diameter (2 (m), as shown in Fig. 1.15. This ensures that the retinal image is always changing even if the scene being viewed does not. If the retinal image is artificially stabilized then indeed the scene does fade. Also, there are present cells which respond to specific features in the image, such as lines at a particular orientation. This reduces the amount of information that must be transmitted.

Thus it can be seen that the eye is not simply a camera, but a complex device for capturing and processing data. The aim of the data processing is not simply to exploit the data channels available in the visual system with greater efficiency, but also to put the image data into a form suitable for interpretation by the brain. This is discussed in later chapters.

1.9   References

[1.1]   Warwick R., Williams P.L., eds. *Gray's Anatomy.* 35th ed. Edinburgh: Longman; 1973: 1097.

[1.2]   American Academy of Ophthalmology Basic and Clinical Science Course 1988–1989. *Section 2: Optics, Refraction and Contact Lenses.* San Francisco: American Academy of Ophthalmology; 1988: 90.

[1.3]   Abrams D., ed. *Duke-Elder's Practice of Refraction.* 9th ed. Edinburgh: Churchill Livingstone; 1978: 23–30.

[1.4]   Koretz J.F., Handelman G.H. How the human eye focuses. *Sci. Am.* 1988; 259(1): 64–71.

[1.5]   Elkington A.R., Frank H.J. *Clinical Optics.* Oxford: Blackwell; 1984: 66–74.

[1.6]   Stiles W.H., Crawford B.H. The luminous efficiency of rays entering the eye pupil at different points. *Proc. R. Soc. London B* 1933; 112: 428–450.

[1.7]   Cornsweet T.N. *Visual Perception.* New York: Academic; 1970: 137.

[1.8]   Moses R.A., ed. *Adler's Physiology of the Eye.* St. Louis: Mosby; 1978: 410.

[1.9]   Masland R.H. The functional architecture of the retina. *Sci. Am.* 1986; 225(6): 90–99.

[1.10] Warwick R., ed. *Eugene Wolff's Anatomy of the Eye and Orbit.* London: Lewis; 1976: 382.

[1.11] Moses R.A., ed. *Adler's Physiology of the Eye.* St Louis: Mosby; 1978: 420.

[1.12] Livingstone M., Hubel D. Segregation of form, color, movement, and depth: Anatomy, physiology, and perception. *Science* 1988; 240: 740–749.

[1.13] Glaser J.S., ed. *Neuro-ophthalmology.* Hagerstown: Harper and Row; 1978: 49.

[1.14] Warwick R., Williams P.L., eds. *Gray's Anatomy.* 35th ed. Edinburgh: Longman; 1973: 961.

[1.15] Zeki S., Shipp S. The functional logic of cortical connections. *Nature* 1988; 335: 311–317.

[1.16] Zeki S. The visual image in mind and brain. *Sci. Am.* 1992; 267(3): 43–50.

[1.17] de Jong B.M., Shipp S., Skidmore B., Frackowiak R.S.J., Zeki S. The cerebral activity related to the visual perception of forward motion in depth. *Brain* 1994; 117: 1039–1054.

[1.18] Allison T., McCarthy G., Nobre A., Puce A., Belger A. Human extrastriate visual cortex and the perception of faces, words, numbers and colours. *Cerebral Cortex* 1994; 5: 544–554.

[1.19] Burgess A.E. Personal Communication.

[1.20] Moses R.A., ed. *Adler's Physiology of the Eye.* St. Louis: Mosby; 1978: 520–521.

[1.21] Warwick R., Williams P.D., eds. *Gray's Anatomy.* 35th ed. Edinburgh: Longman; 1973: 1123.

[1.22] Glaser J.S. ed. *Neuro-ophthalmology.* Hagerstown: Harper and Row; 1978: 205.

[1.23] Boynton R.M. *Human Color Vision.* New York: Holt, Rinehart, and Winston; 1979.

[1.24] Hering E. *Outlines of a Theory of Light Sense, 1877.* Cambridge: Harvard University Press; 1964.

[1.25] De Valois R.L., De Valois K.K. Neural coding of color. In: Carterette E.C., Friedmann M.P., eds. *Handbook of Perception.* New York: Academic; 1975: 117–166.

[1.26] Land E.H. The retinex theory of vision. *Sci. Am.* 1977; 237(6): 108–128.

[1.27] Boettner E.A., Wolter J.R. Transmission of the ocular media. *Invest. Ophth.* 1962; 1: 776–783.

[1.28] Hart W.M. Color vision. In: *Adler's Physiology of the Eye.* St. Louis: Mosby; 1992: 709–710.

[1.29] Blackwell H.R. Contrast thresholds of the human eye. *J. Opt. Soc. Am.* 1946; 36: 624–643.

[1.30] Cornsweet T.N., Pinsker H.M. Luminance discrimination of brief flashes under various conditions of adaptation. *J. Physiol. (London)* 1969; 176: 294–310.

[1.31] Campbell F.W., Kulikowski J.J., Levinson J. The effect of orientation on the visual resolution of gratings. *J. Physiol.* 1966; 187: 427–436.

[1.32] Campbell F.W., Maffei L. Contrast and spatial frequency. *Sci. Am.* 1974; 231(5): 106–115.

[1.33] Davidson M.L. Perturbation approach to spatial brightness interaction in human vision. *J. Opt. Soc. Am.* 1968; 58: 1300–1308.

[1.34] De Palma J.J., Lowry E.M. Sine-wave response of the visual system. II. Sine-wave and square-wave contrast sensitivity. *J. Opt. Soc. Am.* 1962; 52: 328–335.

[1.35] Bryngdahl L. Characteristics of the visual system: Psychophysical measurement of the response to spatial sine-wave stimuli in the mesopic region. *J. Opt. Soc. Am.* 1964; 54: 1152–1160.

[1.36] van der Horst C.J., de Weert C.M., Bourmann M.A. Transfer of spatial chromaticity contrast at threshold in the human eye. *J. Opt. Soc. Am.* 1967; 57: 1260–1266.

[1.37] Campbell F.W., Green D.G. Optical and retinal factors affecting visual resolution. *J. Physiol.* 1956; 181: 576–593.

[1.38] Ratliff F. Contour and contrast. In: *Image, Object and Illusion. Readings from Scientific American.* San Francisco: Freeman; 1974: 21.

[1.39] Kelly D.H. Visual responses to time dependent stimuli. I. Amplitude sensitivity measurements. *J. Opt. Soc. Am.* 1961; 51: 422–429.

[1.40] Cornsweet T.N. *Visual Perception.* New York: Academic; 1970; 146.

[1.41] Moses R.A., ed. *Adler's Physiology of the Eye.* St. Louis: Mosby; 1978: 623–627.

2
Detection of Vision Information

Charles A. Kelsey

2.1   Introduction

The purpose of this chapter is to review briefly some of the early theories of vision, to discuss some simple visual experiments that vividly demonstrate properties of the eye, and to discuss several current models of vision.

One of the problems with studying how we detect and process information visually is that everyone performs this task millions of times every day. Any theory or model must satisfy our intuitive knowledge of how the process works, a process that works rapidly and extremely well.

The eye is often compared to a camera, and in many ways the similarities are very strong. There is, however, one important difference: the visual process projects images outside the body. The lenses of both the camera and the eye focus light on a detector. In the camera, photochemical changes produce a latent image on film. In the eye, photochemical changes produce nerve impulses that are transmitted to the brain. The brain interprets these impulses as arising from objects outside the body; it is impossible to perceive images as lying on the retina. Even nonimages are projected outside the body. For example, the bright after-images that appear after looking briefly at a bright light or a camera bulb are inevitably projected outside the body even with the eyelids closed. It's doubtful whether film "thinks" to project the image outside the camera. This ability to project the image outside the body is so ingrained we never think about it, but we cannot turn it off. It is representative of image processing and interpretation at a very elementary level.

2.2   Early Theories of Vision

Two opposing theories of vision arose in early Greece.[2.1-2.5] One, called the emanation theory, proposed that the viewer gives off a visual spirit or pneuma that originates in the brain and passes along the hollow optic nerve into the eye and then is emanated into space as a cone of linear rays (see Fig. 2.1). The visual spirit was then reflected back into the sender's eye by objects in the world. This emanation theory was supported by Aristotle and the strength of his reputation carried its

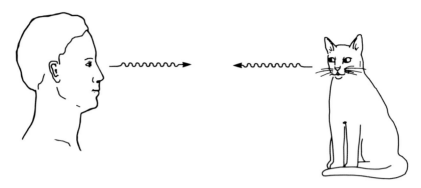

FIGURE 2.1. The emanation theory postulated that the eye emits a visual spirit to illuminate external scenes. The visual spirit was reflected back into the eye to produce vision.

acceptance into the Middle Ages. One argument in favor of the emanation theory was the fact that the eyes of cats and dogs can be seen to glow at night, proving they were giving off light. We know now that this is light reflected from the retina.

2.2.1   The Evil Eye

The emanation theory is reflected in the belief in the "evil eye." It was widely believed (and still is today, in some parts of the world) that some individuals were able to project maleficent essences onto a victim. The emanations, when accompanied by hostile, evil intentions, were believed to cause harm to the recipient. Such phrases as "if looks could kill," "the eye is the window to the soul," and "caught my eye" also reflect the emanation theory.

An opposing explanation of vision was known as the emission theory. In its earliest form, this theory stated that light consists of ethereal substances given off from luminous or illuminated bodies. The tiny ethereal substances were called atoms and were believed to be exact replicas of objects. That is, dogs gave off tiny dog particles, houses gave off tiny house particles, and cats gave off tiny cat particles (see Fig. 2.2). Democritus was a proponent of this theory. Plato tried to combine the emanation and emission theories and spoke of an inner light emitted from the eyes and an outer light that originated from luminous bodies.

The first description of eye physiology as we know it now was put forth by Felix Platter of Basle (1536–1640) who said the lens is the optic and the retina the receptor. Johanas Kepler of astronomy fame (1571–1630) showed theoretically that a reduced and inverted image would be formed by the lens on the retina. Réné Descartes (1596–1650) demonstrated experimentally the formation of this image when he placed a screen in the location of the retina in the excised eye of a bull. The image on the screen was reduced and inverted.

Sir Isaac Newton (1642–1727) believed that light was corpuscular in nature, was emitted from illuminated objects, and traveled in straight lines. He believed the lens focused the light on the retina and the retina was excited by the light in a

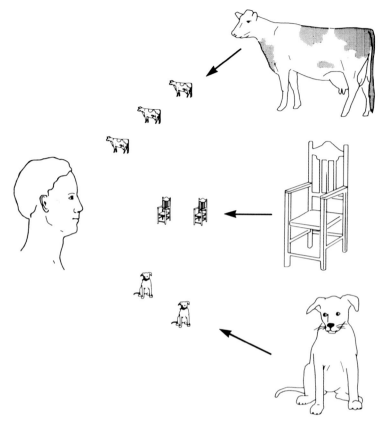

FIGURE 2.2. The emission theory postulated that each illuminated object gives off tiny replicas of itself. These replicas were called "atoms."

manner similar to the way that sound waves excite the ear drum. These excitations were then carried along the optic nerve to the brain.

Current studies indicate that eyes have developed in at least 40 separate evolutionary tracks.[2.6] Hill and Davidson report striking similarities between the developmental genes in the mouse and the fruit fly drosophila.[2.7]

2.3 Simple Experiments

2.3.1 Inverted Image on the Retina

This experiment is designed to demonstrate that the upside down image on the retina is inverted by the brain. The experiment consists of two parts; the first demonstrates that the image projected by the eye's lens onto the retina is upside down, the second demonstrates that the upside down image is inverted by the brain.

FIGURE 2.3. Experiment to demonstrate the blind spot. Cover your left eye and hold this figure about 30 cm from your eye. Focus on the "+" and move the figure close or farther from your eye until the spot disappears.

Part 1. To demonstrate that a simple lens produces an inverted image, set an empty camera on a table or window sill. Focus it on a distant scene. Open the back of the camera and place a sheet of thin typing paper against the back of the open camera. Set the time to the longest exposure time (or *B* or *T* if it is available) and open the shutter. The upside down scene will appear on the typing paper. All simple lenses, including that of the eye, invert the image.

Part 2. To demonstrate that this upside down scene is inverted by the perception interpretation apparatus in the brain, follow these steps: (1) Punch a small hole in a 3" x 5" card with a pin; (2) with one eye, stare at a relatively bright open area such as a blank wall or open sky through the hole with the card held about 30 cm from your eye; (3) focus on the hole, which will appear as a bright spot in the card; (4) with the other hand hold the pin (with pin head up) as close as possible to your eye (just touching your eyelashes); (5) move the pin up into your visual field from below. What you will see is the shadow of the pinhead descending into the bright spot from above. The pin is too close to the eye surface to be focused by the lens, so the shadow of the pinhead is projected onto the retina directly. The brain inverts the entire image including the pin head shadow.

To demonstrate that this shadow is actually on the retina and is superimposed by the brain onto the perceptual field of view, punch a second hole in the card about 5 mm from the first hole and repeat the experiment. Notice that now you see a pin head in each of the two spots. The pinhead images are again inverted by the perception mechanism. The pinheads appear to be in the center of each hole. The perception mechanism places the retinal pinhead shadow on each of the hole images. The brain superimposes the pinhead shadow signal on the rest of the visual field signal to produce an inverted image of the pinhead in each of the bright spots.

2.3.2   Blind Spot

The region where the nerves and blood vessels enter and exit the eyeball is known as the optic disc. It contains no receptors and so objects projected by the lens onto the optic disc are not detected.[2.8–2.9] For this reason the optic disc is also known as the blind spot. A simple experiment using Fig. 2.3 can demonstrate the presence of the blind spot. Hold Fig. 2.3 approximately 30 cm from your right eye and cover your left eye. Focus on the "+" and move the figure closer or farther away

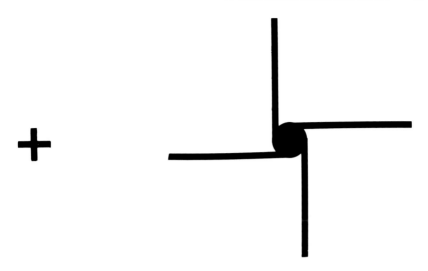

FIGURE 2.4. Experiment to demonstrate that the sensitivity and resolution of the eye differs for horizontal and vertical figures. Repeat the blind spot experiment of Fig. 2.3 with this figure and notice that the two vertical lines merge into a continuous line, whereas the two horizontal lines remain separate.

while keeping your eye focused on the "+" until the spot disappears. If you use the left eye instead of the right eye, the figure must be inverted to make the spot disappear.

2.3.3 Horizontal and Vertical Detectors

The eye does not detect horizontal and vertical lines or features with equal sensitivity. This difference between horizontal and vertical detectors is demonstrated with the aid of Fig. 2.4. Cover your left eye. Place the spot with spokes at the blind spot by focusing on the "+" and adjusting the distance to the figure until the central spot disappears. You will notice that the vertical lines appear to merge into a single continuous line, whereas the horizontal lines are perceived as two separate lines displaced in space. This difference in vertical and horizontal discrimination is not often appreciated and is direct evidence for some sort of feature detection mechanism in the visual process.

  The fact that this discrimination is for lines and not curves is demonstrated in Fig. 2.5, where wavy lines have been substituted for the straight vertical and horizontal lines. Repeat the blind spot experiment from above by placing the central spot in Fig. 2.5 on the blind spot to demonstrate that both the vertical and horizontal wavy lines are perceived as separate lines and separated in space. This shows that the feature detection process is specific for straight lines and not wavy lines. This difference in feature detection sensitivity is discussed further in Sec. 2.8.

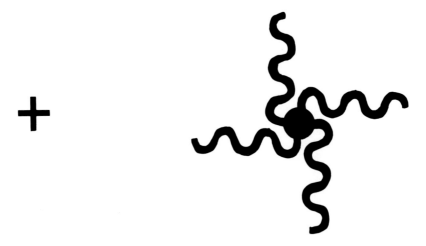

FIGURE 2.5. Experiment to show that the differences in horizontal and vertical sensitivity are limited to straight lines. Repeat the experiment of Fig. 2.3 with Fig. 2.5 and notice that the vertical and horizontal lines appear separate.

2.3.4   Mach Bands

Mach bands are the perceived brighter and darker regions near an edge.[2.10-2.13] Figure 2.6 demonstrates the appearance of Mach bands on either edge of the center line of the figure. Although the area to either side of the center line is uniform as measured by an optical densitometer, the image appears to have a dark band parallel to the center line on the dark side, and a light band parallel on the light side. The apparent light and dark bands are called Mach bands and are named after Ernst Mach (1836–1916), who described them in 1865. Mach bands can appear wherever there is a sharp change in brightness. Mach band patterns often appear in images presented for interpretation or analysis such as in chest X rays near the edges of the lungs or in aerial surveillance images.

2.3.5   The Craik, Cornsweet, O'Brien Illusion

The Craik, Cornsweet, O'Brien illusion is similar to the Mach band illusion except that it extends all the way to the edges of an image.[2.2] See Fig. 2.7. The perception is that the left side of the image is brighter than the right side, but the brightness on either side is actually the same. One explanation for this illusion is that edges are the prime perception target for the eye-brain system. The eye-brain detects the abrupt change at the center and extends it to the edges, transforming the perceived image to be lighter on one side and darker on the other. The fact that the two sides are actually the same shade of gray can be demonstrated by covering the central portion Fig. 2.7 (about 2.5 cm wide) of the figure.

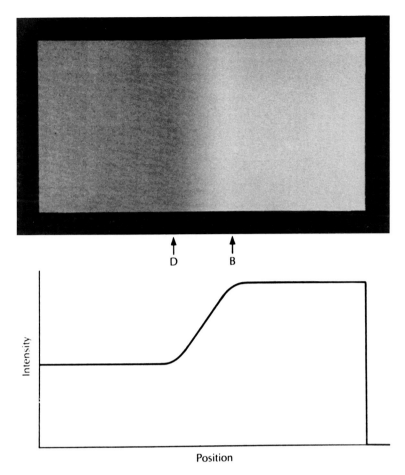

FIGURE 2.6. Mach bands appear as dark (*D*) and bright (*B*) bands on either side of the center line. The lower half of the figure is an optical densitometer plot of the image showing that there is no bright or dark band in the actual image; they exist only in the perceived image. (Image from reference 2.2.)

2.3.6   Herman Herring Grid

The Herman Herring grid,[2.14] shown in Fig. 2.8, is one easy way to demonstrate the presence of lateral inhibition caused by a central excitory field surrounded by an off-center annular area that is inhibitory. The gray spots that appear at the intersections of the white lines of Fig. 2.8 are caused by the firing of the lateral, off-center inhibitory cells.

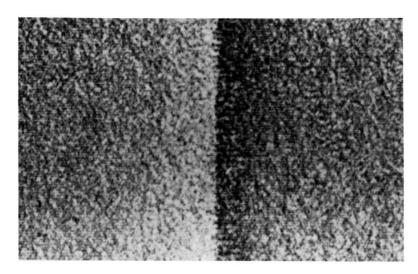

FIGURE 2.7. Craik, Cornsweet, O'Brien illusion. In this image there are dark and bright bands parallel to the gradient in the middle of the figure. The rest of the figure is the same shade of gray. To show that the apparent dark and light sides are an illusion, cover the center line area (about 2.5 cm wide). (Image from reference 2.2.)

2.3.7   The Moon Illusion

From the time of cave dwellers, people have noticed that the moon on the horizon appears larger than the moon overhead. A great deal of nonsense—pure unadulterated and scientific—has been written about the moon illusion. Because the distance to the moon does not change, the actual size of the moon and the size of the moon's image on the retina do not change. Thus, this is a perception phenomenon. One explanation that has been put forth claims that, because the light from the moon passes through more of the earth's atmosphere when the moon is near the horizon, the rays are refracted in a way that makes the moon appear larger.

Another explanation is that, as the moon rises above the horizon, the angle of the eye observing the moon changes from nearly perpendicular to the body axis to steeply elevated upward and that this angle is used by the brain to interpret size, making things overhead appear smaller than things at the horizon. Interestingly enough, even though this explanation is not correct, one of the tests for the angle hypothesis was to view a horizon moon upside down by looking through your legs. When you do this, part of the moon illusion does disappear; that is, the moon does appear to get smaller. The reason will be discussed below.

One of the more plausible explanations for the moon illusion is based on the fact that in distance and depth perception the brain compensates for distance by interpreting distant objects as being larger.[2.14] Thus two bicycles, one close and one several blocks away, are interpreted as being the same size even though the retinal image of the distant bicycle is significantly smaller. This comparison hypothesis

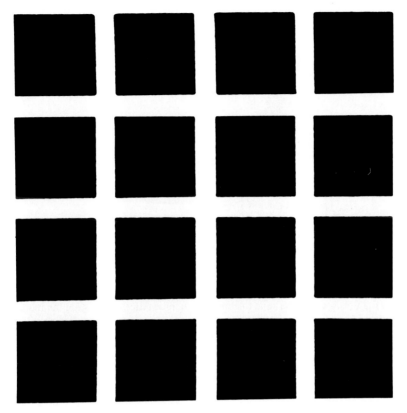

FIGURE 2.8. Herman Herring grid demonstrates the effect of lateral inhibition in the visual chain. The gray spots appearing at the intersection of the white bars are due to lateral inhibition in the visual processing system.

explains the moon illusion by stating that things on the horizon are compared with the moon and the brain interprets these differences as related to distance and size and therefore makes the moon appear large. This explanation seems rational except that the moon appears large even when viewed over an ocean where there are no objects on the horizon to provide these comparisons.

The theory that provides the most reasonable explanation for the moon illusion is based on the receding horizon perception. Even in an image without any landmarks, such as a flat desert or an ocean or large lake, the brain perceives and interprets the horizon as receding into the distance. The confirmation of this explanation of the moon illusion is relatively simple. View the moon when it is near the horizon through a paper tube such as a core of a paper towel or toilet paper roll as illustrated in Fig. 2.9. The illusion immediately disappears. This is not due to viewing the moon through only one eye, as can easily be verified by looking at a moon on the horizon through both eyes and then either one or the other eye individually. The moon illusion remains until the horizon is hidden by the tube.

FIGURE 2.9. Use of a paper tube to eliminate the moon illusion. Viewing the moon near the horizon through a paper tube eliminates the appearance of an extra large moon.

Viewing the moon upside down through your legs does remove part of the illusion (i.e., it doesn't appear as large); this is probably because of the brain's different interpretation of the upside down horizon.

2.4   Adaptation and After Images

Anyone who has sat in a hot tub is familiar with adaptation, the phenomenon that the sensation of heat diminishes with time. The water that at first seems so hot as to be hardly bearable quickly becomes very comfortable. Most studies show that the rate of nerve firing from a constant stimulus decreases as the time of stimulus increases.[2.11] When the stimulus is first presented, the nerves' firing rate depends on the intensity of the stimulation. As time progresses, the firing rate decreases as the inhibitory process come into play. The individual is beginning to "adapt." As the stimulus continues, the inhibitory synapses fire until the total system output drops to approximately the spontaneous or "no stimulus" rate.

When the stimulus is removed, the excitation stops but the inhibitory firing requires time to decay. During this time the nerve firing rate decreases below the spontaneous rate. This is the cause of an afterimage. To illustrate this phenomenon, put a 2 cm diameter red spot on a white sheet using a red pen or crayon. Stare at the red spot, illuminated by a bright light, for about 20 seconds and then look away. The afterimage appears aqua blue and is caused by the inhibitory firings. Blue is the inhibitory opposite of red.

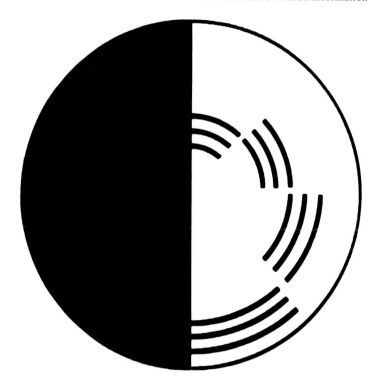

FIGURE 2.10. Black-and-white disk which produces a perception of color when spun rapidly. See text for explanation.

The afterimage of extremely bright lights such as flash bulbs is not negative and is believed to be caused by overstimulation in the photochemical receptors, which continue to excite the nerves and send out firing signals even after the stimulus is removed.

Cornsweet[2.11] and Rock[2.14] discuss how the after-effect arguments can be extended to more complex after-effect phenomena such as the sensation of motion in the opposite direction after actual motion stops. All such discussions include an excitatory synapse and an inhibitory synapse. After-effects occur in the perception of color, motion, detection of angles, and many other features. Cornsweet postulates that the adaptation phenomenon is only one aspect of a higher-level feature detector, which is designed to detect "change." Aspects of these hypotheses will be discussed in more detail in Sec. 2.8.

One other phenomenon that demonstrates the complexity of color vision can be demonstrated by the black and white disc shown in Fig. 2.10. Color vision is based on receptors in the eye that are sensitive to red, green, and blue light. However, it is possible to produce a perception of color from a black and white image. Figure 2.10 provides demonstration of how a spinning black-and-white image can produce the sensation of color.

Reproduce Fig. 2.10 on a 3" x 5" card. Be sure to use black ink or a black marker pen. Push a thumbtack through the center of the circle and tape the thumbtack in place with transparent tape. Attach the shaft of the thumbtack to a variable-speed drill and rotate the figure under a bright white light or in bright sunlight. The sensation of colors will appear. The particular colors perceived depend on the speed and direction of rotation.

The explanation for this phenomenon lies in the fact that the processing and transmission time from the retina to the brain are different for the different colors. (The physiology of color vision has been discussed in Chapter 1.) The white areas reflected from the spinning figure send all colors to the eye; the black areas send none. This leads to a perception of color. The processing and transmission times for the three primary colors are different and so the "cutoff time" for the three colors arriving at the brain varies depending on speed and direction of rotation. The brain interprets the differences in red, blue, and green signals reaching it as shades of color, rather than black or white.

2.4.1   Stimulus and Sensation

It is important in studying vision and perception to remember that stimulus and sensation are not the same. Sensation depends on stimulus but there is an important difference. A sensation can be described as a light sense, a color sense, form sense, or motion sense. A stimulus is the physical process that produces the nerve firings that result in a sensation. For vision, the stimulus is most often light, but it can be pressure or radiation. To demonstrate how pressure can produce sensations, close your eye, wait about one minute until any afterimages fade. Gently press on your eyelid. You will see a bright field from the eye you have pressed on. Many persons report this brightness perception as red or yellow in appearance.

2.5   Three-Dimensional Vision

Learning to see is one of the earliest skills a baby acquires. This incredibly complex ability to follow and focus on moving objects requires the use of all the head and neck stabilization muscles in addition to seven sets of eye muscles. Involuntary saccades shift the fixation point several times per second, (see Sec. 1.5),[2.15] so visual pursuit of a moving object requires the combination and coordination of saccade and pursuit muscles. This ability to follow moving objects and the perception of near and far as evidenced by depth perception[2.15–2.17] are usually developed in the first few months after birth. In viewing actual three-dimensional objects, we use many clues to get the relation between objects, including size, perspective, and shadowing between the objects. Size is important because we soon

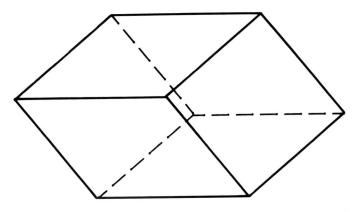

FIGURE 2.11. Two-dimensional line figure which produces an impression of a three-dimensional figure.

learn that things far away appear smaller. We also use perspective, noting that distant objects appear higher on the horizon. Shadowing—the fact that an object in front of a second object will cut off a portion of the more distant object—also helps us perceive three-dimensions.

Artists throughout history have used these characteristics to impart a three-dimensional perspective to a flat two-dimensional image. Figure 2.11 shows how a two-dimensional group of lines can be interpreted as a three-dimensional box. Figure 2.12 illustrates how we combine shading, perspective, and size in a two-dimensional image to convey three-dimensional information. For example, the airplane in the picture is perceived to be in the far distance, and the man to be located between the two women.

2.6   Stereoscopic Viewing

Figure 2.13 shows two different views of a steel bridge. Careful examination will show that these two images are not exactly identical. They are shifted slightly so the two images can produce a three-dimensional image when viewed properly. A simple way to view these three-dimensionally is to place a postcard or a 3" x 5" card between the images and hold it in place at the end of your nose. Gaze out into the distance and focus on a distant object. The two images will form a single image that appears three-dimensional. The eye-brain system combines the image from the left eye and the image from the right eye to form an image perceived in three dimensions. This clearly demonstrates that three-dimensional images are produced by the brain from two individual images. This is the principle upon which the stereopticon viewer was based.

FIGURE 2.12. Two-dimensional illustration demonstrating how perspective, light, shading, and overlapping contours combine to produce an appearance of a three-dimensional image.

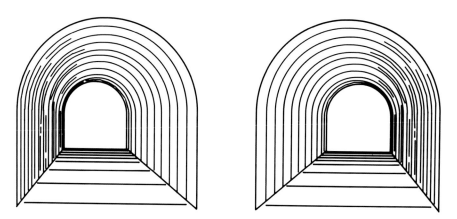

FIGURE 2.13. Two images of a bridge. To view a three-dimensional image, use either the card or cross-eyed techniques described in the text.

2.7   Cross-Eyed Technique of Three-Dimensional Viewing

A second way to form a three-dimensional image from these two images involves crossing your eyes, so that the right eye views the left image and the left eye views the right image. Hold Fig. 2.13 about 18 cm in front of your face and place your thumb between the two images. Focus your eyes on your thumb and gradually bring your thumb towards your nose, keeping your eyes focused on your thumb. Notice that in the background an additional pair of images appears. As you bring your thumb closer to your nose, the two inner images will blend into a single image so you see three images. With practice, the center one will have a three-dimensional appearance and be in sharp focus. With a little experience you can learn to ignore the two outer images and concentrate on the central three-dimensional image. Radiologists use this "trick" to view stereoscopic radiographs without stereopticon devices.

Julesz[2.18–2.19] has demonstrated that the brain can perceive three-dimensional information without any of the usual perceptual clues. He produced a random dot image and then shifted it slightly to produce a second image, such that when viewed either through a stereopticon viewer or using the card or the crossed-eye techniques mentioned above, the random dot pattern appears as a three-dimensional image. He then surrounded these left and right random dot patterns with an identical random dot pattern. When viewed individually, the dot patterns appear identical. When viewed stereoscopically, the central square pattern seems to float in space. Figure 2.14, which reproduces these dot patterns, demonstrates that size, perspective, and shadowing clues are not required for three-dimensional viewing.

2.8   Models of the Visual System

The purpose of a visual system model is to set up a simplified version of parts of the vision process to study. Different models of the visual system are designed to explain different visual phenomena or to concentrate on one aspect of vision to be studied more thoroughly. Eventually we hope to have a single model to explain all parts of vision. This section will discuss several current models explaining the evidence on which these models are based.

2.8.1   Feature Detection

Experimental studies in animals such as the horseshoe crab and other test systems have shown that there are specific nerves that fire only when excited by a particular color, a particular form, a vertical line, an angled line, a circular blob, or upon perception of motion in a specific direction. The discovery of these specialized groups of nerves lead naturally to the concept of "feature detectors." Feature

FIGURE 2.14. Random dot images shifted to produce a three-dimensional image without size, perspective, or shadowing clues. From reference 2.14.

detectors are nature's way of selecting certain specific features (e.g., things moving toward us) out of the flood of information detected by our eyes.[2.20–2.21]

Through studies on horseshoe crabs, cats, and monkeys, evidence has accumulated showing that there are nerves that fire only under specific conditions.[2.22–2.25] There are some nerves that fire only when presented with vertical lines, other nerves that fire only when presented with straight lines at some specific angle to the vertical. There are other nerves that fire when motion is detected from left to right, or when the motion is in another direction.

The fact that some nerves fire only when presented with specific features leads to the feature-detection model, in which a scene is made up of individual features. These relatively large features are detected simultaneously, and the finer details are filled in by slower scanning of the scene, usually with foveal vision. This known as the bottom-up model.

2.8.2   The Bottom-Up and Top-Down Models

According to advocates of the bottom-up school, a perceived image is sequentially formed by building up individual features of the image until the entire scene is recognized. The opposite view is known as the top-down approach, in which an immediate overall impression, a "gestalt" of the entire scene, is initially formed and the individual features are filled in later.

The problem with the bottom-up model is that recognition of a scene requires too much time. We're all familiar with the experience of walking into a crowded room and immediately picking out one or two familiar faces. Flash presentation studies in which a subject is briefly presented with an image demonstrate that the perception of a familiar face, or the differentiation of a cat from a cow, occur too rapidly to have been accomplished through the buildup of individual features.[2.14] On the other hand, the presence of feature detectors in the visual system is incontrovertible. These filters are in the visual system and must be used for something.

2.8.3   Ideal Observer Studies

One series of studies, called ideal observer studies, compares human observers with the best any observer (the ideal observer) could do.[2.26-2.28] In an ideal observer experiment, the observer, either human or "ideal," is presented with two images, one of which contains a target. The observer is asked to choose which image contains the target, a two-alternative forced choice (TAFC). Questions concerning the effect of different amounts of contrast, different signal-to-noise ratios, size, etc., can be experimentally determined using TAFC experiments. The target signal and its possible location are indicated for both images. Figure 2.15 shows a representative TAFC image including the target example above the two squares, one of which contains the target. In this example the target is in the left image.

The ideal observer (usually a computer program) compares the signal from the known target with the signal from each of the images in the location where the target would be if it were present. The ideal observer then picks the image with the best match. In mathematical terms this is a calculation of the correlation between the image signal and the test signal. Because the image signals also have noise, it is possible for the random fluctuations of noise to give a higher correlation signal and thus cause the "ideal observer" to choose incorrectly.

One of the criticisms of ideal observer experiments is that in real life human observers seldom, if ever, know exactly where the target will be. This is true, but ideal observer experiments are designed not to study how human observers operate in real-world conditions, but rather to shed light on some of the properties of the visual system.

In one of the more interesting results from TAFC experiments, Burgess et al. showed that the human observer, under some conditions, is only about 70 percent as efficient as the ideal observer. Many things have been read into this result, including the idea that, because the human observer is so close to an ideal

FIGURE 2.15. Two-alternative forced-choice images. The target, whose form is shown above the two test squares, is located in one of the two test squares together with the background noise. The marks on the edges indicate where the target is located, if present, in the test square.

observer, mechanical or computer image processing can't improve the detection of targets very much. The fact remains, however, that under real-world conditions the observer does not know where the target may be and often doesn't even know what type of target might be present. Thus, it's too soon to count out computerized image processing as a way to improve the detection of information and visual perception in real-life situations.

2.8.4   Computational Models

Computational models use computers to model the human visual system. There are two reasons for studying detection of visual information with computational models. The first is that the human system is very efficient and can be used as an example to improve computer recognition of images; the second is that the visual process can be broken down into a series of steps including detection, processing, and recognition. These steps can be divorced from the hardware involved, i.e., they don't depend on how the light is detected, or what circuits or nerves do the processing. Developing the computer processing approach may help explain what is happening in the human system.

   Poggio et al. have attempted to use computer models to detect and process three-dimensional images.[2.29–2.30] At first glance, stereoscopic image processing appears to be a simple exercise in geometry. On closer examination, however, it turns out to be an extremely complicated problem that can be solved only by making many assumptions. For instance, the assumption that Poggio and others make is that real objects are usually smooth, continuous, and contain discontinuities only at their edges.[2.31–2.32] This is supported by the Craik, Cornsweet, and O'Brien illusion in Fig. 2.7, which demonstrates that edge detectors have major effects on the human visual system. The results of Poggio and others working with computer models have helped our understanding of how the visual system operates.

2.8.5   Preattentive and Attentive Processing Textons

Neisser suggested in 1967 that the speed of visual perception implies some processing early in the perception chain (at what is known as the preattentive level) followed by focused attentive processing.[2.33] This approach combines the top-down and bottom-up models. Current theories state that the preattentive processing operates automatically and employs parallel processing. Some simple features, such as color, line orientation, or other features involving line intersections, are detected at the preattentive level. These features are called "textons" by Julesz.[2.34–2.35] The time required to detect a target from similar objects that differ in one texton (e.g.,color) is almost independent of the number of nontargets in the field. This is self-evident and obvious when an observer is assigned the task of, say, picking a red target out of a field of yellow distractors. It doesn't matter much how many yellow distractors are present; the time to find the red target is almost constant.

   Julesz classifies terminators of line segments and crossing of line segments as textons. Figure 2.16 illustrates preattentively discriminable texton pairs. The Xs are easy to discriminate from the background Ls because the textons are different. The Xs have a crossing terminator, the Ls do not. Notice that it requires a more detailed search to locate the area containing Ts in the background of Ls. This is because the textons of T and L are the same. Figure 2.17 illustrates discrimination of texton pairs at the preattentive level. It contains one X and one T in a background of Ls. To distinguish and locate the Ts from the Ls requires an attentive search. It is easy to distinguish the Xs from the background Ls because of the crossing texton absent in the Ls. It is not, however, easy to distinguish the Ts without an attentive, individual search.

   Fiorentini has discussed the question of whether processing in foveal vision is different than in nonfoveal vision.[2.36] Texton differences permit a target to be rapidly recognized, and give rise to eye-head movement to bring the target image onto the fovea. Foveal vision is typically used to scan with attentive search. Fiorentini has studied the time required to detect and count the number of targets under foveal or parafoveal vision. Her results indicate that the time required for counting colored spots in a different-colored background is the same for foveal

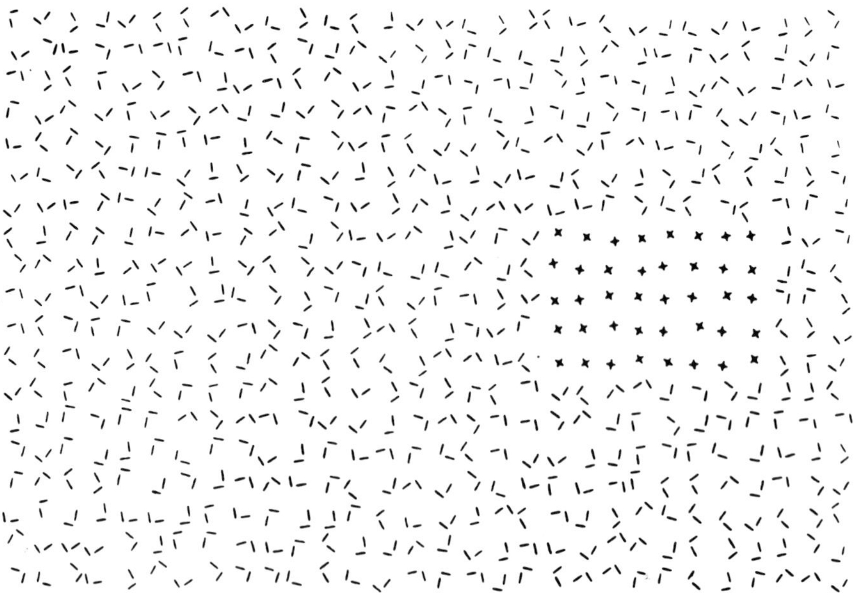

FIGURE 2.16. Example of texton discrimination. It is easy to rapidly pick out the locus of the Xs in the background of Ls. It requires a careful, attentive search to locate the Ts in the background Ls. This is because X has a different texton that L, but the T texton does not differ from L. From reference 2.27.

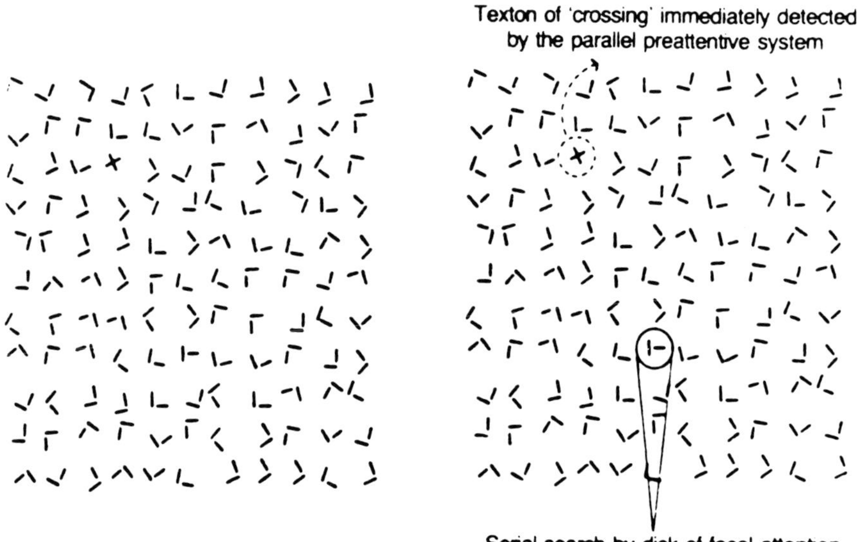

Texton of 'crossing' immediately detected by the parallel preattentive system

Serial search by disk of focal attention

FIGURE 2.17. Second example of texton discrimination. The T must be found by foveal attentive search.

and nonfoveal vision. This implies that the task of counting requires attentive search and not merely texton detection.

The detection of visual information is an incredibly complex process. Several levels of processing are required in order to present the information to our brain in a form that can be used in a timely fashion.

2.9   References

[2.1]   Duke Elder S., Weale R.A. The physiology of the eye and vision. In: *System of Ophthalmology*. Duke Edler S., ed. St. Louis: Mosby; 1985: 435–436.

[2.2]   Wertenbaker L. *The Eye—Window to the World*. Washington: U.S. News Books; 1981.

[2.3]   Lindberg DC. *Theories of Vision from Al-Kindi to Kepler*. Chicago: University of Chicago Press; 1976.

[2.4]   Pastore N. *Selective History of Theories of Visual Perception: 1650–1950*. New York: Oxford University Press; 1971.

[2.5]   Jaffe C.C. Medical imaging, vision and visual psychophysics. *Med. Radiogr. Photogr.* 1984; 60: 2–48.

[2.6]   Salvini-Plawan L.V., Mayr, E. On the evolution of photoreceptors and eyes. In: *Evolutionary Biology* V10. New York: Plenum; 1977.

[2.7]   Hill R.E., Davidson D.R. Seeing eye to eye. *Current Biology* 1994; 41155–41157.

[2.8]   Luckiesh M. *Visual Illusions*. New York: Dover; 1965.

[2.9]   Livingstone M. S. Art, illusion, and the visual system. *Sci Am.* 1988; 258:78–85.

[2.10] Ross J., Morrone C, Burr D. The conditions under which Mach bands are visible. *Vision Res.* 1989; 6: 699–715.

[2.11] Cornsweet T.N. *Visual Perception*. New York: Academic Press; 1970.

[2.12] von Bekesy G. *Sensory Inhibition*. Princeton: Princeton University Press; 1967.

[2.13] Ratliff F. *Mach Bands: Quantitative Studies on Neural Networks in the Retina*. Oakland: Holden Day; 1965.

[2.14] Rock I. *An Introduction to Perception*. New York: MacMillan; 1975.

[2.15] Schall J.D. Neural basis of saccade target selection. *Reviews in the Neurosciences* 1995; 6: 63–85.

[2.16] Avanzini G., Villani F. Occular movements. *Curr. Opin. Neurol.* 1994; 7: 74–80.

[2.17] Buttner U., Fuhry L. Eye Movements. *Curr. Opin. Neurol.* 1995; 8: 77–82.

[2.18] Julesz B. *Foundation of Cyclopean Perception.* Chicago: University Chicago Press; 1971.

[2.19] Julesz B. Stereoscopic vision. *Vision Res.* 1986; 26: 1601–12.

[2.20] Georgeson M.A. From filters to features: location, orientation, contrast and blur. *Ciba Found. Symp.* 1994; 184: 147–65.

[2.21] Burr D.C., Morrone M.C. The role of features in structuring visual images. *Ciba Found. Symp.* 1994; 184: 129–41.

[2.22] Bowne S.F. Contrast discrimination cannot explain spatial frequency, orientation, or temporal frequency discrimination. *Vision Res.* 1990; 30: 449–461.

[2.23] Burveck C.A., Yap Y.L. Two mechanisms for localization? Evidence for separation dependent and separation independent processing of position information. *Vision Res.* 1990; 30: 739–750.

[2.24] Burr D.C., Morrone C., Spinelli D. Evidence for edge and bar detectors in human vision. *Vision Res.* 1989; 29: 419–431.

[2.25] Derrington A.M., Goddard P.A. Failure of motion discrimination at high contrasts: Evidence for saturation. *Vision Res.* 1989; 29: 1767–1776.

[2.26] Burgess A.E., Wagner R.F., Jennings R.J. Efficiency of human visual signal discrimination. *Science* 1981; 214: 93–94.

[2.27] Burgess A.E., Ghandeharian H. Visual signal detection. II. Signal-location identification. *J. Opt. Soc. Am.* 1984; 1: 906–910.

[2.28] Burgess A.E., Colborne B. Visual signal detection. IV. Observer inconsistency. *J. Opt. Soc. Am.* 1988; 5: 617–627.

[2.29] Poggio T., Gamble E.B., Little J.J. Parallel integration of vision modules. *Science* 1988; 242: 436–440.

[2.30] Poggio T. Vision by man and machine. *Sci Am.* 1984; 250: 106–116.

[2.31] Davidson M., Whiteside J.A. Human brightness perception near sharp contours. *J. Opt. Soc. Am.* 1971; 61: 530–536.

[2.32] Ratliff F.L. Contour and contrast. *Sci Am.* 1972; 226: 90–101.

[2.33] Neisser U. *Cognitive Psychology.* New York: Appleton-Century-Crofts; 1967.

[2.34] Julesz B. A brief outline of the texton theory of human vision. *Trends Neurosci.* 1986; 7: 41–45.

[2.35] Julesz B. Textons, the elements of texture perception and their interaction. *Nature* 1981; 290: 95–97.

[2.36] Fiorentini A. Differences between fovea and parafovea in visual search processes. *Vision Res.* 1989; 29: 1153–1164.

3
Quantification of Visual Capability

Arthur P. Ginsburg and William R. Hendee

3.1 Introduction

The interpretation of visual images is characterized by ambiguity. The ability of an individual to extract information from an image is difficult to quantify. One approach to this challenge has been to define the visual capability of an individual in terms of metrics derived from an understanding of how individuals perceive spatial information.

This understanding has been expressed in terms of models of the human visual system. A particularly useful model of perception is one that includes the selection and processing of certain types of visual information and the suppression of others. This model is termed "selective filtering." Although perception is certainly more complex than simple selective filtering, this model at least offers an initial explanation of how people detect and recognize images. The explanation leads to the next challenge of how to quantify how well an observer sees images. This challenge is approached from the standpoint of contrast sensitivity.

This chapter discusses certain physical characteristics and filtering properties of the human visual system and how these characteristics and properties help to explain the ways that an observer uses visual information. The filtering properties of the observer are presented in terms of how they can be quantified through contrast sensitivity measurements obtained with a vision test chart. Advantages of this approach to the quantification of visual capability are compared with those of alternative methods of vision testing.

3.2 Visual Acuity

Visual acuity refers to the ability of the observer clearly to see high contrast spatial information in an image. Visual acuity may be limited by optical blur in the eye, light scatter due to media opacities of the lens or by loss of neural sensitivity in the retina-brain system. A blurred image may be caused by any one or a combination of problems that may occur in the cornea, lens, or pupil size. The increased coarseness of retinal sampling towards the periphery of the eye, compared with that in the fovea, also limits visual acuity.

Visual acuity is generally evaluated by having the observer detect and identify high contrast targets of different sizes placed at a prescribed distance. Visual acuity has been the cornerstone of vision assessment, dating back to 1862 with the creation of the *Snellen visual acuity chart* (Fig. 3.1). The longstanding acceptance of the Snellen chart, and the need to create vision standards for the military in World War I, are responsible for the contemporary notion that a result of "20/20" with the Snellen chart denotes good vision. Snellen visual acuity is an acceptable measure of the optical quality of the eye in the sense of how well an image is focused. It is primarily a quantitative, not a qualitative, measure of visual capability. Assessment of the patient's ability to resolve over the visual angle defined by projecting a dark letter on a light background (Snellen visual acuity) does not correlate very well with routine visual tasks such as detecting and recognizing objects of different sizes and contrasts when, e.g., driving an automobile under conditions of poor visibility, such as at night.

Snellen-type visual acuity charts are currently used for three purposes. First, they are used to assess visual acuity as an indicator of visual health. Many patients with ocular disease, however, display 20/20 vision as measured with a high contrast Snellen visual acuity chart. Testing the ability of these patients to see low contrast information often reveals a functional loss indicative of ocular or neurological disease. Second, acuity testing is used as a tool for evaluation of refraction disorders. Judgements of letters are subjective and refractive overcorrection can occur as a consequence of difficult and unreliable testing situations. Third, acuity testing is used to reveal whether a person can see well enough to pilot vehicles safely, even though measurements of Snellen acuity may not correlate well with visual performance under conditions encountered in the outside world.

In addition to fundamental limitations in the use of the Snellen test to evaluate functional vision, several more problems are associated with the test itself. Among these problems are:

1. There are different numbers of letters per line in most Snellen acuity charts. Therefore, allowing one or two mistakes per line has a different meaning for different levels of visual acuity.

2. There is not a regular progression in letter size, which makes reporting a loss of two lines of vision misleading.

3. Various Snellen letters differ in perceptual difficulty. An A or L is easier to perceive than an E. Different lines do not have letters of equal difficulty.

4. Test luminance conditions are not standardized for use with the Snellen acuity chart.

The standard Snellen visual acuity chart presents only a single high level of black-and-white contrast. Consequently, it does not provide a full assessment of a person's visual performance, defined as the ability to see varying sized objects that differ in contrast. When comparing the quality of vision corrected with soft

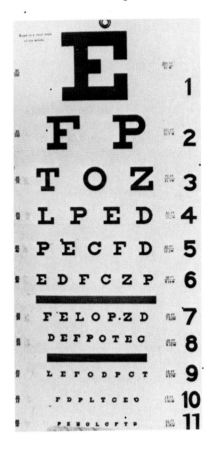

FIGURE 3.1. The Snellen visual acuity test chart.

contact lenses to that corrected with eyeglasses, e.g., a patient may see 20/20 with either but often comments that vision is much crisper with the glasses. In this instance, the poorer quality of vision with the contact lenses may be caused by flexing of the soft lenses. Flexing produces variations in optical focus and requires continuously changing accommodation. Evaluation of the patient's sensitivity to low contrast information assists in demonstrating the subtle differences in vision quality between the glasses and contact lenses. As another example, patients with contact lenses (hard or soft lenses) can develop a diffuse proteinaceous film on the surface of the lenses. These patients often complain of poor vision, even though their vision may actually be 20/20 or reduced to no more than 20/25 or 20/30 as measured with the Snellen visual acuity chart. A contrast sensitivity test often reveals that the overall quality of vision is, in fact, reduced much more sharply than indicated by the Snellen acuity chart. Similarly, the visual acuity of cataract patients may be decreased only slightly, whereas contrast sensitivity may disclose a much greater loss of vision quality.

3.3   Contrast Sensitivity

Contrast sensitivity describes the abilty of the observer to discern subtle differences in shades of gray present in an image. After 25 years of scientific and
clinical development, contrast sensitivity is finally becoming accepted as a more
comprehensive approach to describing vision, compared with the Snellen visual
acuity test. In addition to identifying visual defects due to optical causes, contrast
sensitivity can also be used to evaluate the quality of contrast perception at the
retina-brain level. At this interface, the retinal image is converted into a neural
code that is based primarily on the shape and contrast of the image. In the everyday
world, images are presented in a wide variety of sizes, shapes, and contrasts. The
sensitivity of the visual system to these images ideally should be tested with a set
of visual targets that represent all target sizes, shapes, and contrasts. Sine-wave
gratings (bar gratings that gradually become light and dark) work well as such
targets.

Instead of using sine-wave gratings to test vision, we might ask why letters,
disks, or more complex characters of different sizes, shapes, and contrasts should
not be used. In other words, why not use targets that appear more natural than sine-
wave gratings? The answer is that just as complex sounds such as those of speech
and music can be decomposed into a combination of single frequency sound waves
of different amplitudes and phases, any visual image can be mathematically analyzed as a composite of individual waves of singular spatial frequency, amplitude,
and phase. Stated in the reverse, a prescribed number of sine waves of specific
frequencies, amplitutdes, and phases can be combined to create the image of any
specific object. The process of decomposing a complex signal into its frequency
components is known as Fourier analysis.

3.4   Visual Physiology

The visual system consists of various receptor fields, including the retina, lateral geniculate body, and visual cortex; see Chap. 1. Within the visual cortex
are different types of cells that respond to structures of different sizes (spatial
frequency), contrasts, and orientations. Some cells (large size cells) respond to
(i.e., are "tuned" to) large size objects of low spatial frequency. Other cells (small
size cells) respond to smaller size objects of higher spatial frequency. Some cells
respond to high contrast and others to low contrast. Cells that are tuned to low
spatial frequency stimuli provide visual information about the gross features of
an object. Cells tuned to high spatial frequency features provide information on
the fine details of an object. Thus, when stimulated by the same object, some
cells provide information on gross features and others provide information on fine
details.

Many researchers believe that the human visual system functions as a crude
Fourier analyzer. Discrete "channels" in the system process visual information

within restricted ranges of spatial frequencies. These information channels represent the activity of the functionally independent, size-selective cells just discussed. The channels are thought to play a major role in filtering relevant information about the visual target, including characteristics such as contrast, size, shape, and orientation. Measurements of contrast sensitivity with sine-wave gratings permit generalizations to more complex visual targets such as letters, aircraft, and road signs.

3.5   Visual Filtering

The visual system offers an immense reduction of data at all stages in the processing of visual information. This data reduction is described as *"filtering,"* defined as the selective processing of certain types of visual information and the suppression of others. The filtering characteristics of the human visual system can be described in part in terms of contrast sensitivity, defined as the reciprocal of threshold contrast. Researchers such as Schade,[3.1] Campbell,[3.2] and others have found that contrast sensitivity measurements with sine-wave gratings can be used to predict the detection of one-dimensional images.

3.5.1   Sine-Wave Gratings

Sine-wave gratings have been shown to be sensitive vision test targets.[3.3-3.5] A sine-wave grating is a repeated series of light and dark bars that can be characterized in terms of three variables: spatial frequency, contrast, and orientation. The visibility of a grating is dependent on the combination of all three variables. Bars that are closely spaced are said to have a high spatial frequency. Usually they are more difficult to visualize than those that are widely spaced. A line pair, composed of a light and a dark bar, makes up one cycle or period of a grating. Spatial frequency is the number of cycles of the grating per unit length, sometimes described as the number of cycles per degree of visual angle (cpd). A fingernail at arm's length subtends about one degree of visual angle. The ligth and dark bars of a grating that occupies the same width represent about 1 cpd on the retina. Examples of sine-wave gratings having low, medium, and high spatial frequencies are shown in Fig. 3.2.

In general, sine-wave gratings can be interpreted in terms of Snellen visual acuity by specifying the number of cycles per degree equivalent to a specific Snellen acuity level. For example, one light or one dark bar (one-half cycle) of a 30 cpd grating subtends 1 min of arc. This is approximately the same visual angle as the subtended by each bar of a 20/20 Snellen letter. That is, the E of the 20/20 line on a Snellen chart is equivalent to 30 cpd. The E of the 20/200 line is equivalent to 3 cpd.

In order to determine the number of cycles per degree for a Snellen letter E, divide 600 by the Snellen line number (SLN). For example, the E of the 20/40

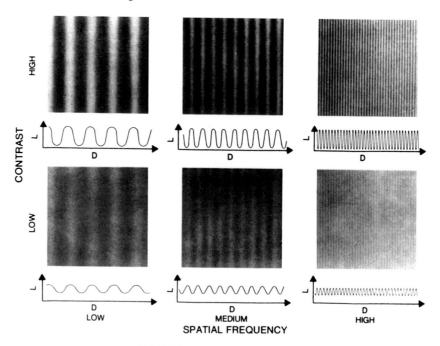

FIGURE 3.2. Sine-wave gratings.

line is $600/40 = 15$cpd. This generalization is accurate only for the letter E. The letter L of similar size on the 20/20 line of the Snellen chart requires only 18 cpd for identification. This difference occurs because the general shape necessary to identify the letter L requires less resolution of the bars than does recognition of the more complex letter E. This difference is also why an L is more easily visible than an E on the same Snellen line.

In addition to spatial frequency, sine-wave gratings permit the evaluation of contrast discrimination. Each bar has a measurable luminance distribution. The luminance difference between light and dark bars determines the contrast of the grating. To define the difference between "figure and field," or contrast C, the Michelson definition is generally used:

$$C = \frac{L_{max} - L_{min}}{L_{max} + L_{min}}, \tag{3.1}$$

where L_{max} and L_{min} are, respectively, the maximum and minimum luminances of the bars of the gratings.

The spatial frequency and contrast of a sine-wave grating can be depicted as a luminance profile for the grating. Examples of these profiles are illustrated beneath each grating in Fig. 3.2. Gratings with greater contrast have higher amplitudes in the luminance profiles.

In addition to spatial frequency and contrast, a sine-wave grating is also described in terms of orientation (i.e., whether the bars are vertical, horizontal, or

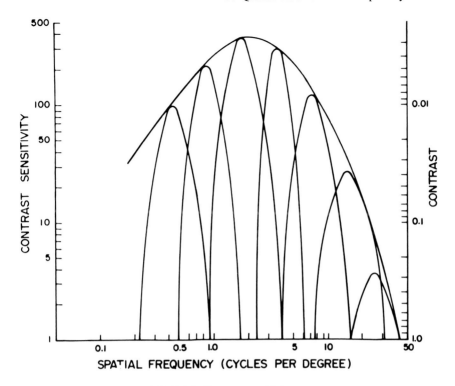

FIGURE 3.3. Contrast sensitivity curve.

tilted to the right or the left). In general, vertically oriented bars are easier to visualize than those that are positioned horizontally or at an angle.

A grating with contrast that varies across a range from invisiblity to visibility is said to encompass threshold contrast. Gratings of different spatial frequencies require different amounts of contrast to reach threshold. Contrast sensitivity, the reciprocal of threshold contrast, is often plotted as a function of spatial frequency to create a contrast sensitivity or contrast response function.

3.5.2   Contrast Sensitivity Function

A typical contrast sensitivity function of the human visual system is shown in Fig. 3.3. The broad, inverted, U-shaped curve describes the visual "window" that depicts the limit in the range of object sizes (acuity) that are visible at threshold contrast. The area above the curve is the region below threshold contrast, where the visual system does not respond to visual stimuli. The area below the curve is the region of detectable contrast where objects are perceptible. Contrast sensitivity decreases (i.e., greater amounts of contrast are required for perception) for spatial frequencies above and below the spatial frequency of about 2–6 cpd where contrast sensitivity is maximum.

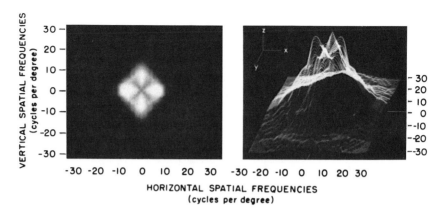

FIGURE 3.4. Two-dimensional (left) and three-dimensional (right) pictorial representations of a modulation transfer function.

Narrow curves shown within the contrast sensitivity function represent relatively narrow bandwidth mechanisms referred to earlier as spatial frequency channels. These channels combine to comprise the overall contrast sensitivity function of the observer.[3.6,3.7] Frequency channels encompass the activity of functionally independent, size-selective cells in the visual system that are considered to play a major role in "filtering" relevant target information such as contrast, size, and basic shape.

A contrast sensitivity curve based on sine-wave gratings can be used to compile a *modulation transfer function* (MFT).[3.8] The modulation transfer function describes how well spatial details in an object are reproduced in an image. Illustrated in Fig. 3.4 is a pictorial representation of a modulation transfer function assembled from contrast sensitivity data. Figure 3.4 (left) represents contrast sensitivity in terms of intensity, where the lighter and darker regions, respectively, correspond to higher and lower contrast sensitivities. The function is asymmetric, with sensitivity higher for linear patterns in the vertical and horizontal directions than in oblique directions. These characteristics suggest that objects with the same size and contrast are more or less visible depending on their orientations. This asymmetry is more obvious in the three-dimensional picture in figure 3.4 (right). The peak contrast sensitivity occurs at about 2 cpd and image details in this range yield higher amplitudes than do lower or higher frequencies. Thus, the filtering characteristics of the observer attenuate the coarse and fine details of an image more than those expressed at intermediate spatial frequencies.

3.5.3   Channel Filter Images

As with auditory processing, only a limited range of spatial frequency information can be processed by the visual system. The system is most sensitive to sine-wave gratings at its peak sensitivity of about 2–6 cpd. Recognition of objects with relevant spatial frequencies above or below the region of peak sensitivity

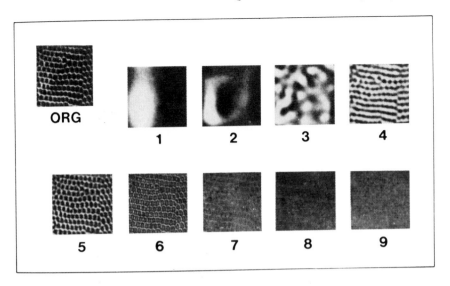

FIGURE 3.5. Channel filtering of visual information from a snakeskin.

requires more contrast. The physiological limit of visual acuity at high contrast is about 60 cpd, and may be significantly less for older observers and under suboptimal viewing conditions. Vision assessment requires evaluations at more than one contrast level. A patient who is 20/20 with the Snellen acuity test, but has impaired contrast vision for broad bars or a low number of cycles per degree, may be unable to see a truck in the fog or to read a low contrast X-ray image. A patient with refractive error might well have contrast losses that are especially pronounced at higher spatial frequencies.

Figure 3.5 shows a series of pictures that depict the operation of the channel filters of the human visual system presented with a natural texture pattern (snakeskin).[3.5,3.6] The first three channels contain average brightness information. Channels 1 and 2 show brightness gradients due partly to uneven illumination of the photograph during digitization; channel 3 contains enough information to identify the presence of texture. In channel 4, the horizontal directionality is clearly revealed by the light and dark bands across the image. The irregularity caused by the ridges also appears clearly as discontinuities in the horizontal bands. The cellular nature of the texture is extracted in channel 5. This channel demonstrates that the segregation of cellular textures, a complex operation for statistical-structural procedures, is a natural byproduct of spatial frequency channel filtering. Smaller cell details appear in channels 6–8. Although edge detection can be performed by numerous statistical-structural methods, this analysis shows that edge detection is a natural byproduct of the same channel-filtering procedure that isolates brightness gradients, directional tendencies, and cellular structures.

Since spatial information about objects can be described in terms of filtering, the same information can be quantified in terms of spatial frequency. For example, the number of cycles necessary for recognition of the Snellen letters E and L has

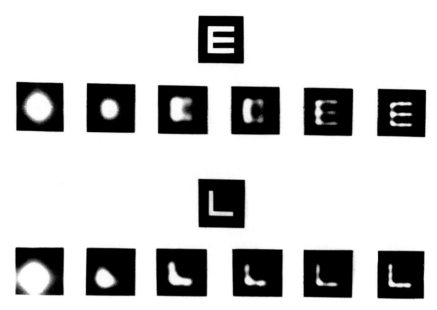

FIGURE 3.6. Snellen letters E and L displayed at 0.5 cpo increments (from left to right).

been determined by Fourier synthesis in steps of 0.5 cycles per object (cpo).[3.9,3.10] These letters were chosen because E is more difficult to resolve than L (see Fig. 3.6). The information contained in frequencies below 1 cpo allows detection, but not recognition, to occur. For recognition, 2.5 cycles are required for E, but only 1.5 cycles are required for L. This is the reason that L is recognized at a greater distance or smaller subtended angle than E. These results suggest that a bandwidth (the relevant number of spatial frequencies for a particular task) of 1.5–2.5 cpo is required for recognition of Snellen letters.

A comparison of vision and hearing tests demonstrates why contrast sensitivity is a more sensitive measure of perception than are high contrast acuity tests. Audiometric testing evaluates the listener's ability to hear pure tones of different temporal frequencies at sound thresholds. Analogous to hearing tests, contrast sensitivity measurements assess the observer's ability to see pure frequency waves (i.e., sine-wave gratings of different spatial frequencies) at various contrast levels. The auditory equivalent of a standard high contrast Snellen eye chart is a hearing test with only one high level of loudness for all sound frequencies tested. Clearly this would be neither a meaningful nor a complete test. Acuity determinations at one high contrast level do not permit statements to be made about other aspects of visual processing. Snellen visual acuity measurements do not relate well to visual performance and degradation of visual quality may not be detectable. Even though some patients can see 20/20 letters of the Snellen chart, they may complain about haziness, glare, and poor night vision. Functional vision requires the capability for optical and physiological visual information processing of visual information that includes integration of size, orientation, and contrast data. The ability of contrast sensitivity tests to measure differences in functional vision under conditions where Snellen visual acuity measurements are useless has been demonstrated in the laboratory, in simulators, and in field trials.[3.11,3.12]

Contrast sensitivity measurements encompass both the optical and neural components of the visual system. These two aspects of vision function independently of one another. The optical aspect is affected by clarity imperfections that produce blurring and light scattering, resulting in a degradation of the retinal image. Blurring affects only the higher spatial frequencies. Examples of such clarity imperfections include contact lenses that are coated with deposits, corneal diseases, lens opacities such as cataracts, and vitreous opacities.

3.6   Causes of Vision Loss

The neural aspects of vision are important in contrast perception. These mechanisms transmit the retinal image to the brain and process the neural signals. Certain neuro-ophthalmological abnormalities can reduce contrast sensitivity at the low and middle spatial frequencies, while higher frequency contrast information remains intact. Under these conditions, Snellen visual acuity is often recorded as normal. Other neuro-ophthalmological abnormalities may reduce sensitivity over all spatial frequencies, often with the middle or low frequencies affected more severely. This situation differs from that of blurring, which preferentially degrades the perception of higher spatial frequencies. Vision loss at high spatial frequencies, with the middle and low frequencies remaining normal, usually indicates an optical (i.e., blurring) cause. Loss at the middle and low frequencies, with the higher frequencies remaining intact, suggests a neuro-ophthalmological abnormality. Examples of neuro-ophthalmological problems include cystoid macular edema, senile macular degeneration, glaucomatous loss and atrophy of the optic nerve, multiple sclerosis, cortical lesions, and amblyopia.

Findings from contrast sensitivity tests can overlap and different optical and neurological problems can generate similar contrast sensitivity curves. A loss of the middle spatial frequencies may indicate optic atrophy, glaucoma, or optic neuritis, whereas a high spatial frequency loss may be due to refractive error, corneal edema, cataracts, macular degeneration, or amblyopia. Impaired contrast sensitivity suggests the need for further testing. Contrast sensitivity is also important in monitoring disease progression and treatment.

3.7   Detection and Identification of Visual Signals

The concepts of contrast sensitivity and visual filtering, when coupled with Fourier analysis, facilitate an understanding of how the visibility of objects changes as a function of size and contrast. Figure 3.7 depicts contrast sensitivity for the detection and identification of at least 50% of the letters on each line of a typical

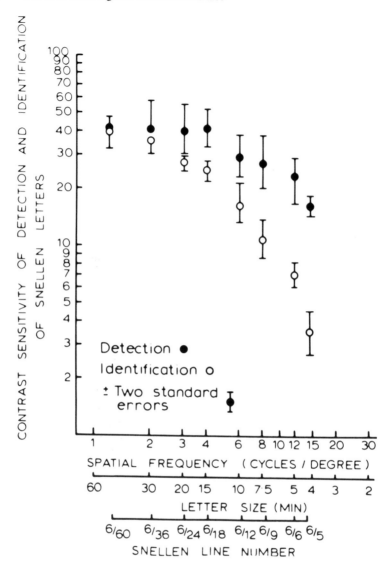

FIGURE 3.7. Contrast sensitivity chart for the detection and recognition of Snellen letters.

Snellen chart.[3.13] In this illustration, contrast sensitivity is defined as the reciprocal of the threshold contrast needed for detection and identification of the letters. Letter sizes 6/60 and 6/6 are identical to Snellen lines 20/200 and 20/20. The measured contrast sensitivity values change somewhat with field size and average luminance and should not be taken as absolutes.

The two curves in Fig. 3.7 describe the contrast required for the human observer to detect and identify Snellen letters whose sizes range from about 60 to 4 min of arc or 1 to 15 cycles per degree. For large letters [line 6/60 (20/200)], detection and identification thresholds are similar, for small letters [line 6/5 (20/15)], however, contrast must be increased by a factor of 5 over the detection threshold before the letters can be identified. These results can be explained by assuming that the detection and identification of larger letters are based upon low frequency components, whereas high frequency components are required for the small letters. Since the contrast sensitivity function decreases rapidly with increasing spatial frequency, much higher contrast is required for identification of small Snellen letters after they have been detected. Thus, the contrast required for identification of an object after detection depends primarily upon the bandwidth between the overall size and the particular features of an object required for identification. It should be noted, however, that high pass filtered letters can be created that permit detection and identification almost simultaneously at any viewing distance.[3.7]

These results further support the premise that objects can be correctly identified from a narrow range of relatively low spatial frequencies. Furthermore, it appears that the particular range of relevant spatial frequencies (i.e., the spatial frequency bandwidth used for identification by the visual system) is determined by the two-dimensional structure of the object. This hypothesis has been tested by predicting Snellen letter acuity from contrast sensitivity functions of patients with abnormal vision caused by amblyopia and multiple sclerosis. The contrast sensitivity functions obtained for these observers had reduced high, middle, or low spatial frequencies and, in one case, a very narrow band loss in sensitivity at middle spatial frequencies.[3.5] The corresponding Snellen acuity of these subjects ranged from 6/4 to 6/200. From knowledge that Snellen acuity requires a narrow band of low spatial frequencies to achieve a certain level of contrast, the Snellen acuities of 17 of 22 subject eyes were predicted by contrast discrimination tests to within one Snellen line. The acuities of the other five eyes were predicted within two Snellen lines. It appears that identification of complex objects (e.g., Snellen letters) can be predicted, at least to a first approximation, from evaluation of the individual filtering characteristics of the visual system as quantified by contrast sensitivity tests.

3.7.1 Measuring Contrast Sensitivity

Three techniques have been used routinely to measure contrast sensitivity with sine-wave gratings. These techniques are computer-controlled video displays, optical projectors, and photographic plates.

Computer-controlled video displays provide the most flexible contrast sensitivity test because they permit a wide variety of test parameters to be specified. Several years ago, Nicolet introduced a sophisticated contrast sensitivity test device that overcame problems encountered with previous methods.[3.14] With this device, bars are projected on a high quality television screen. Hence, the starting point for the test can be randomized. Results are printed out automatically and time-consuming tasks such as luminance calibration, data collection, and plotting are also done automatically. The main drawback to the Nicolet instrument is its cost and the time required for testing (at least 10 minutes per eye).

Optical projection slides for testing contrast sensitivity overcome some of the limitations of computer-generated video systems. The test results are difficult to standardize, however, because of variations in projector bulb luminance and differences in the reflective properties of the image screens. These drawbacks reduce the usefulness of optical projection slides for monitoring contrast changes accurately over time. The slides are useful for contact lens evaluation and fitting, refraction tests, and detection of pathological changes between eyes of the same patient.

Arden photographic plates[3.14] were one of the first attempts to bring contrast sensitivity tests into the clinical arena. Each Arden plate has a different-sized bar grating that gradually becomes fainter (lower contrast) from the bottom to the top of the plate. Arden plates are difficult to reproduce photographically and the resultant data are very subjective. There is no way to verify the accuracy of test results. With a subjective contrast threshold, a motivated individual can easily deceive the examiner.

The contrast sensitivity test system (CSTS) is a contrast sensitivity test chart developed to overcome the limitations of Arden plates and the Nicolet computer-video system.[3.11,3.12,3.15] The CSTS photographic plate is shown in Fig. 3.8. It has spatial frequencies and orientations that reflect the filtering characteristics of visual channels. With all the advantages of optical projection slides, the CSTS also allows standardization of data by utilizing a calibrated light meter to maintain constant test luminance. The CSTS is useful for documenting visual changes over time. It is a quick and accurate method for assessing contrast sensitivity, with the data comparing favorably to values obtained by a computer-video system. It provides standardized contrast sensitivity data by evaluating the eye's ability to distinguish between an object and its background under varying contrast conditions.

Today the CSTS test is the most widely used type of contrast sensitivity test. Its use proceeds according to the following steps. First, the patient is instructed to report the orientation of the faintest perceptible patch of sine-wave grating patches in each of five rows, labeled $A-E$ for spatial frequencies from 1.5 to 18 cpd. A response that correctly identifies the stripe orientation provides reasonable certainty that the stripes are actually seen. The patch number of least contrast correctly identified in each row is recorded on a special evaluation form. After testing is completed, these data points are connected to form the contrast sensitivity function. An area on the evaluation form allows immediate comparison of the patient's contrast sensitivity function to that of a normal population. The contrast sensitivity data can also be converted to Snellen visual acuity values.

Illustrated in Fig. 3.9 is the difference in contrast sensitivity functions for a patient with 15-month-old extended wear soft contact lenses coated with debris, compared with spectacles. New contact lenses gave this patient a contrast sensitivity function equal to that with spectacles correction. Different contact lens brands and materials have also been shown to have varying effects on contrast sensitivity.[3.16]

Contrast sensitivity testing using sine-wave gratings has been used to describe cataract-related vision loss.[3.17-3.19] An example that relates to contrast sensitivity

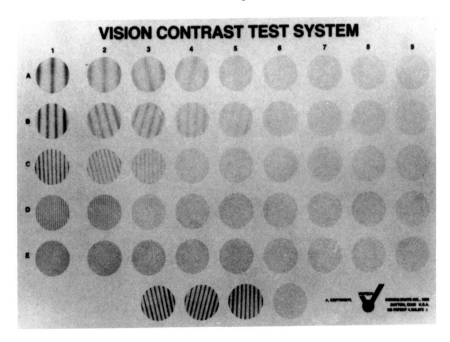

FIGURE 3.8. The contrast sensitivity test system.

and cataracts is shown in Fig. 3.10. The contrast sensitivity of the 55-year-old patient's left eye is depressed due to a posterior subcapsular cataract. Although the patient's standard letter acuity is 20/20, the contrast sensitivity for the affected eye is significantly below that of his right eye and of the normal population. In reality, the patient's acuity in the left eye is equivalent to having a letter acuity of 20/50 due to refractive error.

Individual differences exist in contrast sensitivity functions. In addition to comparing a subject's contrast sensitivity function to population norms, a base line contrast sensitivity function should be established for comparison with subsequent tests. Especially when contrast sensitivity is used to determine the efficacy of a particular imaging medium, the normal value must be considered. The contrast sensitivity function has been shown to predict the detection and identification of letters,[3.20] faces,[3.5] aircraft in simulators[3.21] and field trials,[3.22] and road signs.[3.23]

3.7.2 Sine-Wave Grating and Low Contrast Letter Charts

There is now a new generation of Variable Control Test System-type charts that use improved sine-wave gratings: the Sine-Wave Contrast Test (SWCT) (Stereo Optical Company) is similar to the VCTS and the Functional Acuity Contrast Test (FACT) (Stereo Optical Company).

The FACT, shown in Fig. 3.11, improves the sensitivity and quality of the VCTS and removes the possibility of aliasing resulting in spurious phantom grat-

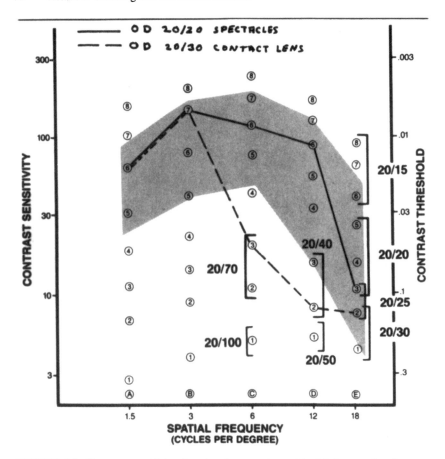

FIGURE 3.9. Contrast sensitivity function for a patient with debris-coated soft contact lenses, compared with spectacles.

ings caused by the high contrast between the white background and the circular gray grating patches.[3.24] The solution for aliasing, previously addressed,[3.25] was implemented by smoothing the grating patch edges into the average gray background of the grating patches. The FACT is further improved over the VCTS to include equal 0.15 log unit contrast steps for greater sensitivity to contrast change, increased patch size to test a larger retinal area, and digitally produced gratings for high quality sine waves.

Since the introduction of sine-wave grating charts, several low contrast letter charts have been developed. The Regan chart simply reduces the contrast of a standard Snellen-type letter acuity chart to furnish several charts, each with a different contrast level.[3.26] The Pelli-Robson chart is a single chart composed of a triplet of letters in 0.15 log unit contrast steps from high to low contrast.[3.27] Although low contrast acuity charts are routinely described as contrast sensitivity tests, they do not provide a contrast sensitivity curve similar to that of sine-

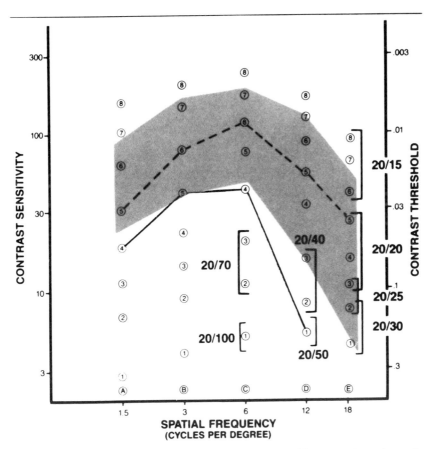

FIGURE 3.10. Contrast sensitivity functions of a patient with a posterior subcapsular cataract in the left eye. Although this patient's standard Snellen visual acuity was 20/20, the contrast sensitivity curve falls in the 20/50 range of the Equivalent Acuity Guide.

wave contrast sensitivity charts. Indeed, Leguire points out important differences between sine-wave grating and low contrast letter acuity charts.[3.28]

Researchers have confused important differences between contrast sensitivity and low contrast letter charts, including the use of the high "reliability" of 0.98 of the Pelli-Robson low contrast letter acuity chart as a "standard."[3.27] Although low contrast letter acuity charts provide high test-retest reliability, their sensitivity to clinically important contrast losses may be low, as discussed later in connection with the early presence of cataracts. The test-retest reliability of a computer-based sine-wave grating system using a forced-choice procedure can be as low as 0.77. Similar VCTS test-retest reliabilities averaging 0.78 across five spatial frequencies from children have been reported by Leguire.[3.28]

The relevance of gratings and letters to clinical and functional vision, and their sensitivity and specificity to visual mechanisms, can be compared simply by plot-

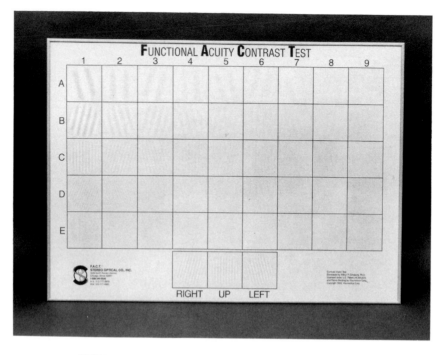

FIGURE 3.11. The Functional Acuity Contrast Test (FACT).

ting the grating and letter spatial frequency (and size) and contrast sensitivity on the same graph. By using the general relationship of threshold letter identification requiring about 2 cycles per letter, the contrast sensitivity of gratings and letters can be compared, as shown in Fig. 3.12.

The relationship between spatial frequency and Snellen acuity of letters is shown along the abscissa of Fig. 3.12. Snellen acuity from 100 to 10 is shown above the spatial frequency notation of 1.5 to 48 cpd. The Snellen acuity value is simply the numerator of the usual "20/xx" notation. The ordinate shows the contrast threshold and contrast sensitivity for both gratings and letters. The graph thus compares the average contrast sensitivities of a sine-wave grating chart (the VCTS), a single size low contrast letter chart (the Pelli-Robson chart), a multi-size low contrast letter chart system (the Regan charts), and a curve of the identification contrast threshold for a Snellen letter chart.

The letter identification contrast threshold shows the upper limit of sensitivity that can be obtained using low contrast letters. The sensitivity of the letters fails to reach the sensitivity of the gratings. Note that each contrast test measures a different range of spatial frequencies and contrast levels.

The VCTS tests contrast levels ranging from 0.3 to 0.0038 over a range of 1.5, 3, 6, 12, and 18 cpd. The Pelli-Robson chart tests only one size letter, 20/360 Snellen or 0.72 cpd (at the suggested test distance) over a contrast range of 0.91 to 0.005. The Regan charts test four levels of letter contrast, 0.96, 0.5, 0.25 and

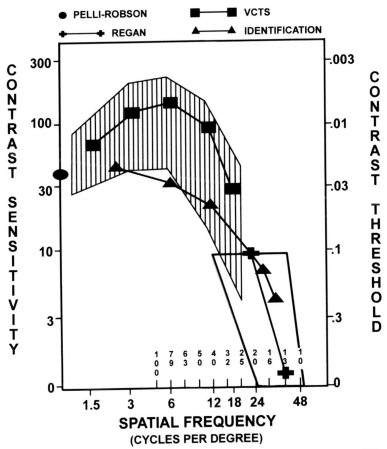

FIGURE 3.12. This chart shows the relationship between spatial frequency and Snellen acuity of letters.

0.11 over a Snellen acuity range of 20/200 to 20/10, or about 5 to 48 cpd. Note that the grating chart test can test almost 30 times more contrast sensitivity at the peak (6 cpd) than can the Regan chart. The grating chart at 18 cpd is about three times more sensitive to contrast than the letter charts.

The three contrast tests measure three different ranges of spatial frequency or size, and three ranges of contrast, as shown by the average population norms of each test in Fig. 3.12. The Pelli-Robson chart tests very large sizes (low spatial frequencies)—over four to eight times larger than the region of peak sensitivity at 3 to 6 cpd. The VCTS (SWCT and FACT) tests a similar range of spatial frequencies around the peak spatial frequencies. The Regan charts test spatial frequencies from the peak and higher, but with less sensitivity. The Regan normal population space (open trapezoid) is generally a continuation of the VCTS normal population space (shaded area), which illustrates the usefulness of this approach.

It may be difficult for the lay person to translate these test spaces into everyday visual experience even when these three tests are graphed on the same spatial frequency and contrast test space. A pictorial analysis can be used to show the functional significance of these test regions. By spatially filtering the street scene in Fig. 3.13(a) using the full spatial frequency ranges of the three tests shown in Fig. 3.12, the resulting filtered images allow visual comparison to the original image of the size and contrast information being tested. The results are shown in the VCTS/SWCT/FACT, Pelli-Robson, and Regan images of Fig. 3.13(b), (c) and (d). It is difficult to know what range of spatial frequencies the Pelli-Robson chart tests because letter identification depends on the relevant letter spatial frequencies reaching threshold. These can vary considerably from one person to another because of individual differences in contrast sensitivity. A generous range of 0.0 to 3 cpd was used for the Pelli-Robson filtered image.

The Pelli-Robson chart tests a size too large to be functionally relevant to this scene (Fig. 3.13(b)). This chart may be useful in predicting the threshold visibility of large objects in the fog, but is not useful in determining the presence of small objects such as the little girl in the street. The VCTS tests a size and contrast range relevant to the complete scene information, especially the little girl and other important scene detail (Fig. 3.13(c)). The Regan letter charts test the contrast and size range most relevant to the sharp edges of the scene. This test reveals little of the image quality and the little girl (Fig. 3.13(d)). The grating chart measures the information most relevant to evaluating the ability to clearly view this scene.

Traveling at 30 mph, under normal weather and visibility conditions on a dry road, the critical stopping distance is 158 ft from the little girl. A driver would have 2.5 seconds to see and avoid hitting the girl. Sine-wave grating technology provides relevant information about whether the driver has sufficient contrast sensitivity to see the girl in time to stop safely.

The important differences between the ability of sine-wave gratings and low contrast letter acuity to measure contrast sensitivity in a clinically meaningful manner is exemplified in a recent paper on early cataracts by Adamsons, et al.[3.29]

The sine-wave grating test reveals statistically significant losses from different types of early cataract at the higher spatial frequencies, 12 to 18 cpd, as shown from other previous research. However, the Pelli-Robson low contrast letter acuity test shows similar contrast from cataract to clear lenses with the exception of posterior subcapsular cataracts. In addition, the Pelli-Robson chart results reveal that cortical and nuclear cataracts have *higher* contrast sensitivity than the clear lens. Clearly, the Pelli-Robson chart provides questionable contrast measures of early cataract and can underestimate their contrast loss.

Although illustrations in the Adamsons, et al. paper imply similar contrast sensitivity measurements between letters and gratings, Fig. 3.12 demonstrates why the grating chart, but not the Pelli-Robson chart, is able to detect contrast loss due to an early cataract. Early cataract causes contrast loss primarily at the higher spatial frequencies. Gratings measuring contrast loss at the higher spatial frequencies detect that loss. Because the Pelli-Robson chart measures very low spatial frequencies, 0.72 cpd (16.7 times larger than the significant 12 cpd grating

Contrast Test Image Space

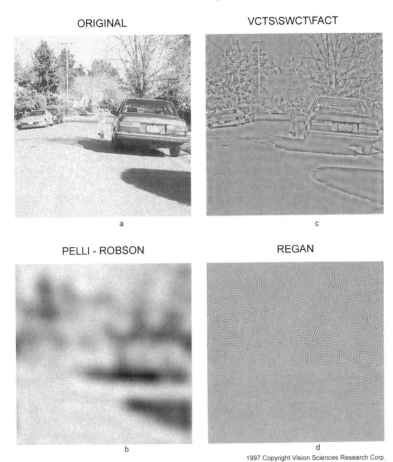

FIGURE 3.13(a). Spatially filtered street scene using full spatial frequency ranges of the three tests shown in Fig. 3.12. (b) The Pelli-Robson chart is testing a size too large to be relevant to this scene. (c) The VCTS tests a size and contrast range relevant to the complete scene information, especially the little girl. (d) The Regan chart tests the contrast and size range most relevant to the sharp edges of the scene, which reveal little of the image quality and the little girl.

contrast losses reported by Adamsons, et al.), detection of cataracts in the early stages is not possible, as demonstated by Adamsons, et al.

A commonly cited advantage of the Pelli-Robson chart is that it has higher reliability compared with grating charts. However, Adamsons, et al. show that this advantage is questionable because the Pelli-Robson chart limits contrast loss measurement to low spatial frequencies and measures higher contrast loss from the clear lens when compared with cortical and nuclear cataract.

The opinion that grating charts have poor reliability has been voiced by Rubin.[3.30] However, Rubin used only one grating chart in his study rather than the three different charts available. The use of three charts can provide repeated measures and allow the determination of means and standard deviations, a more statistically powerful measure then the Pelli-Robson two-of-three-letters-correct rule.[3.31] Others have found the grating charts to be highly reliable, including Leguire, et al. in their research on amblyopia in children.[3.28]

Ravalico, et al. compared the grating and Pelli-Robson charts to visually evoked potential (VEP) measurements, a more objective measure of visual function.[3.32] They could not replicate Rubin's results of low reliability of the grating chart. They found that grating chart measurements were repeatable and, unlike the Pelli-Robson letter chart, the grating chart contrast measures correlated that of the VEP.

Pelli, Rubin, and Legge suggest that the Pelli-Robson chart measures the peak of the grating contrast sensitivity function, leading to the notion that a contrast sensitivity function could be obtained by creating a parabolic curve from the peak measure and a standard acuity measure.[3.33] Recently, Rohaly and Owsley demonstrated why that approach is wrong.[3.34] The Pelli-Robson letter chart cannot predict peak contrast sensitivity, and a contrast sensitivity curve cannot be described by a single parametric curve. The inability of the Pelli-Robson chart to measure peak contrast sensitivity for use as one of two free parameters to create a contrast sensitivity function is easily explained by considering the significant variability in contrast sensitivity functions of normal observers.

The inability of any large letter or edge contrast test to measure contrast loss at important peak and higher spatial frequencies should warn against its use for testing functional contrast related to optical devices, drugs, or clinical issues.

3.7.3   Test Modalities for Contrast Sensitivity

Contrast sensitivity can be measured using chart, lightbox, view-in tester, and computer/video systems. Sine-wave test systems will be discussed because of the inherent limitations and unknown clinical and functional utility of other contrast test targets such as letters, as previously discussed. This does not mean that other test targets may not be useful for contrast testing; however, there is no evidence that other test targets provide the breadth of contrast sensitivity information obtained with gratings.

The VCTS, SWCT and FACT charts test both near and far contrast sensitivity and use an accompanying light meter to standardize test light levels. Using defined light levels for test charts standardizes the contrast sensitivity, an important consideration not achieved with some light boxes. The test charts have the advantages of low cost, simplicity, rapid results and, for the near tests, portability and convenience. Disadvantages are the need for standardized room light levels and the need for multiple charts for repeated measures.

The CVS-1000 (VectorVision, Inc.) uses gratings in a "standardized" light box. Unfortunately, the CVS-1000, by standardizing the background light to varying light levels of the room, does not standardize the contrast. If the CVS-1000 is used in different lighting environments, the contrast of the gratings changes. Therefore, the room light level still must be standardized in order to standardize the CVS-1000.

An additional problem is that the variable white background of the CVS-1000 light box can create a source of variable glare. As room luminance decreases, the white background of the grating patches increases relative to the dark gray grating patches. Although the light reaching the eye from the white background remains constant, there is considerable difference in contrast between the white background and the grating patch. This creates a glare source that is readily noticeable in afterimages.

Sine-wave grating contrast sensitivity is available in two view-in test systems: Multivision Contrast Tester (MCT) (Vistech Consultants, Inc.) and the OPTEC series (Stereo Optical Co.). The MCT has VCTS gratings under variably-controlled luminance with glare. The OPTECs have SWCT and FACT gratings under controlled luminance conditions with glare option.

The main advantages of the view-in testers are that they require only a small test space and eliminate the need to control room light levels. Some of the OPTEC systems have other vision tests such as color and stereopsis, so the view-in systems offer considerable testing power. Disadvantages are the increased expense over that of charts if only contrast sensitivity testing is desired.

Computer/video grating test, such as the Optronix, BVAT, VisioWorks and VSG2/2. offer the highest degree of test flexibility in terms of test technique, test target configuration, and spatial frequencies. However, these systems are relatively costly and complex and require careful consideration and monitoring of contrast calibration of the video display.

3.7.4   Interpreting Contrast Sensitivity: EyeView

Testing vision with sine-wave gratings results in a contrast sensitivity curve as shown in Fig. 3.11. Interpreting the different possible shapes that this curve can acquire comes with experience. Still it can be difficult to translate these shapes into meaningful descriptions to others. EyeView™ (Visumetrics Corp.) is a tool that helps explain the curve in a simple, straightforward manner.

EyeView is an image processing software program that can create either newsprint or street scene pictures modified by either acuity of sine-wave grating contrast sensitivity data. By comparing pictures processed by EyeView to age-matched norms, acuity to contrast sensitivity, or pre/post treatment, the implications for what the patient can see becomes apparent. Eyeview allows others to "see" through the eyes of the patient.

A functionally important example of EyeView-processed pictures is that of a person with histoplasmosis. This person ran into an overturned truck on a major highway on a rainy night. Although tested at 20/25 acuity, her contrast sensitivity, compared with an age-matched normal shown in Fig. 3.14, was severely depressed at the low and especially at the middle spatial frequencies. Her visual mechanisms

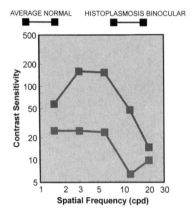

1997 Copyright Vision Sciences Research Corp.

FIGURE 3.14. A street scene is shown with average/normal vision, 20/25 vision and histoplasmosis. The chart graphs contrast sensitivity and spatial frequency for average/normal vision and histoplasmosis binocular vision.

for high spatial frequencies were intact, resulting in the maintenance of good visual acuity.

EyeView pictures of this individual in Fig. 3.14 show the severe consequences of the loss of contrast sensitivity. While the 20/25 acuity suggests normal visual capability, the contrast sensitivity curve shows otherwise. Note the similarity in sharpness of detail of the high contrast picture to that of the 20/25 picture. She has not lost the ability to see high contrast detail as indicated by the 20/25 acuity; her functional vision problem is the loss of the contrast and other picture information. While the 20/25 picture could tolerate a large reduction in contrast before it would look similar to the low contrast picture, a similar reduction in contrast of the low contrast picture would render it virtually featureless. Thus, the 20/25 picture suggests that she has a large contrast reserve while the low contrast sensitivity shows otherwise. Although it can never be proved that her poor vision was the sole cause of the accident, clearly she would be visually handicapped under the poor contrast conditions existing on a rural highway on a rainy night.

In summary, differences between the contrast and spatial frequency and size of any contrast test are easily examined from graphs of the average and normal population test space on the same graph depicting contrast threshold and spatial frequency and size. This graphical representation should become the standard by which to compare contrast tests. The filtered-image technique then can be employed to demonstrate the relative importance of different test spaces to functional vision: the visibility of real-world scenes and objects. Different test instruments for contrast sensitivity testing have different advantages and disadvatages. Finally, EyeView pictorial analysis allows one to begin to "see" the world through the eyes of the patient.

3.7.5   Glare Testing

In most cases an effective contrast sensitivity test without glare can measure vision loss due to cataracts. It sometimes is important, however, to document vision losses cited in glare-related patient complaints that may not be obvious using nonglare contrast sensitivity tests.

Glare is the reduction in image contrast caused by light scattering within the eye.[3.35] A cataract increases the debilitating effects of glare by increasing light scatter. Consequently, glare can seriously affect an individual's (especially a cataract patient's) performance of everyday visual tasks. The ability to see objects in the presence of glare, and the time required visually to recover from glare, rank (after contrast discrimination) as the second and third most important visual factors in night driving.[3.25]

There are many ways to interpret the disabling effects of glare in visual tasks. For example, various studies have used the methods of lowest perceptual contrast brightness,[3.36] comfort levels,[3.37] glare disability acuity scores,[3.38] and light scatter factors.[3.39]

One important study investigating individual differences in glare testing used the measure of subjective comfort provided by an illuminated border.[3.40,3.41] Persons over 50 years of age could tolerate only half the glare acceptable to 25-year-olds. Only 28% of the individual differences could be accounted for with that single measure, however, whereas 55% of the results could be explained by individual differences in test performance.

These results suggest that it is not sufficient to create a measure of glare disability. In addition, individual differences in sensitivity to glare must be considered and any sensitive measure of glare should be expressed as a ratio with and without glare. Although many different measures may be useful in characterizing certain aspects of the effects of glare on vision, a direct measure of glare disability employing sine-wave gratings would be especially sensitive and relevant to the documentation of functional vision loss due to cataracts. Examples of a normal individual's and a cataract patient's contrast sensitivity functions with and without glare are shown in Fig. 3.15. Even though both patients have 20/20 acuity, there

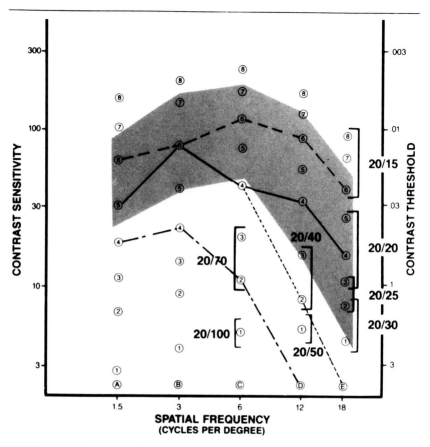

FIGURE 3.15. Comparison of the effects of glare on the vision of two patients with good acuity. One patient is normal and the other has a cataract.

are significant differences in their contrast sensitivity functions, both with and without glare.

3.7.6   Applications to Imaging

The effectiveness of a system for displaying visual images can also be measured using contrast sensitivity.[3.42] After completing the contrast sensitivity test, the subject's contrast sensitivity is again measured with the image projected through the windshield or on a display or video screen. Decreased contrast sensitivity from the first to the second test indicates a display that limits the observer's vision. This knowledge is important in display analysis because it provides feedback on the accuracy and reliability of the image received under specific viewing conditions. In this way, contrast sensitivity permits evaluation of the quality of the display in a quick, simple, inexpensive, and accurate manner.

Both the contrast sensitivity of the observer and the effect of the image display system are important in image detection and recognition. For example, an individual with low contrast sensitivity may be unable to make some judgements with even the best imaging system. Conversely, an individual with normal or high contrast sensitivity may be unable to discriminate fine detail on a less than optimum imaging system.

3.8   Conclusions

This chapter provides a brief overview of the visual system's filtering processes and functions, along with an analysis of how these properties help to explain the visual information used by an observer. Quantification of the human observer's visual filtering characteristics in terms of contrast sensitivity are discussed, along with the importance of the contrast sensitivity function. A new method of measuring contrast sensitivity, using the contrast sensitivity test system (CSTS), provides a quick, simple, inexpensive, and accurate way to measure contrast sensitivity and to evaluate the transmission of information through the observer's visual system. Further research with this system will be useful in testing and improving imaging systems as well as for evaluating the abilities of human observers.

3.9   References

[3.1] Schade O.H. Optical and photoelectric analog of the eye. *J. Opt. Soc. Am.* 1956; 46: 721–739.

[3.2] Campbell F.W. The transmission of spatial information through the visual system. In: Schmitt F.O., Warden F.G., eds. *The Neurosciences Third Study Program.* Cambridge: MIT; 1974: 90–103.

[3.3] McNelis J.F., Guth S.K. Visual performance: Further data on complex test objects. *Illum. Eng.* 1969; 64: 99–102.

[3.4] Ginsburg A.P. Sine-wave gratings are more visually sensitive than disks or letters. *J. Opt. Soc. Am.* 1984; 1(12): 1301.

[3.5] Proenza L., Enoch J., Jampolsky A., eds. *Clinical Applications of Visual Psychophysics.* New York: Cambridge University Press; 1981: 70–106.

[3.6] Ginsburg A.P., Coggins J. Texture analysis based on filtering properties of the human visual system. *Proc. IEEE Int. Conf. Cybern Soc.* 1981; 112–117.

[3.7] Howland B., Ginsburg A.P., Campbell F. High-pass spatial frequency letters as clinical optotypes. *Vision. Res.* 1978; 9: 1063–1064.

[3.8]   Ginsburg A.P. Specifying relevant spatial information for image evalua-
        tion and display design: An explanation of how we see certain objects.
        Proceedings of the Society for Information Displays 1980; 21(3).

[3.9]   Ferris F.L. III, Kassoff A., Bresnick G.H. et al. New visual acuity charts
        for clinical research. *Am. J. Ophthal.* 1982; 94: 91–96.

[3.10] Ginsburg A.P. Need for standard glare, contrast sensitivity tests. *Ocular
        Surgery News* 1988; 6(6): 25–32.

[3.11] Ginsburg A.P. A new contrast sensitivity vision test chart. *Am. J. Opt.
        Physiol. Opt.* 1984; 61(6): 403–407.

[3.12] Corwin T.R., Richman J.E. Three clinical tests of the spatial contrast sen-
        sitivity function: A comparison. *Am. J. Opt. Physiol. Opt.* 1986; 63(6):
        413–418.

[3.13] Ginsburg A.P. The visualization of diagnostic images. *Radiographics* 1987;
        7: 1251–1260.

[3.14] Committee on Vision, National Research Council. *Emergent Techniques
        for Assessment of Visual Performance.* Washington: National Academy;
        1985.

[3.15] Woo G.C., Bohnsack H. Comparison of the distance and near Vistech
        vision contrast test systems (VCTS). *Can. J. Opt.* 1986; 48(1): 12–15.

[3.16] Ginsburg A.P. The evaluation of contact lenses and refractive surgery using
        contrast sensitivity. In: Dabezies O.H., ed. *Contact Lenses: The CLAO
        Guide to Basic Science and Clinical Practice* (update 2). Orlando: Grune
        and Stratton; 1987: 56.1–56.17.

[3.17] Hess R., Woo G.C. Vision through cataracts. *Invest. Ophthal. Vis. Sci.*
        1978; 17: 428–435.

[3.18] Ginsburg A.P., Todesco J. Evaluation of functional vision of cataract and
        YAG posterior capsulatomy patients using Vistech contrast sensitivity
        chart. *Invest. Ophthal. Vis. Sci.* 1986; 27(3) Suppl: 107.

[3.19] Ginsburg A.P. Clinical findings from a new contrast sensitivity test chart.
        In: Fiorentini A., Guyton D.L., Siegel I.M., eds. *Advances in Diagnostic
        Visual Optics.* Berlin: Springer; 1987: 132–140.

[3.20] Ginsburg A.P. *Visual Information Processing Based on Spatial Filters
        Constrained by Biological Data* (AFARML report). Cambridge: University
        of Cambridge; 1978: 78–129.

[3.21] Ginsburg A.P., Evans D., Sekuler R., Harp S.A. Contrast sensitivity predicts
        pilots' performance in aircraft simulators. *Am. J. Opt. Physiol. Opt.* 1982;
        59(1): 105–109.

[3.22] Ginsburg A.P., Easterly J., Evans D. Contrast sensitivity predicts target detection field performance of pilots. *Proc. Hum. Factors Soc.* 1980; 1: 269–273.

[3.23] Evans D., Ginsburg A. Contrast sensitivity predicts age-related differences in highway sign discriminability. *Hum. Factors* 1985; 27: 637–642.

[3.24] Thorn, F. Effects of dioptric blur on the Vistech contrast sensitivity test. *Opt. Vis. Sci.* 1990; 57: 8–12.

[3.25] Ginsburg, A.P. Spatial frequency and contrast sensitivity test chart. Washington: U.S. Patent office, U.S. Patent No. 4,365,873, 1982.

[3.26] Regan D., Neima D. Low-contrast letter charts as a test of visual functions. *Ophthalmology* 1983; 90: 1192–1200.

[3.27] Pelli D.G., Robson J.G., Wilkins A.J. The design of a new letter chart for measuring contrast sensitivity. *Clin. Vision Sci.* 1988; 2: 187–199.

[3.28] Leguire L.E. Do letter charts measure contrast sensitivity. *Clin. Vision Sci.* 1991; 6: 391–400.

[3.29] Adamsons I., Rubin G., Vitale S., Taylor H., Stark W. The effect of early cataracts on glare and contrast sensitivity. *Arch Ophthalmol.* 1992; 110: 1081.

[3.30] Rubin G. Reliability and sensitivity of clinical contrast sensitivity tests, *Clin. Vis. Sci.* 1989; 1: 169–177.

[3.31] Ginsburg A.P. Testing functional vision: important relationships between gratin contrast sensitivity and low-contrast letter acuity tests. Bellingham, WA: Society of Photo-Optical Instrumentation Engineers. SPIE Report No. 2127. 1994: 36–43.

[3.32] Ravalico G., Baccara F., Rinaldi G. Contrast sensitivity in multifocal irtraocular lenses. *J. Catar. Refr. Surg.* 1993; 19: 22–25.

[3.33] Rubin G.S., Legge G.E. The psychophysics of reading. VI. The role of contrast in low vision. *Vision Res.* 1989; 29: 79–91.

[3.34] Rohaly A.M., Owsley C. Modeling the contrast sensitivity functions of older adults. *J. Opt. Soc. Amer.* 1993; 10: 1591–1599.

[3.35] Holliday L.L. The fundamentals of glare and visibility. *J. Opt. Soc. Am.* 1926; 12: 271–332.

[3.36] Lebsensohn J.E. Night driving. *Am. J. Ophthal.* 1949; 32: 860–862.

[3.37] Olsen P.A., Sivak M. Glare from automobile rear view mirrors. *Hum. Factors* 1984; 26: 269–282.

[3.38] Nadler D.J., Jaffe N.S., Clayman A.M. et al. Glare disability in eyes with intraocular lenses. *Am. J. Ophthal.* 1984; 97: 43–47.

[3.39] Paulsson L.E., Sjostrand J. Contrast sensitivity in the presence of a glare light: Theoretical concepts and preliminary clinical studies. *Invest. Ophthal. Vis. Sci.* 1980; 19: 401–406.

[3.40] Bennett C. The demographic variables of discomfort glare. *Lighting Des. Appl.* 1977; January: 22–31.

[3.41] Bennett C. Discomfort glare. Concentrated sources parametric study of angularly small sources. *Illum. Eng.* 1977; 2: 244–246.

[3.42] Ginsburg A.P. Direct performance assessment of HUD display systems using contrast sensitivity. In: *National Aerospace Electronics Conference Mini-Course Notes* (Dayton, Ohio, May 17–19). New York: IEEE; 1983: 33–44.

4

A Multiscale Geometric Model of Human Vision

Bart M. Ter Haar Romeny and Luc Florack

4.1   Introduction

A crucial factor in human perception is that we are able to move around in the three-dimensional world we live in. This induces continuous changes in the structure of the visual world as it is projected onto the retina. Much attention has been paid to the analysis of the "pictorial mode" of perception, the analysis of the retinal images as such. Gibson was one of the pioneers in this field, studying the behavior and perception of aircraft pilots during landing manoeuvres.[4.1] He coined the term "ecological optics" for the study of the natural inflow of information, in which the deformation of structure due to relative movements of objects and observer (or the observer's eyes) is studied.

No matter how we move around, the objects around us are perceived as stable in a three-dimnensional environment. Changes, such as translations, rotations, and deformations are described mathematically by *coordinate transformations* of the space in which the objects are located. So, in one way or another, we are particularly sensitive to features that do not change under translations, rotations, and so on. It is quite natural to use these stable concepts in our perception.

There is a firm mathematical basis for the description of properties invariant under coordinate transformations. The mathematical discipline concerned with these transformations of coordinates and induced changes of objects is called *tensor calculus* and the subdiscipline focusing on invariant objects in particular is called *invariants theory.*

Coordinates are used merely as a tool to take measures of an object and have no meaning by themselves. The freedom in picking new coordinates forces us to be careful about them: We have to be able to describe an object in all allowable coordinates in order to get an objective impression and not be misled by any particular coordinate choice. Once achieved, such a description allows us to study *invariants,* i.e., quantities that do not change, whatever coordinates are used to express them. To begin, then, we have to elaborate on some basic results from tensor calculus and invariants theory.[4.2–4.3] Before we come to that, however, it is even more important to explain what we mean by the *structure* of an object or a scene. This is certainly a nontrivial matter, which requires the introduction of the physical concept of *scale*. The structure of an object or a scene can be operationally defined through observations and to each observable there belongs an intrinsic scale set by the resolution of the measuring device, as well as an outer scale, which is determined by the finite size of the observable. When interested in *scene structure*, all we can do is to make an image and investigate *image structure* as completely as possible, which tells us something about a minor (but hopefully, to us most interesting) part of the scene. Making lots of images on different scales increases our knowledge about the scene, but it should be obvious that the knowledge of scene structure is always inherently incomplete.

Therefore, at each point in the image we need to describe the structure of the image in terms of local spatial derivatives of the brightness. Image structure can be studied at each point separately by looking at how the brightness varies in a small neighborhood around that point. We do this by using a truncated Taylor expansion at each point of the image. To that end we need to know all spatial derivatives up to a certain order.

Furthermore, the outside world has a heirarchically ordered structure; objects exist in many sizes, and larger objects (or structures) are made up of smaller objects. It is much like an atlas: We can look at the coarse structure to have an overview of the structure of the visual field, but we can also "zoom in" and look at finer details if we want to study lower scale structure. We call the size of the smallest, still recognizable features in an image the *inner scale*. Inner scale is a property of the imaging device. The largest scale represented in an image is called *outer scale* and is determined by the size of the field of view. In between the inner and outer scales an image (or retinal image) usually represents a scene on a scale range of a few orders of magnitude. Each object has its own scale; thus, for example, a tree is viewed at a different scale than are its leaves. Each object has its own scale simultaneously by looking at many simultaneous levels of resolution. The totality of these images at different scales, when stacked on top of each other as a one-parameter family, is called a *scale space*. If we understand this internal scale space representation of an image, we can also form an understanding of how we recognize objects in the image, and how the intrinsic hierarchy of nested visual structures is organized. Finally, we may even be able to understand what is learned by our "neural nets,"[4,4] or to perform a kind of "image algebra."

Lowering the resolution of an image, i.e., blurring, means that the visual system in one way or another performs a spatial averaging. This spatial averaging can be done in a variety of ways. We see, however, how certain natural requirements pose strong restrictions on the way the blurring should be done. In fact these requirements lead to a unique blurring strategy based on a Gaussian blurring kernel.

The use of scale space, combined with a description image structure by means of local spatial derivatives, leads to a nice mathematical model, where we have available all the tools of differential geometry,[4,4] which is a natural basis to study invariant behavior under coordinate transformations. Much of the theory presented in this chapter relies heavily on the pioneering work of Koenderink and Van Doorn.

There seems to be neurophysiological and psychophysical evidence that the visual system indeed determines spatial derivatives, probably up to at least the fourth order.[4.5–4.6] This is likely to be done at any location in the visual field. In this chapter we elucidate this model in greater detail. It is based on solid, established mathematics. Our intention in the chapter is twofold: first, to get a well-founded intuitive feeling for the analysis that is likely to be performed by our visual system from the physical and mathematical point of view, and second, to show that the proposed multiscale geometric model of human vision has an important impact on research areas such as computer vision. As a consequence, the perceptual and computer vision research areas gradually show an increasing overlap.[4.7] We try to explain the physical and mathematical aspects in simple terms and we present many clarifying examples. See also. [4.8–4.10]

There is a direct relation with neurophysiological findings in animals and psychophysics that supports this theory. We review these results in some detail in Sec. 4.5.

From everyday experience we know that perception does not significantly change when we put on out sunglasses, or when we distort the grayscale intensity mapping when looking at scenes on a TV screen. This implies that underlying perceptual descriptors exhibit an invariance for scaling of the luminance. Geometrical invariants are defined as those quantities that do not change under a certain group of coordinate transformations. The kind of features that are invariant depends on the type of transformation group. For example, if the group consists of all rotations of a regular polygon, then a concept such as "angle" has an objective meaning, since it does not change after a rotation. Thus the geometrical notion of an "angle" is invariant under the rotation group. Concepts such as "horizontal," on the other hand, are not invariant. Weyl's theorem says that "any invariant must have a geometrical meaning."[4.11] We encounter numerous examples of invariants in due course.

Closely related to the theory of invariants is tensor calculus. We devote some effort to explaining the basics of tensor calculus, in order to provide the background necessary to appreciate the meaning and construction of differential invariants.

We are concerned only with the static perception of structure and mainly deal with a limited set of allowable coordinate tranformations: translations, rotations, and mirroring. Nevertheless, we should like to stress that the model can be generalized to cope with the perception of time-varying structure. We then enter the popular field of *optic flow*. A crucial point is to realize that all structure described in this way is determined locally: at each point we have all the spatial (and, in optic flow theories, also temporal) derivatives, but it is still a largely unsolved problem how to derive global structures. The solution must be in the *deep structure* of scale space (the relations between structural features of the image on different scales). The image structure on a fixed level of scale is denoted as *horizontal structure*. If we blur, more and more structure is lost; none is ever created. This destruction of detail is described by *catastrophes,* which take place when the scale is increased in a causal and topologically well-defined way. For example, an intensity maximum always annihilates with a saddle point. Knowledge of these catastrophes can help us determine hierarchical relations among image structural properties. With the knowledge of local geometrical properties, such as elliptic or hyperbolic (this is a second order property), up to a given order, we classify "patches" of the visual field, i.e., sets of points that are connected by the common value of a certain geometric classifier.

The challenge is to understand the image on all levels of resolution simultaneously and not as an unrelated set of derived images. Then, this theoretical insight may be the basis for a practical implementation in image processing and image understanding, e.g., for image segmentation. This is intrinsically related to across-levels information and is a problem which forms one of the major difficulties in many contemporary computer-visioin applications.

In summary, the idea of modeling the visual system as a "geometry engine," and determining image structure by measuring partial derivatives new approach. The mathematics turn out to be well understood and can be extended from two dimensions to three dimensions. The retinal image is two-dimensional, but many medical image datasets, such as those from computed tomography (CT) or magnetic resonance imaging (MRI) scanners, deliver dozens of consecutive slices, forming three-dimensional dataset. We focus in our theory on two dimensions and sometimes generalize the expressions to three or D dimensions. There has been much concern about taking derivatives to a high degree.[4.12] Noise, always present in images, may rapidly become a limiting factor. It turns out that simply the combination with scale space, where we have a spatial averaging at each point, more than compensates for this effect. Differential geometry is especially attractive and feasible in scale-space in blurred images.

4.2   Scale-Space

To obtain a full understanding of the concept of scale, we look at an image as a representation of a physical scene. The only way to obtain structural information about a physical scene is to extract *observables* with the help of some measuring apparatus. Because of this operational definition of scene structure we inevitably face the problem of *fixing the proper scale,* because observables are always characterized by an intrinsic, finite scale range. The lower bound is determined by the sampling characteristics of the device (*inner scale*), whereas the upper bound is limited by the scope of the field of view (*outer scale*). What happens when scaling the dimensions of an object up and down is a highly nontrivial question. Observables on different scales may lead to completely incomparable descriptions. Nowadays we are familiar with this hidden scale phenomenon of physical quantities. It is implicit in the dimensionality of physical quantities, by means of which we incorporate the inner scale (or several inner scales) in dimensional units.

So by taking an image of some scene (or even simply by looking at it), we already lose all information on scales beyond the image's range. Usually, however, this does not bother us, since we may not be interested in those scales anyway. But if we are, we can alwyas zoom in or out so as to reach the proper scale range we are looking for, e.g., by using a microscope or a telescope.

Because of the scale limitations of observations, we appreciate the fact that, whereas *image analysis* may yield a *complete* description of image structure, *scene analysis* makes sense only within a given scale range and thus is always *imcomplete.* Thus it makes no sense to try to "recover a scene" from an image. In most practical cases, however, a given image does represent a scene over a scale range of typically a few orders of magnitude.

Summarizing the above observations, we reach the following conclusion. *Scene structure can only be operationally defined; any description can only be based on observables, which suffer from the intrinsic artifact that they represent the scene on a finite scale range only. Therefore, inferring (part of) scene structure necessarily requres a multiscale interpretation of image structure.* These observations naturally lead to the necessity for a multiscale description of image structure.

It is our purpose in this chapter to combine the tools that were specifically developed to handle *structural* problems in mathematics, viz., *differential geometry*, with the physical aspect of scaling. We show that this is in fact a harmonious combination that leads to a complete and robust description of the image in a certain well-defined way.

The operator that transforms the image when we increase scale, say from $s \rightarrow s + \delta s$, must be *shift-invariant,* (i.e., not affected by the shift in scale). Therefore, the operation has to be a *convolution* of $L(\mathbf{r}, s)$ with some kernel $G(\mathbf{r}, s \rightarrow s + \delta s)$. A *scale space,* by definition, is a one-dimensional family of images generated in a *continuous* manner from a given image L_0 by means of convolution with a suitable spatial kernel.[4.13-4.16] The family represents that image on various levels of resolution, parameterized by a scale or resolution parameter s.

How do we know what is the best kernel? We can find the expression for the optimal kernel $G(\mathbf{r}, s)$ by considering a number of fundamental requirements:

1. *Spatial Isotropy.* There should be no *a priori* preferred direction in the spatial domain. This means that $G(\mathbf{r}, s) = G(r, s)$.

2. *Separability.* We can make independent length measurements in each of the D independent spatial directions. Therefore, scale applies to each independent direction separately, so we need separability of the kernel in Cartesian coordinates: $G(r, s) = \prod_{i=1}^{D} G^{(i)}(x_i, s)$.

3. *Normalization.* The kernel should be normalized to unit weight.

4. *Kernel Dimension.* Convolving an image with the kernel should lead to a scalar of zero spatial dimension, so the dimension of the operation should be that of an inverse D-dimensional spatial volume. This is so because the convolution operation itself is a spatial integration and hence brings in a spatial volume factor.

5. *Self-Similarity.* There should be no preferred scale. Blursteps between different levels should all be performed with filters that look the same; they

must exhibit *relative size invariance* or *self-similarity*. In other words, the kernel should retain its shape although both size and height may vary in a way consistent with its physical dimension (i.e., without destroying its normalization).

6. *Positivity.* The kernel must be non-negative.

In fact this set of quite natural requirements is largely redundant (but fortunately consistent!). For example, the first three requirements already uniquely suffice to establish the kernel. But let us take the last two requirements as an Ansatz and calculate the actual kernel's shape. Changing scale from $s_b \to s_e$ is linear shift-invariant, so we use the convolution

$$L(s_e) = L(s_b) \otimes G(s_b \to s_e).$$

Going through an intermediate scale $4s_i$ yields

$$L(s_e) = L(s_b) \otimes [G(s_b \to s_i) \otimes G(s_i - s_e)].$$

So

$$g(s_b \to s_i) \otimes g(s_i \to s_e) = g(s_b \to s_e).$$

Because there is no preferred scale, $g(s_2 \to s_1)$ can only be a function of $s_2 - s_1 = s$. We may as well take many intermediate steps:

$$G(s_n - s_1 = G(s_2 - s_1) \otimes G(s_3 - s_2)\dots G(s_n - s_{n-1}). \tag{4.1}$$

Because $G(s_1)$ must have the same form as $G(s_2)$, except for the resolution, G has to be (normalized) Gaussian:

$$G(s) = \frac{1}{\sqrt{4\pi s^D}} \exp\left(-\frac{r \cdot r}{4s}\right). \tag{4.2}$$

Another way of looking at this result is to consider that the repetitive convolution in Eq. (4.1) leads to the Gaussian kernel through the *central limit theorem*. But then each term on the right-hand side has to be a Gaussian as well because of the self-similarity constraint! Note that all these requirements are indeed consistent.

We now see that scale space is the solution of the *isotropic diffusion equation* with the input image L_0 taken as the initial condition $L(s = 0)$ (yet another starting point for the derivation of the convolution filter [4.14]):

$$\begin{cases} \dfrac{\partial L}{\partial s} = \Delta L \quad (r, s) \in \mathbf{R}^D \times (0, \infty) \\ L(s = 0) = L_0 \\ L = \to 0 \text{ as } \|x\| \to \infty \end{cases} \tag{4.3}$$

which states that L should increase (or decrease) with s by an amount proportional to the Laplacian ΔL at level s. This holds in D dimensions. The Laplacian is defined

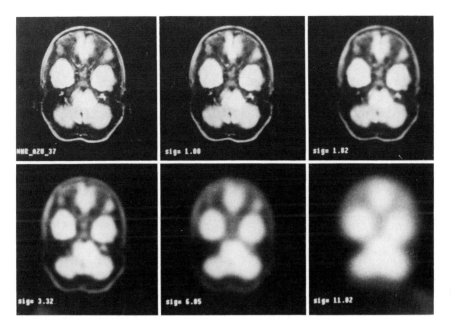

FIGURE 4.1. Several levels of the scale space generated from a magnetic resonance image, presented in order of increasing scale, $\delta = 1.0, 1.8, 3.3, 6.1, 11.0$ pixels, respectively (δ is the kernel width, which is related to the parameter s by $s = \delta/2$).

as the sum of the pure second-order derivatives, so in two dimensions and three dimensions we find:

$$\Delta_{2D}L = \frac{\partial^2 L}{\partial x^2} + \frac{\partial^2 L}{\partial y^2}, \qquad \Delta_{3D}L = \frac{\partial^2 L}{\partial x^2} + \frac{\partial^2 L}{\partial y^2} + \frac{\partial^2 L}{\partial z^2}.$$

The Gaussian kernel is called the *Green's function* of the diffusion equation. In Fig. 4.1, images from the scale space of an MRI scan (sometimes referred to as a *stack*) are shown.

Diffusion in a given point amounts to performing a weighted average with a Gaussian weight factor of increasing size s. A comparison can be made with diffusion as we find in a nonhomogeneously heated isotropic object: the temperature distribution is the analog of the luminance distribution and, while time passes, the temperature distribution gradually becomes more and more homogeneous, leading to an equilibrium state in which the temperature is constant all over the object. Similarly, all structure high in scale space eventually fades out, with a constant luminance distribution equal to the average luminance in the original image.

The Gaussian kernel has a number of nice properties:

1. It is isotropic; thus it is rotationally invariant.

2. It is separable in Cartesian coordinates: if the blurkernel is denoted by $G(x, y; s)$ (in two dimensions for simplicity), we have $G(x, y; s) = G(x; s)G(y; s)$.

3. It has the same shape in the Fourier domain.

4. It is the only kernel that leads to monotonic destruction of detail under consecutive blurring. Gaussian blurring is a causal process, as it should be; no *spurious resolution* is created (in the one-dimensional case $\mathbf{R}^1$ a valid quantification of detail may be the number of extremes; the fact that $\partial L/\partial s$ is proportional to $\Delta L = L''$ (as we see later) implies that this number does not increase). Causality means that "dales" increase in luminance and "hills" decrease; see Fig. 4.2.

5. The Gaussian filtered image $L(s)$ at any level $s > 0$ is a smoothed (even analytical!) member of the family generated by the initial condition L_0, no matter how irregular this may be (within certain weak restrictions). This practically implies that Gaussian blurring is very robust against noise.

To consider the *horizontal structure* on different levels of scale appropriately, we need a closer look at the spatial coordinates. Because self-similarity implies that all Gaussians in Eq. (4.1) must look exactly the same except for scale, we need an appropriate rescaling of the length measures to *natural length units.* Keeping in mind that G is itself some kind of inverse volume, it is clear that it has to scale consistently with the spatial coordinates. And indeed it does. Natural spatial coordinates are defined as the *dimensionless* numbers ξ associated with the spatial coordinates r at scale space level $x > 0$ through:

$$\xi = \frac{\mathbf{r}}{\sqrt{4s}}.$$

Having defined natural length units, it seems rather trivial to remark that we now have a *natural distance measure* for the separation of two points r_1 and r_2 on a given level s:

$$\delta_{1,2} \equiv \|\xi_2 - \xi_1\| = \frac{\|\mathbf{r}_2 - \mathbf{r}_1\|}{\sqrt{rs}}. \tag{4.4}$$

Note the singularity of this measure at the highest (fictitious) resolution $s = 0$. Two distinct points at $s = 0$ are always infinitely far apart; e.g., the points may be on different sides of an intensity discontinuity, thus there can be an arbitrarily large amount of structure in between.

The diffusion equation in natural spatial coordinates becomes

$$\left(\frac{1}{4s}\Delta\xi + \frac{1}{2s}\xi \cdot \Delta_\xi - \frac{\partial}{\partial s} \right) L(\xi, s) = 0 \tag{4.5}$$

which clearly reveals the artifact at $s = 0$. The problem in fact does not reside in this singularity (there is nothing wrong with the mathematics), but in our specification of the initial condition at $s = 0$. Because of the shift invariance with respect to s, this is an indisputable choice from the *mathematical* point of view. We could equally well have specified it at some $s = s_0 > 0$, but there is no mathematical preference for any of these possibilities. We see in a moment that

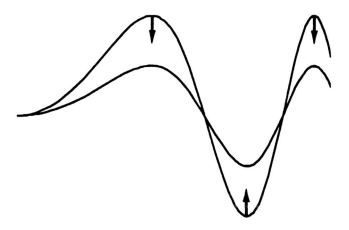

FIGURE 4.2. Causality in blurring with the Gaussian kernel: maxima decrease, minima increase, points where the second derivative is zero do not change.

on *physical* grounds we can pick out a single, natural choice. But first we remove the singularity by taking a scale parameter that corresponds more closely to the physical notion of scale.

Significant changes due to rescaling occur only when we scale the dimensions of an object up or down one or several *orders of magnitude* rather than linearly. Hence it is more natural to reparameterize our scale parameter, thus removing the artificial singularity at $s = 0$. *A natural, linear scale parameter τ is obtained by the following reparameterization of s:*

$$4s = \varepsilon^2 \exp\{2\tau\} \qquad \tau \in (-\infty, +\infty). \tag{4.6}$$

We may invert Eq. (4.6) to obtain

$$\tau = \frac{1}{2} \ln\{4s/\varepsilon^2\} = \ln\{\sqrt{2}\sigma/\varepsilon\}, \tag{4.7}$$

where σ is the standard deviation ("width") of the Gaussian kernel. Note that we are forced to incorporate, on dimensional grounds, a "hidden scale" ε, which carries the dimension of a length! A natural choice for an intrinsic scale inherent in any imaging device is the sampling width of pixel width.

Now we have a dimensionless scale parameter τ that indicates in a continuous manner the order of magnitude of scale relative to ε and that can take on, at least in theory, any real value. (It is interesting to compare this parameterization of scale with the concept of *fractals*. By repetitive application of a certain transformation we can construct mathematical objects that look the same on all orders of scale. These objects exhibit a periodic behavior in τ. They have no inner scale. This they have in common with a physical object, though the latter of course never exhibits this peculiar periodicity property.) If we take ε to be the sampling width, then $\tau = 0$ corresponds to a resolution where the width of the blurring kernel is of the same order of magnitude as the pixel width ε, i.e., the inner scale of the image. This sets a practical lower limit to the kernel widths, at which discretization effects start to contribute to a significant degree. The range $\tau \in (-\infty, 0)$ corresponds to subpixel scales that are not represented in the image and in which all structure has been averaged out. In a continuous scale space, the actual representation of scale is not very important, but when building up a discrete scale space it is most natural to use an *equidistant sampling of τ*, because it is this parameter that, by definition, precisely formalizes the physical notion of scale (it is much as if we zoom in or out at a constant rate of change in scale).

This logical scale reparameterization finally transforms the diffusion equation (4.5) into a representation closest to our physical interpretation in terms of naturally scaled spatial coordinates and linear scale parameter:

$$\left(\frac{1}{2}\Delta_\xi + \xi \cdot \nabla_\xi - \frac{\partial}{\partial \tau}\right) L(\xi, \tau) = 0. \tag{4.8}$$

Because of shift invariance with respect to the scale parameter and because of the absence of small scale structure in the image, it is most natural to specify the initial condition L_0 at $\tau = 0$ instead of $\sigma = 0$.

4.3   Scaled Differential Operators

The diffusion equation (4.3) is linear. This implies that, apart from the Gaussian itself [Eq. (4.2)], *all of its derivatives also satisfy the equation.* In this way we obtain an entire class of *scaled differential operators.* For example, in two dimensions, let $G_{n,m}$ denote the nth and mth order partial derivative of G in x and y, respectively:

$$G_{n,m}(x, y; s) \equiv \frac{\partial^{n+m}}{\partial x^n \partial y^m} G(x, y's) \tag{4.9}$$

or, in natural coordinates and in D dimensions:

$$G_n(\xi; s) \equiv \frac{1}{\sqrt{4s^n}} \frac{\partial^{n_1 + \cdots + n_D}}{\partial \xi_1^{n_1} \cdots \partial \xi_D^{n_d}} G(\xi; s). \tag{4.10}$$

[The *multi=index n* stands for the set of D non-negative integers $(n_1, \ldots, n_D)$.] These Gaussian derivative kernels, sometimes referred to as fuzzy derivatives, play an important role as local neighborhood operators. Their shape in two dimensions is depicted in Fig. 4.3. These kernels, as local neighborhood operators, *are the only sensible way to define derivatives in a scale space!* This is contrary to "pure mathematical derivatives": in mathematics, there is no scaling possibility in a given function (no "physics" in the function). The Gaussian kernel and its derivatives

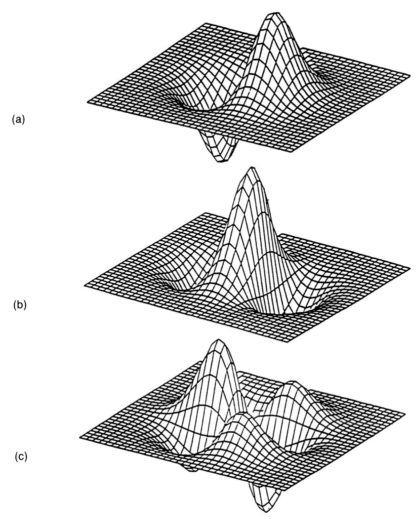

(a)

(b)

(c)

FIGURE 4.3. Some Gaussian derivative profiles. (a) First order in x. (b) Second order in x. (c) Mixed kernel: first order in x, second order in y.

show a remarkable resemblance to depictions of the receptive fields of the human front-end visual system.[4.17–4.18] Consequently, it seems that the retinal image is preprocessed precisely to yield the solutions of the diffusion equation at all levels of scale simultaneously, as well as all of its derivatives up to some order. It is this property in particular that encourages us to use the method of geometry in this scale space induced by diffusion,[4.14] since we know that the human visual system is astonishingly well designed. We are now ready to investigate *image structure* by means of these spatial differential operators.

4.4   Image Structure

Consider a given image $L(x, y; s)$ in two dimensions at a fixed scale, say the original resolution. If we are interested in the geometrical structure of this image in the neighborhood of some fixed point $r = (x, y)$ in $\mathbf{R}^2$ it suffices to consider its Taylor approximation up to a sufficient order N. If we move a very little distance α and β around point (x, y), we can truncate the local Taylor expansion:

$$L(x + \alpha, y + \beta; s) = \sum_{n=0}^{N} \frac{1}{n!} \left(\alpha \frac{\partial}{\partial x} + \beta \frac{\partial}{\partial y} \right)^n L(x, y; s).$$

If we look at local structure, e.g., to the second order, we find

$$\begin{aligned}
L(x + \alpha, y + \beta; s) = L(x, y; s) \ & + L_x(x, y; s)\alpha + L_y(x, y; s)\beta \\
& + \tfrac{1}{2} L_{xx}(x, y; s)\alpha^2 + L_{xy}(x, y; s)\alpha\beta \\
& + \tfrac{1}{2} L_{yy}(x, y; s)\beta^2 + \cdots.
\end{aligned}$$

We denote spatial derivatives as subscripts to L, so, e.g.,

$$L_{xx} = \frac{\partial^2 L}{\partial x^2}.$$

In a realistic model we have to truncate the series to be manageable. It has been proved that the coefficients of the truncated Taylor expansion are the optimal choice for the representation of structure to a certain order. As we see later, there are indications from neurophysiology that the visual system works with derivatives limted to (but including) the fourth order. We define the N-jet of L at $\mathbf{x}$, notation $J^N[L(\mathbf{x}; s)]$, as the set of all functions that share the same Taylor expansion at $\mathbf{x}$ up to order N. Alternatively, we may identify the N-jet of L at $\mathbf{x}$ with the set of derivatives of order less than or equal to N, but only to the extent that the choice of the Cartesian coordinate system is implicit:

$$J^N[L(\mathbf{x}; s)] = \left\{ \frac{\partial^{i+j} L(\mathbf{x}; s)}{\partial x^i \partial y^j} \right\}_{i+j \leq 0 \ldots N}.$$

To infer local structure from this truncated Taylor expression we have to know the coefficients in the expansion, i.e., we need to determine all spatial derivatives up to a given order, the N-jet. The diffusion equation leads to a Gaussian blurred image. We can calculate the N-jet for each blurlevel separately. Note that differentiating a blurred image amounts to filtering the *original* image with a suitable filter, viz., the corresponding *partial derivative of the Gaussian kernel,* since we have

$$\frac{\partial L}{\partial x^i} \otimes G = \frac{\partial G}{\partial x^i} \otimes L$$

and so on.

The scaled partial derivatives of the image facilitate the use of methods from *differential geometry* in a straightforward and elegant way, as we see later. Again we see the intricate relation between derivative operators and scale. It frees us from regularization problems, encountered by many authors.[4.19–4.20] The difference in the differentiation operation in mathematics and physics is best explained by the fact that in physics it is related to a length (in our case, the natural coordinates), which is taken relative to a unit-length measure, which can be of arbitrary scales, but always finite (e.g., Ångstrøms or lightyears).

Several authors have stressed the importance of spatial frequency processing in "bands" in the visual system, with bands allocated to a particular scale. The Gaussian derivative kernels in the Fourier domain indeed do have the shape of band filters, as can be seen in Fig. 4.4.

4.5 Description of the Early Vision System

The fundamental properties of visual perception can be understood by studying the behavioral responses of human observers with different visual tasks (psychophysics) or by studying the neurophysiology of the visual system. Recently our knowledge has increased greatly, particularly due to the studies of Hubel and Wiesel of the macaque monkey visual cortex, for which work they received the Nobel Prize in 1981. Their work is summarized in the highly recommended book by Hubel.[4.22]

The more than 100 million receptors in the retina, the rods and cones, project through an intricate set of retinal layers to about 1 million ganglion cells in the retina whose axons form the optic nerve (see Chap. 1). So there is a considerable convergence of neuronal connections. It appears that on average 100 to 150 receptors, projecting via the bipolar, horizontal, and amacrine cells to one ganglion cell, form a circular area, called a *receptive field* (RF). Receptive fields exist in many sizes; the smallest are found in the central region (the fovea) and diameters increase linearly with eccentricity. Typical diameters of monkey foveal RFs are a few minutes of arc; far out in the periphery they can be one degree or more. Receptive fields overlap substantially (see Fig. 4.5(a)). Studying the shape and structure of RFs gives us insight into brightness and contrast perception. The initial response of the rods and cones that are the primary retinal receptors is approximately logarithmically related to local luminance. Illuminating the retina diffusely over a large area, however, does not produce a response of the retinal ganglion cell because both excitory and inhibitory connections project about evenly on the cell. Most retinal RFs are found to be center-surround, in two classes: "on-center" (with a ring of inhibitory projecting receptors around it) and "off-center" (the reverse) (see Fig. 4.5(b)).

The ganglions in the retina project to the next stage in the visual system, the lateral geniculate nucleus (LGN), in an almost perfect *somatotopic projection*, i.e., neighborhood relations are kept in order. Cells that are neighbors in the retina

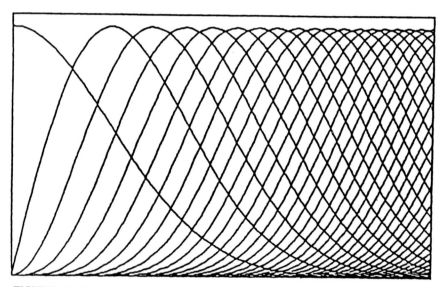

FIGURE 4.4. The normalized spatial frequency spectra of Gaussian derivative filters of different order, showing band-limited properties. The data[4.21] are plotted on linear scales.

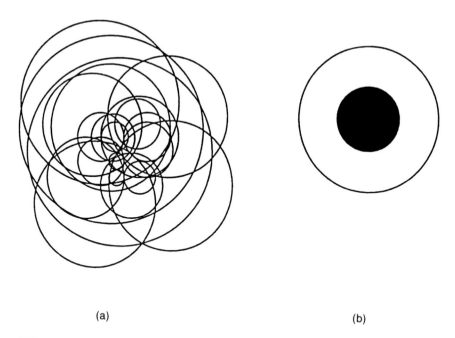

(a)   (b)

FIGURE 4.5. (a) The overlapping structure of receptive fields of many sizes. This suggests that at any location receptive fields of various structures and sizes are present. There is a linear decrease in density of the smaller receptive fields with eccentricity. (b) On-center retinal receptive field.

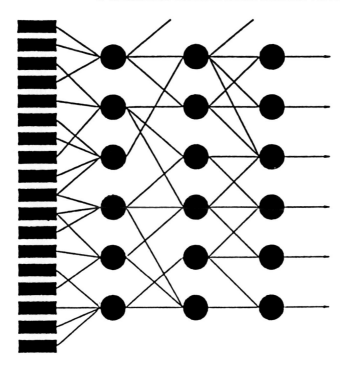

FIGURE 4.6. A schematic representation of the visual pathway with somatotopic, converging projections of the receptive fields from one stage to the next, forming more and more complex structured "compound" receptive fields.

are also neighbors in the LGN and the next few higher centers. A small number of fibers lead off to regulate pupil size regulation and trigger other instinctive responses. In the LGN we again find grouping in receptive fields, now leading to more complex receptive fields in the outer visual field, i.e., with a more complex structure which can be found by stimulation. From the LGN, signals travel further along the geniculo-striate bundle to the visual striate cortex, with its areas $V1$ and $V2$ and beyond. See Fig. 4.6 for a schematic representation. Each layer is topographically perfectly projected and it appears that this scheme is found throughout the central nervous system. There seems to be a strong similarity in our other receptive systems; e.g., the pressure-sensitive Pacini receptors in the skin exhibit a receptive field structure similar to that found in the retina, as do the hair cells in the organ of Corti in the inner ear, etc. There is no mixing of signals from both retinas in the LGN; this binocular interaction occurs later, in the cortical areas.

The human visual cortex is about The human visual cortex is about 2 mm thick, covers an area of 30 cm^2, and contains in the order of 200 million cells. Here "simple cells," "complex cells," and "end-stopped cells" are found, each with their characteristic RF representation on the retina. These are in increasing order more complex than the center-surround cells found in the retina itself. The cortex is

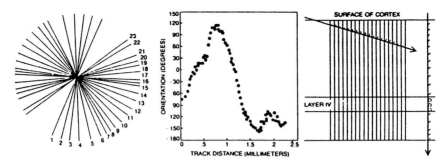

FIGURE 4.7. The strict ordering of orientations of the RFs in the visual field, shown by tracking the electrode parallel to the visual cortex surface at some depth, thereby passing a great number of neighboring cortical columns. After Hubel and Wiesel.[4.23]

organized in so-called *cortical columns*: if we penetrate an electrode perpendicular to the surface, we find many cells with different behavior (expressed in the outer visual field), but at the same location. Most of the RFs in the visual cortex show a more complex response than RFs in the retina. A wide variety of structure is found: *simple cells, complex cells, hypercomplex cells,* and *end-stopped cells.* Simple cells are orientation-sensitive, i.e., they respond only to stimuli orientated in a certain direction and can be figured as RFs with three or two bands ($+ - +$ or $- + -$, or $+ -$ resp $- +$), having the same size (or a little larger) in the outside visual world as retinal RFs have. It is simple to deduce that these cells are responsive to lines or edges oriented in a certain direction. They come in all sizes and directions, so sampling of the world can be complete. Complex cells are orientation-sensitive, but it does not matter where the line-stimulus falls on the RF profile, as is the case with simple cells. End-stopped cells show an RF with many $+$ and $-$ areas.

There is only little indication that RFs of orders higher than four are present.[4.5–4.6] If they are, they are probably distributed much more sparsely. Also intriguing is the fact that the first order Cartesian RF seems to be absent.

There is no (x, y) axis system representation in the visual system. Instead, the neurophysiological studies of the directional sensitivity of the different RFs clearly indicate that all directions are represented in an orderly fashion. If an electrode penetrates the visual cortex along a track horizontal to the surface, thus passing many consecutive cortical columns, a continuous change of direction is found[4.23] (see Fig. 4.7).

When changing from one Cartesian coordinate system to another, the N-jet does not change (it is said to be a *geometrical object*), but of course its coefficients do. The change of bases (e.g., polar to Cartesian), however, imposes well-defined transformation properties on them, expressing the old ones in terms of the new ones. They are said to transform as *tensors* under these orthogonal transformations (CCT: Cartesian coordinate transformations). Koenderink has shown that the representation of the different receptive fields can be represented equally in Cartesian and polar coordinates.[4.24] This leads to a series of basic receptive fields, depicted to the fourth order in Fig. 4.8. The center-surround fields are modeled by

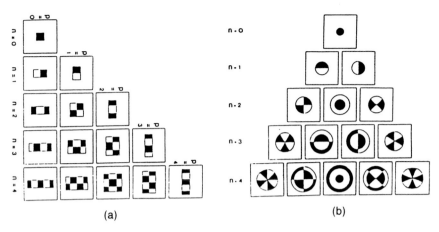

FIGURE 4.8. (a) Receptive field taxonomy in Cartesian coordinates up to the fourth order. The pair (n, p) labels the nth order operator $\delta^n / \delta x^{n-p} \delta y^p$. (b) Receptive field taxonomy in polar coordinates up to the fourth order. Black indicates excitatory projection; white indicates inhibitory projection. After Koenderink and van Doorn.[4.24]

the second order circularly symmetric polar representation, which is equivalent to the Laplacean in the Cartesian representation.

4.6   Differential Invariants

At this stage, we concentrate on an image L viewed on a fixed level of scale; in other words we are interested in the horizontal structure of scale space. We have already introduced the concept of an N-jet extension of the image, which gives us an Nth order approximation in the sense that it represents only those local geometrical features that are completely determined by the first N derivatives at each point. In practice, we hardly ever go further than $n = 2$, since this already gives us a fair amount of freedom in defining all kinds of geometrically meaningful properties, i.e., true image properties.

Once we have calculated the N-jet, we are provided with all partial derivatives of the image up to and including the Nth order. However, one such derivative, say L_x, does not represent any geometrically meaningful property. This is clear when we realize that the choice of the coordinate axes is completely arbitrary. If we restrict ourselves to using an orthonormal basis, then we still have the possibility of rotating a given coordinate frame over any angle we choose. Clearly, such a choice does not have anything to do with the image. But in order to give a mathematical description of image geometry and make actual calculations (e.g., on a computer), we do need coordinates!

To escape from this dilemma, we choose any allowable coordinate system, but at the same time assure ourselves that the result calculated, whatever it may be, is independent of that particular choice. So, although the individual derivatives such as L_x change when we rotate the coordinate system, the functions of interest to us, describing true image properties, should not change their values. This explains why these functions are called "invariants."

The relevant geometrical properties to be discussed later are mathematically described by these *differential invariants*.[4.25] To get a basic understanding of the mathematical discipline concerned with these differential invariants, i.e., *invariants theory*, it is necessary to learn some *tensor calculus* first.[4.2] We show how we can form any function describing a true image property in a systematic way. It shows that we can in fact construct an infinite number of such invariants in each point of the image, but only a few of them are independent. This is to say that, to a given order N, we can build any geometrically meaningful quantity as a function of those (typically very few) independent or *irreducible invariants*.

It is clear that we cannot form an invariant out of a single derivative like L_x. Tensor calculus describes the transformation behavior of quantities like L_x, called *tensor components*. A closed set of tensor components constitutes a *tensor* in the given coordinate system (actually, a tensor has a somewhat more abstract definition, but for our purpose we do not need that level of mathematical rigor). The meaning of the word "closed" in this context is that, after a change of coordinates, each tensor component acquires a new value that can be expressed as some function of the old tensor components. This function depends only on the transformation performed. For example, the partial derivative L_x changes, after a rotation over an angle α, according to the following rule:

$$L_{x'} = \cos \alpha L_x + \sin \alpha L_y.$$

This shows that L_x cannot be the single component of a tensor, we should at least add the component L_y to it. Indeed, this suffices to obtain a two-component tensor $\{L - x, L_y\}$, since the two components do transform in a closed way; for L_y we find

$$L_{y'} = -\sin \alpha L_x + \cos \alpha L_y.$$

The two transformations can be written in matrix form:

$$\begin{pmatrix} L_{x'} \\ L_{y'} \end{pmatrix} = \begin{pmatrix} \cos \alpha & \sin \alpha \\ -\sin \alpha & \cos \alpha \end{pmatrix} \begin{pmatrix} L_x \\ L_y \end{pmatrix}.$$

The relation between the original coordinates x, y and the rotated coordinates x', y' is likewise given in matrix form:

$$\begin{pmatrix} x' \\ y' \end{pmatrix} = \begin{pmatrix} \cos \alpha & \sin \alpha \\ -\sin \alpha & \cos \alpha \end{pmatrix} \begin{pmatrix} L_x \\ L_y \end{pmatrix}.$$

From this it is also clear that the coordinates $\{x, y\}$, taken as a pair, constitute a tensor.

Before we proceed, it is convenient to introduce a more compact notation for both the tensor and the transformation matrix. This is the tensor *index notation* with the following conventions: L_i stands for any component of the one-index tensor L, i.e., $i = 1 \ldots D$ if a *free index* (D is the dimension of the image; most frequently we use $D = 2$, but medical images often form three-dimensional datasets, so we then have $D = 3$). Let R_{ij} stand for the (i, j)th element of the transformation matrix; then the above-mentioned transformation can be conveniently written in condensed form as

$$L'_i = R_{ij} L_j.$$

In this notation, the so=called *Einstein summation convention* is in effect, which states that we should sum over indices that occur more than once in each term, i.e., j. So, by this convention, the above is equivalent to

$$L'_i = \sum_{j=1}^{D} R_{ij} L_j.$$

This is particularly useful when more than one index is repeated. Such a summation over repeated indices is usually called a *contraction of indices*. We also often encounter the term *trace* to describe it.

Because L_i has only one index, it is called a 1-tensor, but we are more familiar with the term *vector*. But we can also consider 2-tensors, 3-tensors, etc. and, in general, *n-tensors*. An example of a 2-tensor is $L_i L_j$, which evidently transforms as

$$L'_i L'_j = R_{ik} R_{jl} L_k L - l.$$

A product of several tensors such as $L_i L_j$ is called a *tensor product*. A less trivial example of a 2-tensor is obtained by considering all second-order partial derivatives. Its transformation is the same as in the above example:

$$L'_{ij} = R_{ik} R_{ji} L_{kl}.$$

It has exactly three *essential components*, i.e., L_{xx}, $L_{xy} = L_{yx}$, L_{yy}. This means that, because of the symmetry of the tensor, we cannot choose all its components independently, as this example shows.

By now it is obvious that all partial derivatives of a given order, say n, form the components of an n-tensor. For each of its free indices its transformation law contains a transformation matrix with one free and one contracted index:

$$L'_{i_1 \ldots i_n} = R_{i_1 j_1} \ldots R_{i_n j_n} L_{j_1 \ldots j_n}.$$

These derivative tensors share the extra property of being *symmetric*, i.e., we can freely interchange indices without any effect, e.g., $L_{ij} = L_{ji}$.

Also of great importance are the following two simple tensors: the symmetric *Kronecker tensor* δ_{ij}, which is always a 2-tensor, and the *Lévi-Civita tensor* $\varepsilon_{i_1 \ldots i_D}$, which is a D-tensor in D dimensions. These are very special because they have *constant components,* independent of the choice of the coordinate axes. They are defined as follows:

$$\delta_{ij} = \left\{ \begin{array}{ll} 1 & \text{if } i = j \\ 0 & \text{otherwise} \end{array} \right\}$$

and

$$
\varepsilon_{i_1 \ldots i_D} = \begin{cases} 1 & \text{if } (i_1 \ldots i_d) \text{ is an even permutation of } 1 \cdots D \\ -1 & \text{if } (i_1 \ldots i_d) \text{ is an odd permutation of } 1 \cdots D \\ 0 & \text{otherwise} \end{cases}
$$

The Lévi-Civita tensor is *antisymmetric*; it changes sign upon an interchange of indices. To show that $\varepsilon_{i_1 \ldots i_D}$ is indeed a tensor, consider the transformation of its single essential component $\varepsilon_{1 \ldots D} = 1$:

$$
\varepsilon'_{1 \ldots D} \equiv R_{1 i_1} \ldots R_{D i_D} \varepsilon_{i_1 \ldots i_D} = \det R = 1.
$$

It can be shown that (anti)symmetry properties are conserved after a transformation. This implies that all components $\varepsilon_{i_1 \ldots i_D}$ are invariant and thus that the Lévi-Civita tensor is well defined. Clearly it is crucial here that we restrict ourselves to rotations for which $\det R = 1$ and exclude reflections ($\det R = -1$), for then $\varepsilon_{i_1 \ldots i_D}$ would not be well defined as a tensor. In the formal mathematical language, the group of all rotations in D dimensions is called $SO(D)$, the *special orthogonal group*, while the *full orthogonal group* of both rotations and reflections is called $O(D)$. When we include reflections into the transformation group (or take an even larger group), the significance of the ε "tensor" still remains as a so-called *relative* or *pseudotensor*. Its transformation law is then slightly modified so as to render its components invariant again:

$$
\varepsilon'_{i_1 \ldots i_D} \equiv (\det R)^{-1} R_{i_1 j_1} \ldots R_{i_D j_D} \varepsilon_{j_1 \ldots j_D} = 1.
$$

The definition of the Kronecker tensor is consistent and holds in any coordinate system. We leave out such issues as covariant and contravariant tensors, etc., that can be found in any competent book on tensor calculus.[4.2] Since we are considering only orthogonal transformations of the coordinate frame, these issues are really irrelevant, since the two concepts then happen to coincide.

Now that we fully understand the transformation behavior of the derivatives under any allowable transformation, we can try to combine them into invariant combinations. This is in fact very easy. Given a set of tensors (in our case these are the image derivatives and the two constant tensors), the way to form an invariant is by means of *full contractions and alternations of indices* in a tensor product. Of course, functions of several invariants are themselves invariants. An alteration of D tensor indices is defined as a full contraction of these indices onto the D indices of the Lévi-Civita tensor, as, e.g., in

$$
\varepsilon_{ij} L_i L_k L_{jk} = (L_x^2 - L_y^2) L_{xy} + L_x L_y (L_{yy} - L_{xx}).
$$

A good way to gain familiarity with these contractions is to verify the correctness of the right-hand side. As a second exercise, show that $L_i \varepsilon_{ij} L_j = 0$ (try to argue this without explicitly writing out the terms in the summation, using symmetry considerations only). The geometrical meaning of this identity is that L_i, i.e., the image gradient vector (∇L in index-free form), always points in the direction perpendicular to the isophote (line of constant intensity), along which the vector $\varepsilon_{ij} L_j$

lies. This vector [let us denote it in index-free notation by $\nabla_{\perp}L = (L_y, -L_x)$] has the same magnitude as the gradient, but is rotated relative to this clockwise by 90 degrees.

It is clear that any property that is invariant under a coordinate transformation must be connected to the image itself and therefore can be given a geometric interpretation and vice versa. Every local image property can be expressed through a differential invariant or invariant relation. This shows that there is an intimate relation between invariants theory and differential geometry. Although this method of forming an invariant out of partial derivatives is all that is involved, it is often a nontrivial problem to assign the correct interpretation to the invariant, especially when higher order derivatives are involved or complex tensor products are considered.

The most trivial 0-jet example of a differential invariant is L, the local image intensity. Another simple example, taken from the 1-jet approximation, is

$$\sqrt{L_i L - i} = \|\nabla L\|,$$

the image gradient magnitude. It represents the maximum slope of the image intensity "landscape" in each point. It is most pronounced on edges, because these correspond to points where there is a strong change in intensity over a relatively short distance. A third, often misinterpreted, 2-jet example is the familiar Laplacean $L_u = \nabla L$. We return to this later on. Figure 4.9 shows these three examples as calculated for a simple test image.

In the tensor notation the invariance of a function manifests itself through a full contraction of indices in the tensor products that make up the function. For this reason we speak of *manifest invariance* when using this notation. When we write out the terms in a contraction in an arbitrary (x, y) system, we lose this manifest invariance, although of course the function is still an invariant.

There is another way to form differential invariants in a straightforward and systematic way, viz., by using *directional derivatives* along geometrically meaningful, mutually perpendicular directions. There are several ways to set up a coordinate frame with geometrical significance. One way is to require, at each point of the image separately, one axis to coincide with the image gradient direction, i.e., the direction of "steepest ascent." The other axis is then automatically directed tangentially along the isophote. Figure 4.10 shows one such coordinate system set up at one point of the image.

Because of the rotational freedom that we have in choosing an orthonormal basis, this kind of requirement is always allowed as long as the image gradient does not vanish and uniquely establishes the desired coordinate frame. We call such a requirement a *gauge condition*, and the resulting coordinates *gauge coordinates*. We should always check whether a gauge condition is *admissible*, i.e., realizable through a suitable transformation provided by the transformation group at hand.

To clarify the problem of gauge fixing, let us set up the special frame alluded to in the previous paragraph. Suppose that we are looking at a given point in the image. We are interested in the local image structure at a given point in the image.

FIGURE 4.9. Some simple examples of invariants calculated for a test image. Top left: the original image. Top right: a blurred version of the original image. Bottom left: the magnitude of the (blurred) image gradient or "edginess." Bottom right: the Laplacean.

We are interested in the local image structure near that particular point. To that end it is convenient to use any local coordinate system in which the given point is the origin ($x = 0$, $y = 0$). To fix the gauge we may rotate the (x, y) system in such a way that in the new coordinate system, say (v, w), the w-axis is directed along the gradient: $L_w = \|\nabla L\|$ and, consequently, $L_v = 0$ (here and subsequently it is implicitly understood that all invariants, in this case L_2 and L_v, are to be taken at the point of interest, i.e., at the origin of the local frame). The relation between the arbitrary coordinates (x, y) and the new, invariant gauge coordinates (v, w), is then given by the following projections:

$$v = \frac{\varepsilon_{kj} L_j x_k}{\sqrt{L_i L_i}} = \frac{\nabla_\perp L \cdot x}{\|\nabla_\perp L\|}$$

i.e., the projection of $\mathbf{x}$ onto the isophote directions $\nabla_\perp L / \|\nabla_\perp L\|$) and

$$w = \frac{\delta_{kj} L_j x_k}{\sqrt{L_i L_i}} = \frac{\nabla L \cdot \mathbf{x}}{\|\nabla L\|}$$

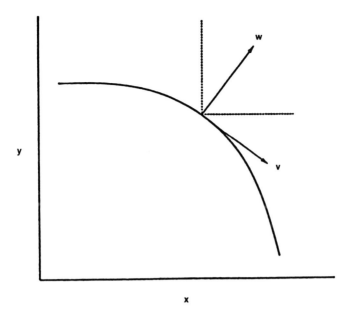

FIGURE 4.10. The (v, w) gauge; the w axis points in the direction of the local gradient.

(the projection of x onto the gradient direction $\nabla L/\nabla L\|$).

This gauge is ill-defined in points with vanishing gradient, but these points form a countable set, at least in *generic images* (images that are topologically "stable" against very small local distortions). An image with a constant intensity subdomain of finite size is unstable, since a small local distortion eliminates almost all zeros and completely changes its topological properties. Blurring a nontrivial image to a certain level of resolution always yields a generic image.

Other gauges are possible, such as the one that diagonalizes the Hessian L_{ij}, say with coordinates (p, q), which is always well defined. In this case the gauge condition is given by $L_{pq} = 0$. Indeed, this gauge condition is admissible, because any symmetric matrix can always be brought into diagonal form by a mere rotation, which is an allowable transformation from our point of view of rotational symmetry.

The directional derivative operators applied to the image yield invariants. Manifest invariance is then an obvious result of writing differential invariants using these invariants differential operators only. We have (check it!):

$$\frac{\partial}{\partial v} = \frac{\varepsilon_{kj} L_j \partial_k}{\sqrt{L_i L - i}} = \frac{\nabla_\perp L \cdot \nabla}{\|\nabla_\perp L\|}$$

and

$$\frac{\partial}{\partial w} = \frac{\delta_{kj} L_j \partial_k}{\sqrt{L_i L - i}} = \frac{\nabla L \cdot \nabla}{\|\nabla L\|}$$

We see that (by a previous example) L_v vanishes identically. This is precisely the motivation for this particular gauge and its power is best illustrated by means of some examples. Consider, e.g., the following two invariants, both in obvious tensor notation and in gauge coordinates:

$$\kappa = \frac{L_i L - ij L - j - L_i L_i L_{jj}}{(L_k L_k)^{3/2}} = -\frac{L_{vv}}{L_w},$$

$$\mu = \frac{L_i \varepsilon_{ij} L_{jk} L_k}{(L_l L_l)^{3/2}} = -\frac{L_{vw}}{L_w}.$$

Because these invariants have an interpretation that implicitly involves the image gradient, they greatly reduce in complexity when written in this particular gauge. The first of these invariants, κ, is the *isophote curvature,* while the second one, μ, is the curvature of the flow lines generated by the image gradient vector field (*gradient flow-line curvature*, in short). The meaning of curvature of a planar curve may be intuitively clear; it is a measure for the local deviation from a straight line. The exact definition is as follows:

$$\kappa = \frac{d\tau}{ds},$$

where κ is the curvature at the point P of interest and τ is the infinitesimal angle the tangent line at P makes with the tangent line of a neighboring point Q an infinitesimal distance ds away from P along the curve.

An equivalent, often more practical definition, is as follows. Put a coordinate frame with its origin at the point P of interest on the curve. The x-axis should be a tangent to the curve. The curve can then be locally described by a function $y(x)$ on an open interval around $x = 0$. In this system the curvature at the origin is defined as the second derivative $y''(0)$. We may check the correctness of both expressions for κ and μ above most easily with this definition and preferably directly in gauge coordinates.

To illustrate the use of differential geometry and at the same time show the power of gauge coordinates that are turned to a particular problem, let us turn to the proof of the first formula above, the one for the isophote curvature. In the (v, w) system we have, by definition, $\kappa = w''(P)$, in which $w(v)$ denotes the function describing the isophote locally near $P(v = 0, w = 0)$. Now the isophote passing through the point P is implicitly given by the equation $L = L(P)$. Taking the total differential of this equation yields $dL = 0$, or

$$L_v + L_w w' = 0.$$

In P we have, by our suitable choice of gauge, $L_v(P) = 0$, hence also $w'(P) = 0$. One more differentiation of the above equation gives us

$$L_{vv} + 2L_{vw} w' + L_{ww} w'^2 + L_w w'' = 0,$$

so in P we finally have $\kappa \equiv w''(0) = -L_{vv}/L_w$, which completes the proof.[4.26] It may be a surprise to learn that, although we have calculated the isophote curvature in a very special, simplifying coordinate system, viz., the most convenient system we could think of, it is a nearly trivial matter to find the expression in any arbitrary coordinate system; we do not even have to perform any rotations! The method is as follows. Write down an invariant in manifest index notation that reduces to the simple expression when evaluated in gauge coordinates. The above index notation for κ can be easily justified this way: its denominator is the third power of the gradient magnitude L_w^3, while its numerator can be rewritten from index to gauge notation as follows:

$$
\begin{aligned}
\kappa L_w^3 &= L - i L_{ij} L_j - L_i L_i L_{jj} \\
&= -L_{xx} L_y^2 + 2 L_x L_y L_{xy} - L_{yy} L_x^2 \\
&= -L_{vv} L_w^2 + 2 L_v L - w L_{vw} - L_{ww} L_v^2 \\
&= L_{vv} L_w^2
\end{aligned}
$$

(the last equality followed from $L_v = 0$). By this simple argument we can now also directly write down the isophote curvature expression in the (p, q) gauge without more ado! We simply replace x by p and y by q in the second expression above and then put the mixed derivative to zero:

$$
\kappa (L_p^2 + L_q^2)^{3/2} = -L_{pp} L_q^2 - L_{qq} L_p^2.
$$

The proof of the flow-line curvature expression is omitted. We only state here the precise definition of gradient flow lines. A flow line of a vector field $\mathbf{v}(\mathbf{x})$, [i.e., the image gradient vector field $\mathbf{v}(\mathbf{x}) = \nabla L(\mathbf{x})$], is defined as the integral curve passing through a given point $P(\mathbf{x}_0)$, which is tangent to the vector field everywhere:

$$
\begin{cases}
\frac{d\mathbf{x}}{ds} = \nabla L(\mathbf{x}) \\
\mathbf{x}(s = 0) = \mathbf{x}_0
\end{cases}.
$$

In order to verify this, it is best to start from yet another curvature expression that is used for *parameterized curves* like the above flow lines. Hence

$$
\mu = \left\| \frac{d^2 \mathbf{x}}{ds^2} \right\|
$$

As another example of manifest invariance using gauge coordinates, consider the well-known Laplacean of the image:

$$
\Delta L = L_{ii} = L_{vv} + L_{ww} = L_{ww} - \kappa L_w
$$

in which the expression for isophote curvature is that which we encountered earlier. From this example we may learn several things. First, it shows that, in general, invariants can be interrelated. More specifically, this example shows the naiveness of edge detection methods based on Laplacean zero crossings often encountered in the literature.[4.26] In fact, these zero crossings have little to do with edges. The relation between zero crossings of ΔL and edges is most obvious when using gauge coordinates. The term L_{ww} is the second derivative of the image along its gradient direction, i.e., normal to the isophote. If we define an edge as the locus of points of inflection defined by the vanishing of this normal derivative, which is quite a natural choice, then the zero crossings of ΔL can accurately describe edges only if the isophotes are sufficiently straight, so that the curvature term can be ignored. It is well known that this condition is certainly not fulfilled near corners and this is one reason why this zero crossings method is too naive. Another reason is related to the notion of *phantom edges,*[4.27] i.e., nonedge points detected by this zero crossings method. A phantom edge is an artifact which arises because of the second-order nature of the Laplacian. Even if the isophote is straight, the second-order normal derivative detects not only local maxima of the gradient (*true edges*) but also local minima, which precisely correspond to points that are the least likely candidate edge points of all!

Now let us give up the restriction on the transformation group and include reflections as well as rotations. After all, these also respect orthonormality of the coordinate basis. So now our ε tensor becomes a relative tensor and invariants containing an odd number of these become *relative invariants,* i.e., quantities that are invariant up to a possible minus sign (which shows up only when the orientation of the coordinate basis is reversed). In fact we can always write a relative invariant using exactly one ε factor. This follows from the fact that any tensor product of an even number of Lévi-Civita tensors can be written in terms of Kronecker tensors:

$$\varepsilon_{ij}\varepsilon_{kl} = \delta_{ik}\delta_{jl} - \delta_{il}\delta_{jk}.$$

It may be evident that we can construct an infinite number of invariants from any finite set of tensors by means of tensor multiplilcations and contractions. For example, an arbitrary *connected polynomial invariant* (an invariant that is not written as a sum or product of other invariants and is a polynomial expression in the image derivatives) in a 2-jet approximation always has one of the following three forms:

$$L$$
$$S_n \equiv L_i L_{ii_1} L_{i_1 i_2} \ldots L_{i_{n-1}j} L_j.$$
$$I_n \equiv L_{ii_1} L_{i_1 i_2} \ldots L_{i_{n-1}j}$$

The index n denotes the number of second-order derivative factors in each product. Additional invariants other than connected polynomial invariants can be constructed as functions of these.

It is also clear, however, that the N-jet has only a finite number of *essential components* (see Sec. 4.2), e.g., in two dimensions the number of essential components of the 2-jet equals five [one for the trivial invariant L, one for the gradient L_i in the (v, w) gauge, and three for the Hessian L_{ij}, or, counted differently, one for L, two for L_i, and two for L_{ij} in the (p, q) gauge). Therefore, we might expect only a finite number of so-called *irreducible polynomial invariants,* i.e., a set of basic polynomial invariants in terms of which all other invariants may be expressed.

However plausible this argument may seem, the proof of it in the general case is far from trivial. It seems that no general proof of this suspicion is available in the mathematical literature. But let us make this statement more precise in the simple context of the 2-jet, for which an irreducible set is available and a proof can be given. Consider the following set of five polynomial 2-jet invariants in two dimensions:

$$S = \{L, S_0 = L_i L_i, S_1 = L_i L_{ij} L_j, I_1 = L_{ii}, I_2 = L_{ij} L_{ji}\}.$$

A proof can be given that this is a complete and irreducible set; we omit the details.
    An example of reducibility is given by the following identity:

$$I_3 = \tfrac{3}{2} I_2 I_1 - \tfrac{1}{2} I_1^3.$$

Algebraically, the reducibility of this example can most easily be seen after diagonalizing the Hessian (which is always possible), say $L_{ij} = \text{diag}(a, b)$. Note that this is just another admissible gauge condition. We have the identity

$$a^3 + b^3 = \tfrac{3}{2}(a^2 + b^2)(a + b) - \tfrac{1}{2}(a + b)^3$$

expressing the trace of the third power of the Hessian in terms of the traces of the Hessian itself and its square.
    The proof and construction of polynomial bases of irreducible invariants is highly nontrivial. For orders N higher than 2, we often encounter sets of polynomial invariants that are complete and irreducible but are also nonminimal in the sense that they contain more invariants than there are degrees of freedom for the N-jet. For these overdetermined sets there then exist nonlinear relations among several members. These kinds of relations are referred to as *syzygies*. The mathematical literature, however, is rather vague and incomplete on this subject.
    There exists, however, another quite simple, complete basis of invariants and relative invariants, although nonpolynomial. This is the complete set of all *directional derivatives* in a gauged coordinate system, such as the (v, w) system:

$$\dot{G} = \{L_v \equiv 0, L_w, L_{vv}, L_{vw}, L_{ww}, L_{vvv}, \ldots\}$$

or the (p, q) system:

$$\dot{G}' = \{L_p, L_q, L_{pp}, L_{pq} \equiv 0, L_{qq}, L_{ppp}, \ldots\}.$$

The proof of this is trivial: just replace each pair of dummy indices in the index notation of an invariant by v and w respectively [or by any other pair of admissible gauge coordinates, such as the ones that diagonalize the Hessian, (p, q)] and add the results. We have already done this once in the discussion on the Laplacean. In the (v, w) system we may put $L_v = 0$ everywhere we encounter it, whereas in the (p, q) system we have, by definition, $L_{pq} = 0$. Here is another example we have met before:

$$L_i L_i L_{jj} - L_i L_{ij} L_j = L_{vv} L_w^2 = L_{pp} L_q^2 + L_{qq} L_p^2.$$

In this way, each polynomial invariant in manifest index notation is decomposed into a sum of terms, each of which is a product of basic, invariant factors. The sets $\dot{G}$ and $\dot{G}'$ are *minimal*, since, up to a given order, they contain precisely as many invariants as there are essential tensor components in the jet.

The set $\dot{G}$ is nonpolynomial, as we see from

$$
\begin{aligned}
L_w &= \|\nabla L\|^{-1}(L_x^2 + L_y^2), \\
L_{vv} &= \|\nabla L\|^{-2}(L_{xx}L_y^2 - 2L_xL_yL_{xy} + L_{yy}L_x^2), \\
L_{vw} &= \|\nabla L\|^{-2}[L_{xy}(L_y^2 - L_x^2) + L_xL_y(L_{xx} - L_{yy})], \\
L_{ww} &= \|\nabla L\|^{-2}(L_x^2L_{xx} + 2L_xL_yL_{xy} + L_y^2L_{yy}), \\
L_{vvv} &= \|\nabla L\|^{-3}(L_{xxx}L_y^3 - 3L_{xxy}L_xL_y^2 + 3L_{xyy}L_x^2L_y - L_{yyy}L_x^3),
\end{aligned}
\tag{4.11}
$$

$$\vdots$$

We can form polynomial invariants from the (v, w) derivatives by multiplying each with a power of the gradient equal to the order of the derivative. So the set

$$
\dot{P} = \{L_vL_w \equiv 0, L_wL_w, L_{vv}L_w^2, L_{vw}L_w^2, L_{ww}L_w^2, L_{vvv}L_w^3, \ldots\}
$$

is polynomial:

$$
\begin{aligned}
L_w^2 &= L_iL_i \\
&= L_x^2 + L_y^2, \\
L_{vv}L_w^2 &= L_iL_iL_{jj} - L_iL_{ij}L_j \\
&= L_{xx}L_y^2 - 2L_xL_yL_{xy} + L_{yy}L_x^2, \\
L_{vw}L_w^2 &= \varepsilon_{ji}L_iL_{jk}L_k \\
&= L_{xy}(L_y^2 - L_x^2) + L_xL_y(L_{xx} - L_{yy}), \\
L_{ww}L_w^2 &= L_iL_{ij}L_j \\
&= L_x^2L_{xx} + 2L_xL_yL_{xy} + L_y^2L_{yy}, \\
L_{vvv}L_w^3 &= \varepsilon_{ij}(L_{jkl}L_iL_kL_l - L_{jkk}L_iL_lL_l) \\
&= (L_{xxx}L_y^3 - 3L_{xxy}L_xL_y^2 + 3L_{xyy}L_x^2L_y - L_{yyy}L_x^3),
\end{aligned}
$$

$$\vdots$$

The price we pay for this nice property is that this set is not complete anymore. Figure 4.13 shows one of the above polynomial invariants, $L_{vv}L_w^2$, which can be aptly called "cornerness."

It may be clear by now that invariance theory provides us with a powerful tool kit for the description of local image properties and that it is intimately related to differential geometry. This gives us an interpretation of the geometrically meaningful quantities that are expressed by the invariants.

4.7   Applications

Computer implementation is most conveniently done in the Fourier domain. Differentiation is done in the Fourier domain simply by multiplication with the frequency coordinates to the power of which we take derivatives. The basic Gaussian zeroth-order kernel G_{00} can be calculated analytically in the Fourier domain, and thus can be implemented directly. A convenient property is that convolution in the spatial domain is equivalent to multiplication in the frequency domain. Thus we get

$$L \otimes G_{nm} = \mathcal{F}^{-1} \mathcal{L}(i\omega_x)^n (i\omega_y)^m \mathcal{G}_{00}\},$$

where $\mathcal{F}^{-1}$ is the inverse Fourier transform (FFT), $\mathcal{L}$ is the Fourier transformed input image L, and G is the Fourier transformed zeroth-order Gaussian kernel G_{00}, respectively. Any polynomial expression of the invariants may then be constructed from these elementary higher order convolutions. The elegance of this representation makes it very suitable for extension to three or even higher dimensions and the use of dedicated FFT hardware. This methodology lends itself well to parallel implementation (e.g., SIMD). After all, this is just what happens in our visual system!

An essential aspect, of course, is that we are dealing with discrete images, either sampled by the receptors on the retina, or represented as discrete pixels in a computer. Discretization issues have been studied extensively.[4.28] Given similar requirements on the blurkernel (e.g., no new structure created in scale space, and isotropic conditions), the best representation is by means of modified Bessel functions of integer order. These functions approach the Gaussian kernels with increasing scale. We have no artifacts with the Gaussian kernel, as can also be seen in the figures. All this theory is continuous and analytical, but the match between the extent of the fuzzy derivative and the intrinsic scale is a natural guarantee for the operations to be performed between the inner and outer scale range.

Applications are best classified by increasing order in the N-jet. We give examples of the invariants on a horizontal structure, i.e., for a fixed scale. First we consider the 0-jet:

$$J^0[L(\mathbf{x}; s)] = \{L(\mathbf{x}; s)\}.$$

This is the set of images that coincide up to zeroth order in $\mathbf{x}$, i.e., all images with the same value $L(\mathbf{x})$. It has only one invariant) component, which is the intensity value atx. The 0-jet extension is a complete description of the image in that it provides all the intensity values, but it does not allow for any comparisons between neighboring points. It gives a local estimate on a $\mathbf{k}(\sqrt{4s})$ scale. So from a geometrical point of view it is not very interesting. To consider *the* intensity at $\mathbf{x}$ is both meaningless and irrelevant. $L(\mathbf{x})$ is meant to refer to $L(\mathbf{x}; s)$ at level s in scale space, and this is the important difference with a mathematically defined function! Next, we consider the 1-jet:

$$J^1[L(\mathbf{x}; s)] = \left\{ L(\mathbf{x}; s), \frac{\partial L(\mathbf{x}; s)}{\partial x^i} \right\}.$$

This is the jet of lowest nontrivial order. It allows us a glimpse into the differences that occur over the neighborhood of $\mathbf{x}$. We can precisely extract one additional independent invariant from it by total contraction of first-order tensors, viz., the *gradient strength* or *edgeness*:

$$\|\nabla\| = \sqrt{L_i L_i} = \sqrt{\left(\frac{\partial L}{\partial x}\right)^2 + \left(\frac{\partial L}{\partial y}\right)^2} = L_w.$$

Edgeness is the geometrical property described by this invariant: it is the slope of the tangent plane to the image at ξ. Edgeness is strongest on edges, so it is basically an edge detection operator. Note that the Canny edge detector[4.29–4.30] for steep edges is equivalent to L_w.

Because of its scaling property under image intensity rescaling, it is said to be *homogeneous of degree 1*:

$$Q(\lambda L_w) = \lambda Q(L_w).$$

As we have seen, only homogeneous quantities are of interest, whatever their degree may be.

The performance in a very noisy image is illustrated in Fig. 4.11. Indeed, we see a fast decrease of the variance of the noise in scale space, despite differentiation.

Now we consider the 2-jet:

$$J^2[L(\mathbf{x}; s)] = \left\{ L(\mathbf{x}; s), \frac{\partial L(\mathbf{x}; s)}{\partial x^i}, \frac{\partial^2 L(\mathbf{x}; s)}{\partial x^i \partial x^j} \right\}.$$

A second order jet already reveals many more details of geometrical significance. Second-order derivatives describe the variation in first-order behavior and this is closely related to the concept of curvature. Some care is needed with this notion. There are many kinds of curvature and careful definition is necessary. Let us examine some examples.

We first study the curvature of the intensity surface, as we move around in our local position. We know that the smallest and largest curvatures are on trajectories perpendicular to each other. These directions are called principal directions, and the curvatures are principal curvatures. From standard geometry texts,[4.4] we find for the principal curvatures K_1 and K_2, the Gaussian curvature G, and the mean curvature M:

$$K_{1,2} = \frac{L_{vv} + L_{ww} \pm \sqrt{\cdots}}{2(1 + L_w^2)^{3/2}}, \tag{4.12}$$

$$G = K_1 K_2 = \frac{\frac{L_{vv}L_{ww} - L_{vw}^2}{(1+L_w^2)^3} = L_{ii}L_{jj} - L_{ij}L_{ji}}{(1 + L_k L_k)^3 = \frac{\det H}{(1+\|\nabla L\|^2)^3}}, \tag{4.13}$$

$$M = \frac{1}{2}(K_1 + K_2) = \frac{L_{vv} + L_{ww}}{2(1 + L_w^2)^{3/2}} = \frac{L_{ii}}{2(1 + L_k L_k)^{3/2}}, \tag{4.14}$$

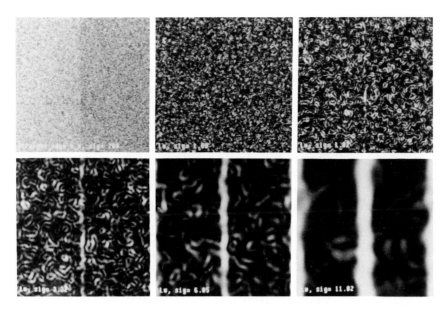

FIGURE 4.11. Edge detection of a vertical straight edge (intensity value 0, 100) drowned in additive Gaussian noise ($\sigma_N = 200$) on different levels of scale space. The edge is most pronounced at its appropriate scale, whereas the noise exists predominantly at small scales. The weaker irregular undulations are the structural properties of the noise *at that particular scale*.

where H is the Hessian matrix of pure second-order derivatives. The (v, w) gauge coordinates are not the appropriate ones to be used in these expressions, because they are related to first-order derivatives, whereas curvatures are intrinsically second-order image properties. In (p, q) the gauge coordinates of the second order, we get a more natural representation (because now $L_{pq} = 0$):

$$K_1 = \frac{L_{pp}}{(1 + L_p^2 + L_q^2)^{3/2}}, \qquad K_2 = \frac{L_{qq}}{(1 + L_p^2 + L_q^2)^{3/2}}. \qquad (4.15)$$

Note that Eqs. (4.12)–(4.15) are not homogeneous and hence of little value. The factor 1 derives from the fact that in the calculation *image intensity is not a spatial parameter.* Note, however, that the denominators of these equations, and hence their signs, *are* homogeneous. So we can sensibly talk of the *signs* of the Gaussian curvature[4.7] or mean curvature, and hence divide each image into *elliptic and hyperbolic patches,*[4.31] see Fig. 4.12. Therefore, we call the mean curvature a *second-order patch classifier.* All extremes are located in elliptic regions and all saddle points in hyperbolic regions. These critical points play a dominant role in catastrophe theory underlying the image's deep structure. Near points where the Gaussian curvature G is zero, the image is either flat or cylindrically shaped. In the latter case these points are referred to as parabolic points. They form closed lines, the parabolic lines in an image. They form the boundaries between the patches. We see now how they can easily be calculated in the visual system. A further

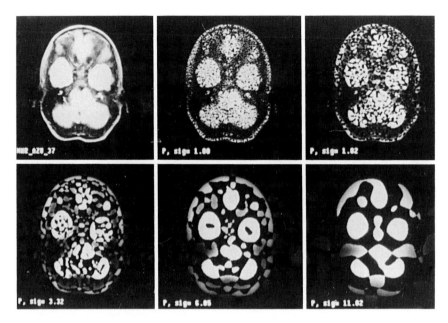

FIGURE 4.12. Second order patch classification in an MRI scan at several levels in scale space (as in Fig. 4.1). Parabolic lines, where the Gaussian curvature is zero, form the natural boundary between elliptic (principal curvatures same sign) and hyperbolic (principal curvatures opposite sign) areas. The hyperbolic areas are suppressed, whereas the elliptic areas are given the same gray values as in the corresponding blurred images. This is achieved by multiplying a binary image (elliptic 1 and hyperbolic 0) by the corresponding scaled original.

classification is not possible due to third and fourth order classification. In three dimensions, we discern three principal curvatures and hence find elliptic and two kinds of hyperbolic volumes.

Isophote curvature has a different geometrical meaning and is homogeneous of degree zero. In fact, it does not change under invertible intensity transformations $L \perp \tilde{L}(L)$. It was discussed as an illustration with the invariants theory in Sec. 4.6. The polynomial invariant $L_{vv}L_w^2$ is a good *corner detector.*

Corners in an image are characterized by points where edges meet at a certain angle. Under blurring, a corner evolves to a (decreasingly string) highly curved collection of isophotes near the original corner position. Moreover, these isophotes are on a steep ramp which remains from the original edges. Isophote curvature is proportional to L_{vv}, but becomes singular for $L_w = 0$. If we inspect Eq. 4.11, we see that L_{vv} is similar, because we divide by L_w^2. We can form a new, nonsingular invariant, *cornerness* C:

$$C \equiv -L_{vv}L_w^2 = \kappa L_w^3. \tag{4.16}$$

A corner detector should be homogeneous of strict positive degree, since it is required to scale with the intensity step. So edginess is combined with isophote curvature to construct an invariant which is homogeneous of degree three and is most pronounced on highly curved edges along steep ramps. In Fig. 4.13, the performance is shown in scale space for corners on a polygon with additive Gaussian noise. Clearly, higher in scale space the larger extents of the Gaussian derivative kernels cancel the noise, and the corners show up *at their own scale*. In the literature a number of corner detectors have been proposed.[4.32] This example shows the robustness of geometric reasoning and no ad hoc choices have to be made. Again, it also shows the stability of the application of differential geometry in scale space.

Geometric properties of the second order may also be understood as variations of the first order. If we consider the Taylor expansion (in D dimensions) in the compact manifest invariant notation:

$$L(\mathbf{x} + \delta\mathbf{x}) = L + L_i \delta x_i + \frac{1}{2!} L_{ij} \delta x_i \delta x_j + \frac{1}{3!} L_{ijk} \delta x_i \delta x_j \delta x_k + O(\delta\mathbf{x}^4)$$

we can define the following properties:

Deviation from flatness $=$

$$D \equiv \sqrt{L_{ij} L_{ji}} = \sqrt{L_{xx}^2 + 2L_{xy}^2 + L_{yy}^2} = \sqrt{\mathrm{Tr} H^2}.$$

The deviation from flatness gives the importance of the mixed terms, indicated by different indices.

Umbilicity $=$

$$U \equiv \frac{L_{ii} L_{jj} - L_{ij} L_{ji}}{L_{lk} L_{kl}} = \frac{2(L_{xx} L_{yy} - L_{xy}^2)}{L_{xx}^2 + 2L_{xy}^2 + L_{yy}^2} = \frac{(\mathrm{Tr} H)^2}{\mathrm{Tr} H^2} - 1.$$

We see that umbilicity is a similar *patch classifier* as the determinant of the Hessian:

$$U = \frac{2\kappa_1 \kappa_2}{\kappa_1^2 + \kappa_2^2} = \begin{cases} U = -1 & \text{pure hyperbolic} \\ -1 < U < 0 & \text{hyperbolic region} \\ U = 0 & \text{parabolic point} \\ 0 < U < -1 & \text{elliptic region} \\ U = 1 & \text{spherical point} \end{cases}$$

Finally, we consider the 3-jet:

$$J^3[L(\mathbf{x}; s)] = \left\{ L(\mathbf{x}; s), \frac{\partial L(\mathbf{x}; s)}{\partial x^i}, \frac{\partial^2 L(\mathbf{x}; s)}{\partial x^i \partial x^j}, \frac{\partial^3 L(\mathbf{x}; s)}{\partial x^i \partial x^j \partial x^k} \right\}.$$

It is straightforward to define all kinds of third-order invariants by performing full contractions and alterations as already explained. It becomes increasingly difficult, however, to explain the geometrical meaning of all these higher-order invariants. But if we start with a well-defined higher order geometrical notion, such as the *gradient of isophote curvature,* then it is not difficult to derive the associated

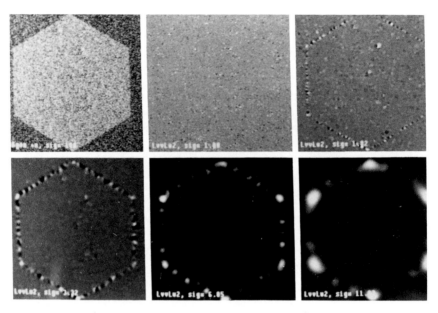

FIGURE 4.13. Performance of the "cornerness detector $-L_{vv}L_w^2$ in scale space for corners on a hexagon with additive Gaussian noise. Scale space levels are ordered from left to right, with σ as in Fig. 4.1. Clearly, higher in scale space the larger extents of the Gaussian derivative kernels cancel the noise and the corners show up at their own scale.

invariant:

$$\nabla\kappa = \left\{ \frac{L_{vv}^2 L_{vw}^2 + L_{vv}^2 L_{ww}^2}{L_w^2} \quad -2\frac{L_{vv}L_{vvv}L_{vw} + L_{vv}L_{vvw}L_{ww}}{L_w^3} \right.$$
$$\left. +\frac{L_{vvv}^2 + L_{vvw}^2}{L_w^2} \right\}^{1/2}.$$

The gradient of isophote curvature is homogeneous of degree zero and may be of some value for a T-junction detector, much like the isophote curvature itself is connected to the corner detector described earlier. Indeed, the isophotes near such a T-junction show a strong change in curvature over a relatively small spatial neighborhood; see Figs. 4.14 and 4.15. An important step in studying the geometric entities in the 3-jet is the choice of gauge coordinates and properties to study along these new coordinates, and the deviations from second-order behavior. This is, however, beyond the detail necessary in this chapter.

4.8   Discussion

In this chapter, a rather mathematical approach (and justification) has been given to the closely related concepts of scale and image structure analysis as proposed in our multiscale geometric model of human visual perception.

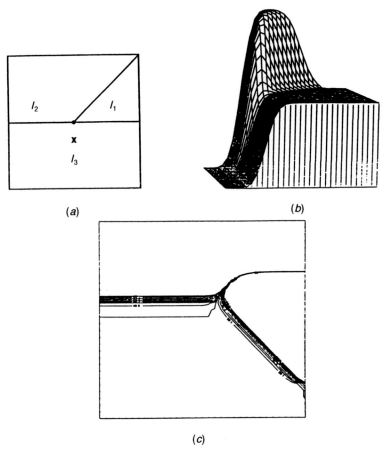

(a)   (b)

(c)

FIGURE 4.14. (a) T junction, formed by three areas with different intensities. (b) Blurred T junction in scale space. (c) Isophotes on a blurred T junction. Clearly the sharp increase in curvature is seen when we change isophotes; therefore the gradient of the isophote curvature is a good geometrical detector for this property of T junctions.

The main themes of this chapter are fivefold:

1. A physical observable is intrinsically related to the concept of *scale*. The natural range of scales within an image is limited by the image resolution, defining the *inner scale* and by its extent, set by the image boundaries, giving the *outer scale*. We have introduced the notion of *natural coordinates* so as to be able to define scale-independent operations.

2. A natural *scale space* is formed by convolution with a kernel, which must have the property of being *self-similar*. We have seen that this requirement uniquely leads to the zeroth-order Gaussian kernel, which is the solution of the isotropic diffusion equaiton. The input image is the initial condition to this equation.

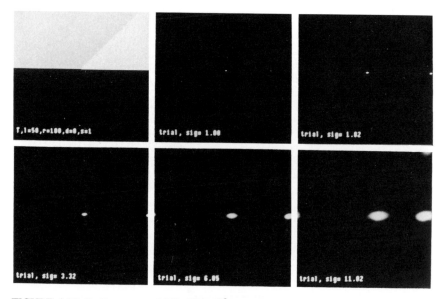

FIGURE 4.15. Performance of $\|\nabla\kappa\|\|\nabla L\|^5$ in scale space. There is a good detection of the T-junction on its appropriate scale. Due to the application of the Fourier domain in the calculations, which "folds" the original image, we get T-junction detection at the boundaries as well, as expected.

3. The Gaussian itself is a solution to this diffusion equation, as well as all of its partial spatial derivatives. These differential operators are the natural operators for the process of differentiation at a certain spatial scale.

4. Properties with an objective, geometrical meaning with a certain transformation group must be invariant with respect to coordinate transformations belonging to that group. We have seen the powerful application of invariants theory and the simplification resulting from the compact manifest invariant notation. Any full contraction of indices automatically leads to a geometrically meaningful image feature. We have also presented another obvious way of forming invariants, viz., through the use of directional derivatives using *gauge coordinates*.

5. In principle, an infinite set of invariants can be constructed to any order. But in the N-jet approximation we find that a finite and, for small N, fairly small set of *irreducible invariants* suffices.

Of crucial importance is the realization that the concepts and methodologies introduced here are all fundamentally intimately related. For example, we cannot differentiate without, at the same time, scaling the image. Another example is that we cannot take derivatives haphazardly and expect the result to make sense; only certain combinations of derivatives are meaningful.

The approach presented has a firm mathematical basis (differential geometry, tensor calculus, and invariants theory to name the most important disciplines). Moreover, it is ideally suited to a parallel computer implementation. The theory can easily be extended to higher dimensions, e.g., three dimensions. Only the extension of differential geometry is not entirely trivial. Many digital acquisition systems of medical tomographic data, such as computed tomography and magnetic resonance imaging, supply data in the form of a three-dimensional dataset with a reasonable resolution in all directions. Here we lose the analogy with the human two-dimensional visual system, but there is no fundamental obstacle from the mathematical point of view.

The models of these combined disciplines to computer-vision applications is relatively new and forms a promising tool for many models, like image segmentation, shape from X, shape description,[4.33] and scene analysis.[4.7] It is intriguing (and indeed very encouraging!) to see how closely this model resembles the receptive field structure of the mammalian visual system.

Data from neurophysiology suggest that receptive fields with a structure higher than the fourth order are rare or may even be absent. It is a bold but interesting conjecture to suggest that there might be a mathematical reason for this. So far, however, we have no indication as to what might determine the highest order to be taken into account. The intimate relation of scale and differential operators leads to a natural taxonomy of the many receptive field profiles that are found.[4.24]

The approach presented in this chapter, based on the intimate relation between invariants theory and scale-related differential operators, gives a good starting point for further research into the *deep structure* of scale space.

It is generally agreed that the issue of *image segmentation* also needs a model-driven approach. The invariants may be good descriptors of such a model, in relation with their behavior in scale space.

The approach given here is essentially local. The evolution of the invariants with increasing scale, especially those forming an irreducible set, may give us a global image description. In scale space we encounter many *catastrophes,* such as annihilation of an extreme and a saddle point with increased blurring, so the application of *catastrophe theory* seems appropriate. Other approaches focus on the structure of *links* between image structural elements in scale space.[4.34–4.35] The theory of differential invariants is presently being developed for the deep structure of scale space. Temporal behavior of invariants is not discussed in this chapter. It is clear, however, that visual perception relies very much on the dynamic changes of the visual scene, induced by object or ego motion, or eye movements. Scale space theory receives much attention in contemporary literature on *computer vision.* Many other approaches are proposed, such as the description with Gabor functions, i.e., complex-valued Gaussians,[4.36] wavelets,[4.37] or the fastly developing field of nonisotropic diffusion.[4.38]

We have only just begun to understand these intricate relationships. There needs to be much more research into the intriguing and astonishingly well-performing system that supports our human visual perception.

4.9 References

[4.1] Gibson J.J. *The Perception of the Visual World.* Boston: Houghton Mifflin; 1952.

[4.2] Kay D.C. *Tensor Calculus* (Schaum's Outline Series). New York: McGraw-Hill; 1988.

[4.3] Gurevich B. *Foundations of the Theory of Algebraic Invariants.* Groningen: Noordhof; 1979.

[4.4] Spivak M. *A Comprehensive Introduction to Differential Geometry* (Vols. I–V). Berkeley: Publish or Perish; 1970.

[4.5] Young R.A. The Gaussian derivative model for machine vision: Visual cortex simulation. *J. Opt. Soc. Am.* 1985; A2(13): 39, 102.

[4.6] Young R.A. Simulation of human retinal function with the Gaussian derivative model. *Proc. IEEE Conf. Comput. Vision Patt. Recogn.* 1986; 564–569.

[4.7] Kanatani K. *Group-Theoretical Methods in Image Understanding* (Springer Series in Information Sciences, Vol. 20). Berlin: Springer; 1990.

[4.8] Ter Haar Romeny B.M., Florack L.M.J., Koenderink J.J., Viergerer M.A. Scale-space: its natural operators and differential invariants. In: Colchester A.C.F., Hawkes D.J., eds. *Information Processing in Medical Imaging* (*Lecture Notes in Computer Science,* 511), Berlin: Springer; 1991: 239–255.

[4.9] Florack L.M.J., Ter Haar Romeny B.M., Koenderink J.J., Viergerer M.A. Scale and the differential structure of images. *Image Vis. Comput.* 1992; 10: 376–388.

[4.10] Florack L.M.J., Ter Haar Romeny B.M., Koenderink J.J., Viergerer M.A. Linear scale-space. *J. Math. Imag. Vis.* 1994; 4: 325–351.

[4.11] Weyl H. *The Classical Groups, their Invariants and Representations.* Princeton: Princeton University Press; 1946.

[4.12] Ter Haar Romeny B.M., Florack L.M.J., Salden A.H., Viergerer M.A. Higher order differential structure of images. *Image Vis. Comput.* 1994; 12: 317–325.

[4.13] Witkin A.P. Scale space filtering. *Proc. Int. Joint Conf. Artificial Intell.* (*Karlsruhe*) 1983; 1019–1021.

[4.14] Koenderink J.J. The structure of images. *Biol. Cybern.* 1984; 50: 363–370.

[4.15] Babaud J., Witkin A., Duda R. Uniqueness of the Gaussian kernel for scale space filtering. *IEEE Trans. Patt. Anal. Mach. Intell.* 1986; PAMI-8: 26–33.

[4.16] Korn A. Toward a symbolic representation of intensity changes in images. *IEEE Trans. Patt. Anal. Mach. Intell.* 1988; PAMI-10: 610–625.

[4.17] Koenderink J.J. Geometrical structures determined by the functional order in nervous nets. *Biol. Cybern.* 1984; 50: 43–50.

[4.18] Koenderink J.J., van Doorn A.J. Representation of local geometry in the visual system. *Biol. Cybern.* 1987; 55: 367–375.

[4.19] Torre V., Poggio T.A. On edge detection. *IEEE Trans. Patt. Anal. Mach. Intell.* 1986; PAMI-8: 147–163.

[4.20] Poggio T., Torre V., Koch C. Computational vision and regularization. *Nature* 1985; 317: 314–319.

[4.21] Koenderink J.J. Image structure. In: Viergever M.A., Todd-Pokropek A., eds. *Mathematics and Computer Science in Medical Imaging* (NATO ASI Series F39. Berlin: Springer; 1988: 67–104.

[4.22] Hubel D.H. *Eye, Brain, and Vision* (Scientific American Library Series 22). San Francisco: Freeman; 1988.

[4.23] Hubel D.H., Wiesel T.N. Brain mechanisms of vision. *Sci. Am.* 1979; 241(3): 130–146.

[4.24] Koenderink J.J., van Doorn A.J. Receptive field families. *Biol Cybern.* 1990; 63: 291–298.

[4.25] Florack L.M.J., Ter Haar Romeny B.M., Koenderink J.J., Viergerer M.A. Cartesian differential invariants in scale-space. *J. Math Imag. Vis.* 1993; 3: 327–348.

[4.26] Marr D., Hildrecht E.C. Theory of edge detection. *Proc. R. Soc. London* 1980; 200: 269–294.

[4.27] Clark J.J. Authenticating edges produced by zero-crossing algorithms. *IEEE Trans. Path. Anal. Mach. Intell.* 1989; PAMI-11: 43–57.

[4.28] Lindeberg T. Scale space for discrete signals. *IEEE Trans. Patt. Anal. Mach. Intell.* 1990; PAMI-12: 234–245.

[4.29] Canny J. A computational approach to edge detection. *IEEE Trans. Patt. Anal. Mach. Intell.* 1987; PAMI-8: 679–698.

[4.30] De Micheli E., Caprile B., Ottonello P., Torre V. Localization and noise in edge detection. *IEEE Trans. Patt. Anal. Mach. Intell.* 1989; PAMI-10: 1106–1117.

[4.31] Koenderink J.J., van Doorn A.J. A description of the structure of visual images in terms of an ordered hierarchy of light and dark blobs. In: Jaeffe S.C., ed. *Proc. 2nd IEEE Int. Conf. Vis. Psychophys. Med. Imag.* (Cat. 81 CH 1676-6) New York: IEEE; 1981: 173–176.

[4.32] Noble J.A. Finding corners. *Image Vis. Comput.* 1988; 6:121–128.

[4.33] Koenderink J.J. *Solid Shape.* Cambridge: MIT; 1990.

[4.34] Lifshitz L.M., Pizer S.M. A multiresolution hierarchical approach to image segmentation based on intensity extrema. *IEEE Trans. Patt. Anal. Mach. Intell.* 1990; PAMI-12: 529–541.

[4.35] Betgholm F. Edge focusing. *IEEE Trans. Patt. Anal. Mach. Intell.* 1987; PAMI-9: 726–741.

[4.36] Bovik A.C., Clark M., Geisler W.S. Multichannel texture analysis using localized spatial filters. *IEEE Trans. Patt. Anal. Mach. Intell.* 1990; PAMI-12: 55–73.

[4.37] Mallat S.G. A theory for multiresolution signal decomposition: The wavelet representation. *IEEE Trans. Patt. Anal. Mach. Intell.* 1989; PAMI-11: 674–694.

[4.38] Ter Haar Romeny B.M., ed. *Geometry-Driven Diffusion in Computer Vision.* Dordrecht: Kluwer; 1994.

5

Human Response to Visual Stimuli

Alastair G. Gale

5.1   Introduction

The human response to a visual stimulus depends upon many factors, not the least of which is the particular task set the observer. At the simplest level we may be interested in the detection (is it present?) or identification (what is it?) of a stimulus which may be just a small spot of light. Alternatively we may be interested in a more complex problem such as the detection of a target in a large visual field containing many possible targets. An example of the latter is the detection of a particular type of military vehicle from satellite imagery.

This chapter briefly reviews techniques employed to measure human responses to simple visual stimuli in the classical psychophysics tradition and then moves on to examine the problem of how we respond to more complex stimuli. This involves the consideration of selective visual attention and the interplay between the visual stimuli and cognitive factors of the observer. The influence of psychological factors on perception is highlighted by consideration of visual search. Although examples are given from several research areas, the domain of medical images is stressed.

5.2   Measuring Human Responses

5.2.1   Low Intensity Signals

The study of the relationship between physical stimuli and the psychological reactions to them is the province of psychophysics. Typically the visual stimulus varies in one dimension. For instance, to investigate the perception of a small spot of light, the intensity of the light—the physical stimulus—can be precisely measured using techniques derived from physics and the investigator's purpose is to measure the way in which it has been perceived.

Often the interest is in measuring the human response to stimuli of very low intensity, i.e., the detection of the stimulus. The classical psychophysics approach to detection involves measuring the *detection threshold*. This is the smallest amount of energy required for the stimulus to be reported correctly half of the time. The lower the detection threshold, the higher the sensitivity to that stimulus. Galanter gives an approximate detection threshold for light as a candle flame seen at 30 miles on a clear dark night.[5.1]

Detection thresholds can be measured by three techniques:

1. *Method of Limits*

   A clearly detectable stimulus is presented to an observer, who is then asked to respond yes/no to detection. This is a two-alternative-forced-choice approach (TAFC). The stimulus intensity is then systematically decreased until the observer cannot detect it. Stimulus intensity is then increased until detection occurs again. This is repeated over a number of trials, so the observer is presented with decreasing and increasing series of intensities. Both series are typically employed together, as different thresholds can be obtained depending upon the direction of the particular series used.

   Two types of error can occur in threshold estimation. These are errors of habituation or preseveration (a tendency to keep making the same response within a series) and errors of anticipation (the observer's assertion that he can detect the stimulus when in fact he cannot). Habituation errors produce higher detection thresholds on an ascending series and lower thresholds on a descending series. Thus by employing both series together these errors should cancel one another out. The same logic applies to overcoming anticipation errors. To overcome the possibility that the observer might simply count the number of presentation trials before altering the response, the series are started at different points.

   Using the two series together is termed the staircase method where, for instance, an increasing series is first presented until the observer changes his response and then the series is decreased. Thus the value of the test stimulus is tracked back and forth across the observer's threshold value. This approach enables the threshold to be monitored even if it is changing, such as during adaptation to background stimuli or after administration of different drugs.[5.2]

   The method can be adapted, e.g., by using only one series, as in the detection of a spot of light by a dark-adapted observer where the use of a decreasing series would affect the research goal. The Snellen eye chart (see Chap. 3) is an example of the use of the descending method of limits coupled with correct identification of the target letters.

2. *Method of Adjustment*

   In this technique, the observer (rather than the experimenter) adjusts the stimulus intensity; such adjustments are often continuous rather than discrete. Observers are much less likely to make errors of habituation and anticipation with the use of this method. With this approach, half the trials are typically started with the stimulus level well below threshold and half with it far above threshold. The technique is useful because it produces a fairly rapid estimate of a threshold. The threshold value obtained is typically less accurate, however, and the values obtained vary between observers.

3. *Method of Constant Stimuli*

A constant set (e.g., five or nine) of visual stimuli, selected to range above and below threshold, are presented to the observer in a random order. The use of a random sequence overcomes errors of habituation and anticipation. Pretesting, for instance by using the previous method of adjustment, is employed to arrive at these final small series of selected values. Each stimulus is typically presented a large number of times which makes this approach somewhat time consuming; however, it does provide the most reliable estimate of a threshold.[5.3]

All three approaches assume a fixed threshold for an intensity of a particular visual stimulus. Unfortunately, thresholds can vary with psychological factors such as expectation and motivation. Thus, such classical psychophysical methods actually measure both the observer's decision-making strategy and his sensitivity to the signal. This point is elaborated in the approach of signal detection theory (covered in more detail in Chap. 9) which is also used to investigate detection and identification of visual stimuli.

5.2.2 Intense Stimuli

With stimuli which can be easily detected, the aim is to determine the smallest detectable change in stimulus that has to be changed in order to be perceived as just noticeably different. The observer's discrimination ability is measured by a *difference threshold*, which is defined as the smallest change in a stimulus that is required to produce a noticeable difference half of the time. Alternatively, a difference threshold can be defined as the amount of change in a physical stimulus required to produce a just noticeable difference (JND) in the psychological sensation. In measuring a difference threshold, a standard stimulus is presented along with a comparison stimulus which is varied. The point where the two stimuli are subjectively judged by the observer to be equal is the point of subjective equality.[5.4] All of the three previously mentioned psychophysical methods can be used to measure discrimination.

Early work by Weber in the 1800s employed the *just noticeable difference* to examine the relationship between the physical stimuli and the observer's responses. No one-to-one correspondence was found between physical stimuli and psychological reactions; instead, Weber showed that the observer's psychological reaction related, not to the absolute size of the change in a visual stimulus, but to the relative size of the change. Weber's law states that the change in intensity (dI) divided by the intensity (I) is a constant $(k$, the Weber fraction):

$$dI/I = k. \qquad (5.1)$$

Weber's law is generally best at predicting discrimination ability in the middle ranges rather than for high or low intensity stimuli and holds for several psychophysical judgments.[5.5]

Fechner subsequently employed Weber's law to produce a scale that related the size of the physical stimulus to the size of the observer's reaction. Fechner's law states that the magnitude of the psychological reaction (R) is equal to the logarithm of the intensity of the stimulus (I) times a constant (w):

$$R = w \log I. \tag{5.2}$$

An alternative relationship between psychological reaction (R) and stimulus intensity (I) is provided by Stevens's[5.6] power law:

$$R = aI^n, \tag{5.3}$$

where a is a constant.

Stevens used the *method of magnitude estimation* in which the observer provides a direct estimate of the magnitude of stimuli based on a specified reference value. In this technique, the observer is instructed that one stimulus (the standard) is assigned a particular value and this is then used in estimating the magnitude of other stimuli.[5.7] Results from magnitude estimation and cross-modality comparison studies support the power law and this function best describes the relationship between stimulus intensity and an observer's reaction. In general, Stevens's power law has better predictive abilities than Fechner's law[5.8] for various visual stimuli.

To measure interstimulus differences in detectability or discriminability with stimuli that well exceed the difference threshold, reaction time is employed. This is the time between the onset of a stimulus and the beginning of an overt response. Simple reation time involves making a single immediate response (e.g., key press) when a stimulus is detected, whereas choice reaction time involves making one of several responses (e.g., press the left-hand key when a blue stimulus is presented or the right-hand key for a red stimulus detection). When simple reaction time is used in detection experiments, it is faster for more intense stimuli. For discrimination tasks, reaction time is shorter for a large change in stimulus intensity. Choice reaction time is used in both discrimination and identification and is a linear function of the amount of information in the stimuli (Hick's law).

5.3   Complex Stimuli

5.3.1   Selective Visual Attention

We have just dealt with the human response to relatively simple stimuli. Our response to more complex stimuli, such as two-dimensional displays containing targets, is more complicated. Knowledge of the physiological structure of the eye suggests that the fovea is used for detailed processing of visual information and that both head and eye movements serve to move this area of high visual acuity around visually to sample the stimulus (see Chap. 1). In addition to this, the psychological concept of visual attention (itself much maligned) plays a very important role. Posner[5.9] described attention as an alignment of sensory input and Norman[5.10] has referred to it as a spotlight with an adjustable beam, which is a useful metaphor. Somewhat similarly, Eriksen[5.11] uses a zoom lens analogy for visual attention. The implication is that we can attend to different areas of visual stimulus, e.g., to all the stimulus or to a possibly varying and smaller selected area. To complicate matters further, this attentional beam may even be split into more than one single area.[5.12] A good overview of the concept of attention is provided by Johnston and Dark.[5.13]

Since the late 1960s, experimental evidence has supported the idea of an essentially dual visual system. Several workers have distinguished between a somewhat parallel processing of visual information across the stimulus and serial processing of selected stimulus information. For instance, Trevarthen[5.14] has distinguished between focal and ambient vision; Held, et al.[15.15] used the terms *identifying* and *locating modes*; Treisman and Gelade[5.16] proposed feature extraction and object identification; Julesz[5.17] identified attentive and preattentive stages; and Breitmeyer and Ganz[5.18] referred to sustained and transient mechanisms. Jonides has further distinguished between covert and overt orienting of attention, where the latter involves the eye-head system to change the visual input.[5.19]

It is now generally accepted that the examination of visual stimuli involves two phases (see Fig. 5.1). The first is an essentially parallel phase which occurs within the first glimpse of the stimulus and provides a global impression. The methodology used to investigate this typically involves presentation of artificial targets using brief (50-200 ms) exposure times. Visual processing in this stage is referred to as being *preattentive* or *distributed*.[5.20] This leads to detection of visual features anywhere over the stimulus. Such detection, termed pop-out, depends on simple local pattern properties, e.g., brightness, size, and slope of lines and blobs.

The second phase is *focal attention*, in which the display is typically serially examined using eye movements. As the eyes scan the display, information is processed in detail not just from the foveal area but from some larger area around the fovea. Mackworth[5.21] has used the term "useful field of view" to refer to this area. Other functionally equivalent terms are "visual lobe"[5.22] and "functional visual field."[5.23] The density of irrelevant display items affects the size of this area[5.24] as does target and nontarget similarity.[5.22] Ikeda and Takeuchi[5.25] have proposed that this area shrinks as foveal load increases, although experience or training with the stimulus may overcome such shrinkage.[5.26]

5.3.2 *Visual Scanning*

Voluntary eye movements are the most massive instrument of selective attention employed in this stage. A perplexing research problem, however, is that accurate measurement of where an observer is looking does not mean he is actually attending there. This fact has been known since the time of von Helmholtz and (as Shepherd, et al.[5.27] point out), whereas it seems impossible for us to make an eye

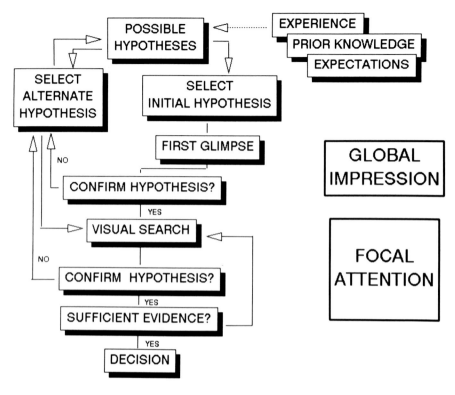

FIGURE 5.1. A schematic outline of the examination of visual display for the possible presence of a real world target (e.g., an abnormality in radiology). Factors such as experience facilitate selection of an initial plausible hypothesis (schema). The first glimpse provides a global impression which is followed by a focal attention stage in which active serial visual search of the stimulus takes place.

movement without producing a change in attention locus, eye movements are not necessary for a change in attention.

What type of eye movements are involved in scanning the visual environment in this sampling fashion? There are several different kinds. The most frequent is the saccade and Cumming[5.28] has pointed out that we make several billion saccades in a lifetime. Saccades are typically used to scan the visual world so as to bring different areas to fall on the fovea. They require about 250 ms to plan their execution and take about 50 ms to execute,[5.29] reaching very high velocities of up to $600°/s^{-1}$. Saccades are typically rapid conjugate movements which are under voluntary control and are ballistic in nature. Lancaster[5.30] pointed out that the majority are less than 15° in amplitude. These eye movements very often overshoot[5.31] or undershoot their target, so producing small corrective saccades.[5.32] They can be curved as well as linear in nature. During a saccade there is a suppression of visual uptake of information.

When we try to accurately maintain a stationary fixation on a target, the eye is still subject to several small motions, generally less than $1°$ in amplitude. These movements are not under conscious control. Ditchburn and Ginsborg distinguish between flicks (miniature or microsaccades), drifts, irregular movements, and high frequency tremors.[5.33] The effect of such movements is to shift the retinal image about the retina over an area larger than the foveal receptors, thus preventing fading of the image.[5.34] This is discussed in Chaps. 1 and 4.

Pursuit eye movements are used to track a moving target and are necessary to maintain a stable image of such a target on the retina. These movements are slow ($30° - 100°/s^{-1}$) and smooth in nature. Related to pursuit movements are compensatory eye movements, which act to compensate for movement of the head and body. These can compensate for a movement rate of about $1° - 30°/s^{-1}$.

In contrast to these types of conjugate movements, vergence movements are disconjugate movements which allow both eyes to focus on the same target. The eyes converge to look at nearby objects and diverge to focus on distant ones. Vergeance movements are slow (around $100°/s^{-1}$) and a typical movement would last approximately one second, much slower than saccades.

A variety of techniques exist to monitor eye movements (see Young and Sheena[5.35]). No single technique is appropriate for every possible situation; choice depends on the task, resolution, and accuracy required, and the sampling rate of the technique employed. For instance, we can very accurately monitor an observer as he reads a single line of text and adroitly alter characters in the next word as he is making saccade towards it. Alternatively, we can record eye movements in the real world as the observer drives, flies, or plays golf. In reading, horizontal saccades and microsaccades would be of interest, whereas in the latter tasks, two-dimensional saccades would be monitored.

In general, research has concentrated on examining how an observer responds to visual information presented in a one- or two-dimensional display. The present discussion concerns two-dimensional displays, in which focal examination of a visual stimulus is considered to be a sequence of saccades and fixations, although the fixations themselves actually encompass miniature movements.

5.3.3 Matching Image and Observer Characteristics

It is important that observer and image characteristics are matched so that the observer has the best opportunity to adequately abstract the necessary information. For instance, the sensitivity of human observers is well known in the spatial frequency domain (see Chap. 3). In a task such as radiological inspection, observation of radiologists demonstrates that they regularly vary their viewing distance and use magnifying (and sometimes minifying) lenses to inspect the display. Such techniques allow better matching of the observer's sensitivity to particular spatial representations in the display. With digital displays the observer can achieve the same goal electronically.

Despite care in designing both visual displays and workstation ergonomics, however, the human observer is still prone to errors. For instance, many visual tasks involve the correct recognition of a target. Here, two types of error can occur; these are *false negatives* (misses) and *false positives* (incorrect detections). In industrial inspection tasks such as the examination of electronic "chips," in one experiment experienced observers missed 23% of anomalous samples.[5.36] In radiology, somewhat similar error rates of 20% to 30% are found. This is true both for experimental investigations (for instance of the accuracy of reporting chest radiographs[5.37]) and in mass surveys of tuberculosis[5.38] or lung cancer.[5.39] Such studies have tended to show large inter- and intraobserver variability and demonstrate that experienced radiologists, when interpreting large numbers of chest films, fail correctly to identify abnormal radiographs even though these do contain demonstable evidence of disease. In radiology, false negatives pose a considerable cost to a patient where the disease may subsequently be more advanced when eventually detected. A false positive report causes unnecessary additional investigations together with considerable psychological trauma.

Earlier we argued that such errors can arise either from a failure to adequately match the visual display to the observer, or from the observer's failure to abstract and interpret relevant visual information.[5.40] Sometimes the sheer size of what is misinterpretated seems amazing. For instance, Tuddenham[5.41] reported failures to notice an amputated female breast on a radiograph and we[5.40] reported similar failure with a scintiscan display of a leg amputee where eye movement recordings clearly showed the experienced observer fixating on where the leg ought to be and then reporting the display as normal!

5.3.4   Focal Attention and Observer Factors

When targets do not preattentively pop out, the observer must focally search the visual display. Saccadic eye movement monitoring can be employed to determine whether targets are consequently missed because they are not looked at or whether they are in fact examined but misinterpreted. From eye movement records it is possible to tease out the reason why a target has been missed. For instance, assuming the useful field of view for the target type is known, then if the observer has failed either to fixate the target or to encompass it within this field of view the error is due to *inadequate search* of the stimulus. If the target is adequately· fixated for a short period of time but still not correctly reported, this is a *detection error* (the observer looked at it but failed to detect it). If the observer has similarly adequately looked at it for a much longer period of time and still fails to report it, then this is termed an *interpretation error* (Fig. 5.2). Kundel, et al. have shown that in chest radiology false negative errors are due more to interpretation (45%) than to search (30%).[5.42] This is not to diminish the role of search, as adequate search is a prerequisite to detection unless targets are detected in the preattentive phase.

Two broad classes of models of eye movement control have been postulated to explain the search process. These can be categorized, depending upon the degree of control postulated, as moment-to-moment control or indirect control models. The former models argue that information from the current fixation controls both

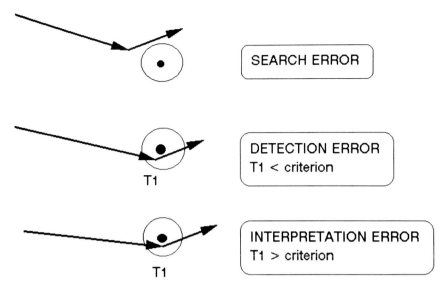

FIGURE 5.2. The three types of errors which can occur during search of a visual stimulus. The circle surrounding the small dark target is the area within which the observer has to fixate in order to detect the target (equivalent to the useful visual field, although this term refers to an area around each fixation). A search error occurs when the observer fails to fixate near the target. Detection and interpretation errors are those where the observer has fixated sufficiently near the target but still fails to detect it. A detection error occurs when fixation time $(T1)$ near the target is less than an empirically derived criterion (e.g., 600 ms) and an interpretation error occurs when this time exceeds the criterion.

the length of the present fixation and where next to move,[5.43–5.45] whereas indirect models invoke cognitive processes.[5.46–5.47]

Thus, viewing a display proceeds mainly as a sequence of saccades and fixations. The main variables of fixation which have been examined are *fixation duration* and *fixation location*. Consider fixation duration. During a fixation the observer is not only encoding information from the current eye position on the visual stimulus but is also programming the subsequent saccade. It appears that only the first 50 ms or so of a fixation (at least in reading) is spent encoding information. Thus, the majority of the fixation time is concerned with preparing the next movement. If an observer could decide on an examination strategy in unit time when presented with a visual stimulus, then he could opt either for using many short fixations or fewer but longer ones. Boynton[5.48] argued that a good searcher is one who makes fewer fixations of longer duration. This is supported by the work of Schoonard, et al.,[5.36] who examined experienced industrial inspectors whose task was to check integrated circuits. It was found that their modal fixation time was 200 ms. These times were shorter than those found in studies of White and Ford[5.49] and Ford et al.[5.50] in somewhat simpler search tasks and where the subjects were less trained. Other work also supports this; for instance, Loftus[5.51] found that there was a higher probability of an observer remembering a picture if he made more fixations on it in a unit time. Fixation time has also been extended to the concept of "gaze" and "dwell time," where individual fixations are grouped together within a spatial threshold value.

What determines which part of a stimulus will be fixated by an observer? Many studies show that fixations tend to occur on areas of high informational content, such as contours, rather than homogenous areas.[5.52-5.55] Mackworth and Morandi[5.56] demonstrated that a group of subjects fixated areas of display which another group independently rated as being highly informative. Much earlier, Buswell[5.57] had found that observers often fixated on faces and hands of people present in the visual scene, these being deemed highly informative. There are some exceptions to this general finding. For instance, Hughes and Cole[5.58] found in a driving task that a large number of fixations fell on an area of the sky, although fixating there presumably contributed no driving information. There is also evidence of some tendency for observers to fixate the visual "center of gravity" of particular displays.

Each individual eye fixation yields some information about the visual stimulus. How are these individual glances integrated together into our percept of the stimulus? This topic has been addressed well and separately by Hochberg[5.59] and Gibson.[5.60] Gibson's approach stressed the role of stimulus invariants (such as the overlap of an object between successive fixations, optical gradients, and discontinuities). In contrast, Hochberg emphasized cognitive factors and proposed that a schematic map is generated which relates these successive inputs from each fixation together. This map is formulated on the basis of previous knowledge and on what the observer has just seen. This relates well to the concept of bottom-up and top-down processing. In the former, stimulus information effectively drives perception, whereas with the latter cognitive factors predominate. Neisser[5.61] related these two opposing views with his "perceptual cycle," in which stimulus information constantly modifies the search plan, itself altered by the sampled stimulus information (see Fig. 5.3).

Therefore, real-world knowledge, such as physical laws, past experience, and expectancies would be expected to affect the way in which successive eye fixation information is amalgamated. This does appear to be the case as the pattern of search is indeed affected by experience. With radiological chest images, Kundel and La Follette[5.62] presented data indicating that naive observers tend to employ what they termed localized central scanning. In contrast, experienced observers tend to use a more circumferential scan. The former consists of clustering of eye fixations interspersed with relatively small saccades, whereas the latter is made up of more widely spaced fixations around significant areas of the image. Radiological experience presumably helps to establish which areas to fixate. Similarly, in industrial inspection, Schoonard, et al.[5.36] found that experienced inspectors tended to fixate areas of integrated circuits where there was likely to be a target and generally ignore areas where there was a low probability of target occurrence. The implication of such findings is that with experience the observer can make more efficient use of peripheral vision, either "attending" to a larger stimulus area at a time and/or determining where best to place the next fixation. Clearly,

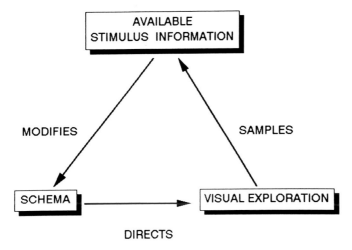

FIGURE 5.3. Perceptual cycle. After Neisser.[5.61]

in searching displays for targets, one role of experience is to produce a set of probable target locations.

Expectancies can be generated on the basis of specific information given to the observer before the stimulus is presented. An often-cited example of how instructions affect the observer's fixation pattern is from Yarbus,[5.53] who monitored an observer's eye movements while examining the same picture under different instructions. The picture was a domestic scene containing several people. For instance, when observers were asked to estimate the material circumstances of the family depicted in the picture, many of their eye fixations fell on furniture and clothing. In contrast, when they were asked to gauge the people's age, fixations fell on the faces of the figures. Effects of instructions on search behavior are found in other situations, such as driving.[5.58–5.63] Search patterns in radiology are affected by prior clinical information given to the observer about the patient.[5.64] Such findings clearly demonstrate that fixation patterns are under voluntary control.

Another example of how intructions affect search can be seen in the area of observer training. Eye movement investigations of how experienced radiologists examine radiographs demonstrate the use of free search rather than the use of a systematic search strategy. This is in direct contrast to how, during training, such individuals are often encouraged to adopt a regular search pattern. The underlying argument for this seems to be that such a systematic approach leads to appreciation of radiological appearance. Indeed, numerous texts specifically advocate this. Somewhat in contradiction, available experimental evidence shows that when a group of observers is taught to adopt a stylized strategy, then although they could adhere to it (albeit with individual variations in eye behavior superimposed) they also evidenced an increase in false positive detections as compared to a matched

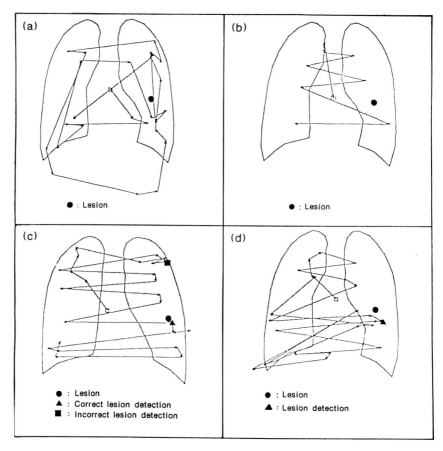

FIGURE 5.4. Eye movement pattern of an experimental subject when examining the same radiograph (a) before and (b) after training, when the scanning strategy was employed. In each case the lesion was missed. Further examples of the same subject examining other radiographs after training are shown in (c) and (d). While in each case the observer uses the taught strategy, large differences in the precise pattern of eye movements are apparent. Reprinted from Gale and Worthington[5.65] with permission.

control group[5.65] (see Fig. 5.4). Likewise, other experimental manipulations which interfere with the normal mode of viewing radiographs produce similar errors. This finding fits well with an explanation of the role of eye movements as resulting from both top-down and bottom-up processing, i.e., from a schematic plan in conjunction with the actual visual information absorbed. The gross omission errors mentioned earlier[5.40] can be explained as instances where the initial hypothesis or schematic map dominates over the actual sampled visual information in a top-down manner.

When examining a visual display, observers do not constantly search out new areas of the stimulus to fixate but instead repeatedly return to the same parts. Early work by Tuddenham and Calvert[5.66] pointed out such non-uniform coverage in radiographic visual search. This finding holds for various classes of stimuli such as faces[5.67] or radiographs.[5.68] Loftus[5.51] found that in 95% of cases, objects in a picture which are subsequently remembered are fixated by the third fixation and are then refixated several times. This implies that increasing the viewing time available for an observer to search a display does not necessarily increase his detection performance. The actual pattern or sequence of saccadic eye movements may be important in how we recognize stimuli. Several researchers have reported a certain regularity in eye movement sequences. For instance, see the classic works of Buswell[5.57] and Yarbus.[5.53] Other experimental evidence was provided by Zusne and Michels.[5.55] The most explicit theoretical proposal invoking the role of eye movement sequence is that put forward by Noton and Stark.[5.69] They proposed that the memory for a picture is composed of an alternating sequence of sensory image and corresponding motor instruction concerning the location of the next fixation (i.e., where next visually to attend). They proposed a scan path hypothesis, which proposes that a new visual stimulus produces a particular fixation sequence, termed the scan path. This pattern recurs when the same stimulus is recognized if it is again presented to that observer. Foveal details, it is argued, are stored together with memory traces of the eye movements. To allow for internalized attention shifts, the internal stimulus representation, a feature ring, is elaborated into a feature network, with the scan path being the preferred sequence in this network. Stark and Ellis[5.70] and Ellis and Smith[5.71] report other work concerning scan paths.

The scan path proposal has not been without its critic, notably Spitz[5.72] and Didday and Arbid.[5.73] Some workers report finding scan paths without ascribing any particular function to them. For instance, using simple random shapes, Locher and Nodine[5.74] found scan paths in about half of their subjects' eye movement patterns but there was no indication that they actually aided recognition. Walker-Smith, et al.[5.67] found some regular scanning patterns in a face recognition task but stressed the role of peripheral information in executing some control over the next fixation in interaction with some internal schema, a theoretical approach later elaborated for ambiguous stimuli.[5.75]

More recently, the scan path hypotheses has been revitalized by Groner, et al.[5.76] who made the useful distinction between local and global scan paths. As Groner pointed out, local scan paths "reflect the spatio-temporal organization of the fixations on a local scale of successive events, while global scan paths reflect the distribution of eye fixations when taking into account the entire inspection process."[5.77] Thus, global scan paths are mediated by search plans in a top-down fashion, whereas local scan paths are dependent on fixation and peripheral information. Groner and Menz[5.78] have also demonstrated that local and global scan paths are subject-specific and that global scan paths are related to the task which the observer has to perform.

5.3.5   Individual Differences

Individual differences, such as detection performance differences found between observers presented with visual information, have led some workers to investigate whether they could measure aptitude for the task and hence to relate this to task performance. The idea is attractive, particularly in the selection of individuals for specific visual tasks, and several psychological tests of ability have been employed in different studies. In general, such work has not been too successful, although in radiology Smoker, et al.[5.79] found some evidence that observer performance could be predicted on the basis of perceptual ability, and there are some initial supportive findings in breast screening.[5.80] Furthermore, individual differences in scanning have often been reported and eye movement analysis has revealed that visual information processing strategies differ between subjects of high and low ability in spatial ability tasks.[5.81]

How we individually respond to visual information may also be affected by the time of day, as performance on several tasks has been shown to vary throughout the day.[5.82] For instance, serial visual search tends to become faster over the day, largely paralleling the variation in body temperature.[5.83] In the past, such performance variations were directly related to changes in physiological arousal. A more complex basis for circadian variation in performance, however, seems plausible.[5.84] In an experimental search task using chest radiographs, we[5.85] demonstrated that a drop in receiver operator characteristic (ROC) sensitivity was obtained after lunch (such a post-lunch dip is found in several tasks). The analysis involved using eye fixation to provide localization of the ROC response (EMROC analysis). Such a finding would be particularly intriguing if it were to translate into the real-world situation.

5.3.6   Improving diagnostic accuracy

Radiologists bring great expertise to the skilled task of interpreting medical images.[5.86] However, human diagnostic performance is largely based on correctly identifying and using appropriate visual information in the complex display and so is ultimately limited by vagaries of human attention as detailed earlier. Approaches to helping observers achieve even greater accuracy are based on identifying salient visual features in the display and highlighting them in some way for the observer.[5.87] Mammography is a good domain to illustrate this. Various research groups are currently developing algorithms for the computer detection of key features indicative of early breast cancer.[5.88] Such methods are fairly successful and lead to systems which can act as diagnostic aids to assist the radiologist. However, a key difficulty with such methods is that they invariably yield both false positive identifications of suspicious areas as well as true positive ones on any image. Just how many false positives a radiologist will tolerate from such approaches before this error rate affects his diagnostic performance is currently a pertinent research topic.[5.89]

Another useful technique (computer-aided perception[5.90]) is to monitor the radiologist's own visual examination of the display and then identify those areas where he looked for more than a predetermined length of time (thus indicating which areas of the display are "informative"). The image is then overlaid with these areas highlighted. Basically such a technique can improve detection and interpretation errors but not affect search errors, whereas prompting based on computer recognition of abnormalities can affect all three error types. An important issue, however, is how best to feed back to the radiologist potential image areas in a way that does not distract him by masking the image features in the process.

The increased understanding of the underlying development of the complex cognitive-perceptual skill[5.91] of interpreting medical images has led to teaching nonradiologists to accomplish some aspects of identifying abnormalities. This enables such personnel to act either as a second reader or as an initial reader, thus increasing the overall system diagnostic accuracy.

5.4   References

[5.01] Galanter E. Contemporary psychophysics. In: Brown R., Galanter E., Hess E.H., Mandler G., eds. *New Directions in Psychology*. New York: Holt, Rinehart, and Winston; 1962:87–156.

[5.02] Jesteadt W., Bacon S.P., Lehman J.R. Forward masking as a function of frequency, masker level, and signal delay. *J. Acoust. Soc. Am.* 1982; 71:950–962.

[5.03] Gescheider G.A. *Psychophysics: Method and Theory*. Hillsdale: Erlbaum; 1985.

[5.04] Baird J.C., Noma E. *Fundamentals of Scaling and Psychophysics*. New York: Wiley; 1978.

[5.05] Laming D. Some principles of sensory analysis. *Psychol. Rev.* 1985; 92:462–485.

[5.06] Stevens S.S. *Psychophysics: Introduction to its Perceptual, Neural, and Social Prospects*. New York: Wiley; 1975.

[5.07] Falmagne J.C. *Elements of Psychophysical Theory*. Cambridge: Cambridge University Press; 1985.

[5.08] Myers A.K. Psychophysical scaling and scales of physical stimulus measurement. *Psychol. Bull.* 1982; 92:203–214.

[5.09] Posner M.I., Snyder C.R.R., Davidson B.J. Attention and the detection of signals. *J. Exp. Psychol.: General* 1980; 109:160–174.

[5.10] Norman D.A. Towards a theory of memory and attention. *Psychol. Rev.* 1968; 86:114–255.

[5.11] Eriksen C. W. Attentional search of the visual field. In: Brogan D., ed. *Visual Search*. London: Taylor and Francis; 1990:3–19.

[5.12] Lambert A.J. Expecting different categories at different location and selective attention. *Quart. J. Exp. Psychol.* 1987; 39A: 61–76.

[5.13] Johnston W.A., Dark V.J. Selective attention. *Annu. Rev. Psychol.* 1986; 37:43–75.

[5.14] Trevarthen C.B. Two mechanisms of vision in primates. *Psychol. Forsch.* 1968; 31:299–337.

[5.15] Held R., Ingle D., Schneider G.E., Trevarthen C.B. Locating and identifying: Two modes of visual processing. *Psychol. Forsch.* 1967; 31:42–62.

[5.16] Treisman A., Gelade G. A feature-integration theory of attention. *Cogn. Psychol.* 1980; 12:97–136.

[5.17] Julesz B. Towards an "axiomatic" theory of preattentive vision. In: Edelman G.M., Gall W.E., Cowan W.M., eds. *Dynamic Aspects of Neocortical Function*. New York: Wiley; 1984:160–187.

[5.18] Breitmeyer B.G., Ganz L. Implications of sustained and transient channels for theories of visual pattern masking, saccadic suppression, and information processing. *Psychol. Rev.* 1976; 83:1–36.

[5.19] Jonides J. Towards a model of the mind's eye's movements. *Can. J. Psychol.* 1980; 34: 103–112.

[5.20] Beck J. Similarity grouping and peripheral dicriminability under uncertainty. *Am. J. Psychol.* 1990; 85:1–19.

[5.21] Mackworth N.H. Stimulus density limits the useful field of view. In: Monty R.A., Senders J.W., eds. *Eye Movements and Psychological Processes*. New York: Wiley; 1976:307–321.

[5.22] Bellamy L.J., Courtney A.J. Development of a search task for the measurement of peripheral visual acuity. *Ergonomics* 1981; 24:597–599.

[5.23] Sanders A.F. Some aspects of the selective process in the functional visual field. *Ergonomic* 1970; 13:101–117.

[5.24] Bartz A.G. Peripheral detection and central task complexity. *Hum. Fact.* 1976; 18:63–70.

[5.25] Ikeda M., Takeuchi T. Influence of foveal load on the functional visual field. *Percept. Psychophys.* 1975; 18(4):255–260.

[5.26] Edwards D.C., Goolkasian P.A. Peripheral vision location and kinds of complex processing. *J. Exp. Psychol.* 1974; 102(2):244–249.

[5.27] Shepherd M., Findlay J.M., Hockey R.J. The relationship between eye movements and spatial attention. *Quart. J. Exp. Psychol.* 1986; 38A:475–491.

[5.28] Cumming G.D. Eye movements and visual perception. In: Carterette E.C., Friedman M.P., eds. *Handbook of Perception.* New York: Academic; 1978:221–255.

[5.29] Hallett P.E. Eye movements. In: Boff K.R., Kaufman L., Thomas J.P., eds. *Handbook of Perception and Human Performance* (Vol. 1). New York: Wiley; 1973:10–112.

[5.30] Lancaster W.B. Fifty years experience in ocular motility. *Am. J. Opthal.* 1941; 24:485–595.

[5.31] Thomas J.G. Subjective analysis of saccadic eye movements. *Nature* 1961; 189:842–843.

[5.32] Ginsborg B.L. Small voluntary movements of the eye. *Br. J. Opthal.* 1953; 37:746–754.

[5.33] Ditchburn R.W., Ginsborg B.L. Involuntary eye movements during fixation. *J. Physiol.* 1953; 119:1–17.

[5.34] Fuchs A.F. The saccadic system. In: Back-y-rita O., Collins C.C., Hyde J.E., eds. *The Control of Eye Movements.* New York: Academic; 1971:343–362.

[5.35] Young L.R., Sheena D. Survey of eye movement recording methods. *Behav. Res. Methods Instrum.* 1975; 7(5):397–429.

[5.36] Schoonard J.W., Gould J.D., Miller L.A. Studies of visual inspection. *Ergonomics* 1973; 16:365–379.

[5.37] Herman P.G., Hessel S.J. Accuracy and its relationship to experience in the interpretation of chest radiographs. *Invest. Radiol.* 1975; 19:62–67.

[5.38] Yerushalmy J. The statistical assessment of the variability in observer perception and description of Roentgenographic pulmonary shadows. *Radiol. Clin. N. Am.* 1969; 1:381–390.

[5.39] Guiss L.W., Kuenstler P. A retrospective view of survey photofluorograms of persons with lung cancer. *Cancer* 1960; 13:91–95.

[5.40] Gale A.G., Johnson F., Worthington B.S. Psychology and radiology. In: Oborne D.J., Gruneberg M.M., Eiser J.R., eds. *Research in Psychology and Medicine* (Vol. 1). London: Academic; 1979:453–460.

[5.41] Tuddenham W.J. Visual search, image organization, and reader error in Roentgen diagnosis. *Radiology* 1962; 78:694–704.

[5.42] Kundel H.L., Nodine C.F., Carmody D. Visual scanning, pattern recognition, and decision making in pulmonary nodule detection. *Invest. Radiol.* 1978; 13:175–181.

[5.43] Rayner K., McConkie G.W. What guides a reader's movements? *Vision Res.* 1976; 16:829–837.

[5.44] O'Regan J.K. Elementary perceptual and eye movement control processes in reading. In: Rayner L., ed. *Eye Movements in Reading: Perceptual and Language Processes.* New York: Academic; 1983:121–139.

[5.45] Carpenter P.A., Just M.A. What your eyes do while your mind is reading. In: Rayner K., ed. *Eye Movements in Reading: Perceptual and Language Processes.* New York: Academic; 1983:275–307.

[5.46] Shebilske W. Reading eye movements from an information processing point of view. In: Massaro D., ed. *Understanding Language.* New York: Academic; 1975: 291–311.

[5.47] Vaughan J. Control of visual fixation duration in search. In: Senders J.W., Fisher D.F., Monty R.A., eds. *Movements and the Higher Psychological Functions.* Hillsdale: Erlbaum; 1979:135–142.

[5.48] Boynton R.M. Summary and discussion. In: Morris A., Horne E.P., eds. *Visual Search.* Washington: National Academy of Science; 1960:231–250.

[5.49] White C.T., Ford A. Eye movements during simulated radar search. *J. Opt. Soc. Am.* 1960; 50:909–913.

[5.50] Ford A., White C.T.., Lichenstein M. Analysis of eye movements during free search. *J. Opt. Soc. Am.* 1959; 49:287–292.

[5.51] Loftus G.R. Eye fixations and recognition memory for pictures. *Cogn. Psychol.* 1972; 3:525–551.

[5.52] Berlyne D.E. The influence of complexity and novelty in visual figures on orienting responses. *J. Exp. Psychol.* 1958; 55:289–296.

[5.53] Yarbus A.L. *Eye Movements and Vision.* New York: Plenum; 1967.

[5.54] Baker M.A., Loeb M. Implications of measurements of eye fixations for a psychophysic of form perception. *Percept. Psychophys.* 1973: 13(2):185–192.

[5.55] Zusne L., Michels K.M. Nonrepresentational shapes and eye movements. *Percept Mot. Skills* 1964; 18:11–20.

[5.56] Mackworth N.H., Morandi A.J. The gaze selects informative details within pictures. *Percept. Psychophys.* 1967; 2(11):547–552.

[5.57] Buswell G. *How People Look at Pictures.* Chicago: University of Chicago Press; 1935.

[5.58] Hughes P.K., Cole B.L. The effect of attentional demand on eye movement behavior when driving. In: Gale A.G., Freeman M.H., Haslegrave C.M. et al., eds. *Vision in Vehicles II.* Amsterdam: North Holland; 1988: 221–230.

[5.59] Hochberg J. In the mind's eye. In: Haber R.N., ed. *Contemporary Theory in Visual Perception.* New York: Holt, Rinehart, and Winston; 1968:309–331.

[5.60] Gibson J.J. *The Senses Considered as Perceptual Systems.* Boston: Houghton Mifflin; 1966.

[5.61] Neisser U. *Conitive Psychology.* New York: Appleton-Century-Crofts; 1967.

[5.62] Kundel H.L., La Follette P.S. Visual search patterns and experience with radiological images. *Radiology* 1972; 103:523–528.

[5.63] Mourant R.R., Rockwell T.H., Rackoff N.J. Driver eye movements and visual workload. *Highway Res. Rec.* 1969; 292:1–10.

[5.64] Kundel H.L., Wright D.J. The influence of prior knowledge of visual search strategies during the viewing of chest radiographs. *Radiology* 1969; 93:315–320.

[5.65] Gale A.G., Worthington B.S. Scanning strategies in radiology. In: Groner R., Menz C., Fisher D.F., Monty R.A., eds. *Eye Movements and Psychological Factors: International View.* Hillsdale: Erlbaum; 1983: 169–191.

[5.66] Tuddenham W.J., Calvert W.F. Visual search patterns in Roentgen diagnosis. *Radiology* 1961; 76:255–256.

[5.67] Walker-Smith G., Gale A.G., Findlay J.M. Eye movements during pattern perception. *Perception* 1977; 6:313–326.

[5.68] Llewelyn-Thomas E. Search behavior. *Radiol. Clin. N. Am.* 1969; 7:403–417.

[5.69] Noton D., Stark L. Scan paths in eye movements during pattern perception. *Science* 1971; 171:308–311.

[5.70] Stark L., Ellis S.R. Scan paths revisited: Cognitive models direct active looking. In: Fisher D.F., Monty R.A., Senders J.W., eds. *Eye Movements: Cognition and Visual Perception.* Hillsdale: Erlbaum; 1981:193–226.

[5.71] Ellis S.R., Smith J.D. Patterns of statistical dependency in visual scanning. In: Groner R., McConkie G.W., Menz C., eds. *Eye Movements and Human Information Processing.* Amsterdam: North Holland; 1985:221–238.

[5.72] Spitz H.H. Scan paths and pattern recognition. *Science* 1971; 173:753.

[5.73] Didday R.L., Arbid M.A. Eye movements and visual perception: A 'two visual system' model. *Int. J. Man-Machine Stud.* 1975; 7:547–569.

[5.74] Locher P.J., Nodine C.F. The role of scan paths in the recognition of random shapes. *Percept. Psychophys.* 1972; 15(2):308–314.

[5.75] Gale A.G., Findlay J.M. Eye movement patterns in viewing ambiguous figures. In: Groner R., Menz C., Fisher D.F., Monty R.A., eds. *Eye Movements and Psychological Functions: International Views.* Hillsdale: Erlbaum; 1983: 145–168.

[5.76] Groner R., Walder F., Groner M. Looking at faces: Local and global aspects of scan paths. In: Gale A.G., Johnson F., eds. *Theoretical and Applied Aspects and Eye Movement Research.* Amsterdam: North Holland; 1984: 523–533.

[5.77] Groner R. Eye movements, attention and visual information processing: Some experimental results and methodological considerations. In: Luer G., Lass U., Shallo-Hoffman J., eds. *Eye Movement Research: Psyiological and Psychological Aspects.* Toronto: C.J. Hogrefe; 1988: 295–319.

[5.78] Groner R., Menz C. The effects of stimulus characteristics, task requirements, and individual differences on scanning patterns. In: Groner R., McConkie G.W., Menz C., eds. *Eye Movements and Human Information Processing. Proceedings of the XXIII International Congress of Psychology.* Amsterdam: North Holland; 1985: 239–250.

[5.79] Smoker W.R.K., Berbaum K.S., Luebke N.H., Jacoby C.G. Spatial perception testing in diagnostic radiology. *Am. J. Roentg.* 1984; 143: 1105–1109.

[5.80] Walker G.E., Gale A.G., Roebuck E.J., Worthington B.S. Training and aptitude for mammographic inspection. In Megaw E.D., ed. *Contemporary Ergonomics.* London: Taylor and Francis; 1989: 456–460.

[5.81] Just M.A., Carpenter P.A. A theeory of reading: From eye fixation to comprehension. *Pychol. Rev.* 1980; 87: 329–354.

[5.82] Colquhoun W.P. Circadian variations in mental efficiency. In: Colquhoun W.P., ed. *Biological Rhythms and Human Performance.* London: Academic; 1971: 39–107.

[5.83] Monk T.H. Temporal effects in visual search. In Clare J.N., Sinclair M.A., eds. *Search and the Human Observer.* London: Taylor and Francis; 1979: 30–39.

[5.84] Folkard S., Monk T.H. Chronopsychology: Circadian rhythms and human performance. In: Gale A., Edwards J., eds. *Physilogical Correlates of Human Behavior.* London: Academic; 1983: 52–78.

[5.85] Gale A.G., Murray D., Millar K., Worthington B.S. Circadian variation in radiology. In: Gale A.G., Johnson F., eds. *Theoretical and Applied Aspects of Eye Movement Research.* Amsterdam: North Holland; 1984: 312–321.

[5.86] Chi M.T.H., Glaser R. Farr M.J. *The Nature of Expertise.* Hillsdale NJ: Lawrence Erlbaum; 1988.

[5.87] Getty D.J. Assisting the radiologists to greater accuracy. In: Kundel H.L., ed. *Medical Imaging 1996*; image perception, Proc SPIE, in press.

[5.88] Gale A.G., Astley S.M., Dance D.R. & Cairns A.Y. *Digital Mammography.* Amsterdam: Elsevier; 1994.

[5.89] Hutt I.W., Astley S.M., & Boggis C.R.M. Prompting as an aid in mammography. In: Gale A.G., Astley S.M., Dance D.R. & Cairns A.Y., eds. *Digital Mammography.* Amsterdam: Elsevier; 1994.

[5.90] Nodine C.F., Kundel H.L. Computer aided perception aids pulmonary nodule detection. In Kundel H.L., ed. *Medical Imaging 1994*; image perception, Proc SPIE 2166; 1994; 55–59.

[5.91] Sharples M. Computer based tutoring of visual concepts: from novice to expert J. *Compt. Assist. Learn.* 1991; 7: 123–132.

6
Cognitive Interpretation of Visual Signals

William R. Hendee

6.1   Early Views of Cognition

How we see, and how we know what we see, are conundrums that have perplexed scientists for decades and intrigued philosophers for centuries. In pre-Socratic times, philosophers attempted to separate interactions with the external world into those that involve the human senses and those that do not involve the senses directly. The theory of Empedocles, typical of the reasoning employed by several early philosophers, suggested that the senses are affected by tiny, particle-like effluences emitted by objects. As the effluences strike the body, they lodge in one set of pores or another depending on their exact size and configuration. Each set of pores gives rise to a particular sensation experienced by the observer. For a stone to be seen as a gray, round object, its effluences would have to be captured by the pores that characterize gray, round objects. These pores are located in the eyes, because the object is seen. Objects that are heard emit effluences that are captured by the ears, and effluences that are captured by the nose give rise to the sensation of smell.

Empedocles' theory presented a materialistic model of perception: An object is sensed by interaction of one physical object with another, through the emission and receipt of physical effluences. In the materialist theory there is no mental imagery or intellectual processing of information. Visual sensation is strictly a physical response and visual detection, recognition and interpretation are all one and the same process. The materialistic model of perception was challenged by many philosophers. They wondered how a round, gray object such as a rock was distinguishable from another round, gray object that is not a rock. The materialists answered that the sensory pores were subdivided in such a manner that effluences from a rock could be distinguished from those emitted by similar but not identical objects. Empedocles said[6.1]: "For by earth we see earth, by water water, by air bright air, and by fire brilliant fire."

Aristotle (384-322 BC) expanded the materialistic theory of perception by suggesting that effluences are a subtle, nonmaterialistic influence of the object that induce a change in the observer. That is, the observer "receives the form of the object without its matter." [6.2] In this manner the observer receives characteristic "sense data" such as color and shape from the object and incorporates them into the appropriate sensory apparatus. The observer then integrates information from each of the affected senses to achieve a unified perception of the object. This unification is accomplished through the exercise of a common faculty, termed "common sense," that is located in the heart. The unified perception is a physical change that remains in the body after the object is removed from the vicinity of the observer. Aristotle referred to these changes as phantasms, and suggested that they constitute the imagination of the observer. The imagination is essential to memory and furnishes the linkage between perception and thought. Thinking depends on the presence of phantasms and therefore perceptual experiences are essential to thinking. In fact, the range of thoughts is restricted by the breadth of experiences. In one brief passage, Aristotle attempted to identify a higher level of thinking, termed "the active intellect," and to distinguish it from the thinking process called "the passive intellect" that is dependent on phantasms. In the remainder of Aristotle's works, however, the importance of phantasms gained from sensory experiences prevails.

6.2   Western Philosophical Speculations on Cognition

Descartes (1596-1650) has been a major influence on philosophical thought from the time of his work in the final century of the Renaissance. He proposed the philosophy of dualism in which the mind and body are two separate and distinct entities. They have some level of interaction, however, because mental events can be "secreted" by the brain. Spinoza (1632-1677) rejected Descartes' dualistic philosophy and replaced it with a theory of "substance monism" in which the mind and body of the individual have no independent existence.[6.3] These entities have no reality either, other than as manifestations of the one and only reality—God. Spinoza believed that the mind and body do not interact, because they are one and the same. He suggested that physical stimulation of the retina by light rays (a modified form of effluences) from an object can be interpreted as an image, but the interpretation is misleading because it is simply a reaction of the body to the light. The reaction is not reliable as a description of any fundamental characteristics of the object itself. In some ways, Spinoza's ideas are compatible with those of the materialistic school of Greek philosophers: the only characteristics detectable about an object are those that can be derived from physical sensations induced in the observer by the object.

Locke (1632-1704) was a major force in early philosophical theories of the mind and thought process.[6.4] In his "casual theory" of perception, Locke proposed that the appearance of an object is an "idea" that is distinct from, but casually related to, the object itself. Thus, ideas such as visual images are the result of the way objects act upon the senses and not a property of the objects themselves. Ideas can be a product of direct actions upon the senses, memories of sensory experiences, or abstract concepts that evolve from experiences. But always there is a casual relationship is based on experiences with the objects. The universe is essentially a system of solid bodies in mechanical interaction, whereas experiences of an observer are a system of "corpuscles" in mechanical interaction. These corpuscles act upon the sense organs to induce "ideas of sensation" in the mind. Locke suggested that these two types of interactions reflect the existence of two categories of properties. One category is termed "primary characteristics" that are properties of objects such as hardness, mass, and extension in space and time. The second category is referred to as "secondary characteristics." These properties do not exist in external reality, but instead are created in the mind. Secondary characteristics include properties such as color, texture, and shape.

Newton (1642-1727) joined Locke as a prominent member of the empiricist school of philosophy. In discussing color in *Opticks*,[6.5] Newton agreed with Locke that the perception of an object is entirely separate from the object itself. For example, an object is perceived as red because it stimulates the sensation of redness in the observer. The mind is connected to the physical world not through direct experiences, but only by fragile threads of nerve that serve as channels for sensory impulses. Consequently, sensations are not direct responses of the individual to properties of an object, but instead are internal properties of the observer that are connected to objects in the most tenuous of manners. By abandoning sensation as direct experience of the physical world, the empiricists opened the way to development of theories of perception as models for bridging the gap between matter and mind.

Leibniz (1646-1716) proposed a radical philosophy of perception in direct opposition to the model of Locke and other empiricists.[6.6] Leibniz suggested that only minds exist. There are many varieties of minds, from those of microbes and insects, to animals and humans, and finally to angels and God. The mind of God is infinite, whereas all other minds are finite and were termed "monads" by Leibniz. Minds differ in the clarity of their perceptions, which include both concepts and sensory responses such as visual experiences. Pleasure is an awareness that perceptions are improving in clarity, wheras pain is caused by a diminution in perceptual clarity. Concepts (ideas in Locke's terminology) are not derived from sensory experiences, however, in opposition to Locke's point of view. Instead, concepts are innate in the mind, as proven by the concept of God for which no sensory experience is possible.

Berkeley (1685-1753), an Anglican bishop, wrote *An Essay Towards a New Theory of Vision* devoted to visual perception.[6.7] He agreed with Locke that perception consists of visual ideas that are distinct from physical objects. He contended that the connection between the two is not casual, however, but rather maintained only through God's pleasure. He suggested that the only connection of the observer with the physical universe is through ideas generated by the senses. Hence, the existence of a physical universe apart from sensory ideas can never be proved or disproved. Berkeley challenged Locke's position that external objects cause and therefore explain the occurrence of ideas by suggesting that only animate beings can be a "cause" of events. Berkeley's philosophy can be interpreted as a negation of all matter and a reliance on sensory impressions as the only reality. Experiences are all simply ideas and it is only through the intercession of God that these ideas yield an apparent unity and coherence of a common world among observers. Berkeley believed that the world is just as it appears because it is the appearance that constitutes the world.

Kant (1724-1804) addressed the connection between sensory experience and conceptual recognition by postulating that the imagination serves as a mediating influence between the two.[6.8] The imagination works to correlate a sensory impression of an obect with past impressions of similar objects, thereby permitting the object to be recognized by connection with previous experience. Imagination works in a transcendental way to conceptualize sensory experiences and serves as the origin of certain general precepts called "categories" that help to organize this experience. Kant also postulated the presence of reason as a higher mental faculty whose exercise leads to complete understanding. Reason must always be connected to experience, however, for otherwise its utilization leads to illusion.

Hume (1711-1776) also addressed the connection between sensory impressions and mental ideas and stated that ideas evolve from a system of beliefs that are not direct products of experience.[6.9] Beliefs are essential because no single object necessarily implies the existence of another object. Generalization of sensory experiences and imposition of cause-and-effect relationships between events require the use of ideas and ideas in turn to require a system of beliefs. Hume recognized the fragile nature of an observer entirely dependent on his model of reasoning of events and was skeptical of the observer's ability to achieve rational knowledge. This skepticism led to the modern school of skeptical philosophy. Kant credited Hume with awakening him from "dogmatic slumbers."

A major influence on the theories of perception and cognition was exercised by Helmholtz (1821-1894), a German physiologist and physicist.[6.10] As a philosopher, Helmholtz was an empiricist who believed that the observer is separated from the external world and can experience it only through neural signals that are interpreted according to assumptions that may or may not be appropriate. Much of the observer's interpretation occurs unconsciously and is not subject to willful reason and influence. How the observer sees and thinks can only be studied experimentally and not through introspection or "thought experiments." Helmholtz was fascinated by visual illusions, because they demonstrate how easily the observer may be misled by incorrect inductive inferences from sensory data. He suggested that the perception of visual images occurs as a consequence of projecting internally organized knowledge upon sensory data. This approach is equivalent to the modern theory of perceptual hypotheses[6.11] in which stored knowledge is imposed on sensory signals according to a "top-down" model of perception and cognition.

Many challenges were leveled at Helmholtz's theories. Among the more significant was the work of Gibson (1904-1979), an American psychologist.[6.12] In his theory of "ecological optics," Gibson regarded perception as "picking up information from the ambient array of light" reaching the observer. This process occurs rather passively and requires no particular processing or analysis by the mind. Gibson's work has been particularly influential in the development of computer vision.

The gestalt theory of vision arose in part in opposition to Helmholtz and other experimentalists.[6.13] Proponents of the gestalt theory were particularly interested in the "figure-ground" problem in which a single object can produce two or more totally different images and the observer can see one or the other image at will, but never both simultaneously. (Examples of this problem are given in Sec. 6.4.1.) To explain this paradox, gestalt theorists suggested that "form" is the fundamental unit of perception and that its origin is mental rather than sensory. Visual stimuli are interpreted by superimposition of a form that is derived from mental fields evolved from configured brain processes. The gestaltists could identify no physiological model for the brain processes that gives rise to mental fields. Instead they proposed several "laws of organization" that are mentally invoked to yield the shapes out of which form is induced by the observer. Objects and events are perceived not because the observer learns how to interpret sensations, but rather because the nervous system has evolved to provide a perceptual organization for superimposition onto sensory stimuli. Although the gestaltist approach to the interpretation of perception raises more questions than it answers, it continues to have adherents who believe it provides a reasonable approach to understanding visual perception and cognition.

Perception and its relation to cognition continue to be of great interest to a small number of physicists, psychologists, and philosophers, and to a larger number of computer scientsts interested in computer vision and artificial intelligence. Marr (1945-1980) has been one of the more influential psychologists working in this area.[6.14] He suggested that the shapes of objects are perceived by a three-stage process. The stages are: formation of a "primal sketch" in which major features and intensity variations of an object are noted; determination of more subtle characteristics such as surface discontinuities and depth referenced to a coordinate frame centered in the viewer; and, finally, mental construction of a three-dimensional model of the object in a coordinated frame centered in the object. The product of this three-stage process is a mental model of the object suspended in object-oriented space to yield an impression that the object is "out there".

The experimental psychologist Gregory[6.11–6.15] has also exerted considerable influence on the subject of visual perception through his analyses of illusions and conjuring. The art of the conjurer is to present a series of events (usually visual) that lead to obvious predictions about future events that ultimately fail to materialize. Optical illusions provide visual impressions that are internally inconsistent and therefore paradoxical, or that conflict with our understanding of the external world as constructed from experience. Illusions and conjuring offer excellent opportunities for studying visual perception and cognition and Gregory is one of the leaders in exploring these opportunities. Rock[6.16] has also been instrumental in improving the insight of psychologists and physicists into the nature of optical illusions. A few optical illusions are examined in some detail in Sec. 6.4.

Current theories of visual perception suggest that the detection and recognition of objects involves the continuous interchange of perception and comprehension of the external world. That is, there is a constant interplay between perception and cognition rather than a single step in which neural signals are integrated into a visual image somewhere in the visual cortex. Hence, it is no longer possible to separate the mechanisms of detection, recognition, and interpretation of visual images. Instead, these processes must be considered as a single interactive process in which the acquisition of visual information is integrated with recognition and interpretation, and even consciousness.[6.17]

6.3   Visual Texture Discrimination

Vision in humans and other animals has evolved in part as a mechanism of protection from adversaries. As any sportsperson can attest, the eye is extremely sensitive and quick in detecting motion in an otherwise stationary landscape. Youngsters who become expert at video games develop amazing abilities to respond almost instantaneously to visual clues and signals provided on a video screen. In similar fashion, images can be examined and internal inconsistencies or inappropriate components identified in a very rapid and accurate fashion. Proofreaders use this characteristic of the eye-brain system to identify misspellings in a text at a much more rapid rate than would be possible if they had to read and think about each word. Radiologists employ the same process quickly to identify abnormalities in X-ray films of patients. This ability is referred to as an "early warning system" because of its relationship to protection and defense. It is also called the pre-attentive or "global" phase of vision. The early warning system permits almost instantaneous (less than 150 ms) detection of changes in the texture of the visual environment that in earlier times were critical to survival and today are important to work and play.

The result of this detection process is an awareness that something in the visual environment is different or anomalous. Identification of what the difference is requires a slower, more thoughtful follow-up process.

The preattentive phase of vision is accompanied by an attentive or local phase of viewing that targets attention on specific features of the visual landscape that have been identified as being different during the preattentive phase. Once motion in a visual scene is detected, e.g., the observer is able to quickly focus attention on the location of the motion to determine its source. During attentive vision, information is processed from an area of the retina that includes but extends beyond the fovea. This area is termed the "useful field of view,"[6.20] "visual lobe,"[6.21], and "functional visual field."[6.22] The proofreader and the radiologist use this efficient two-phase process to detect and then to identify problems present in a manuscript or X-ray film. Preattentive vision represents an unconscious attempt of the observer to quickly extract information from an image through a global search operation that involves active grouping processes.[6.18] These processes are illustrated in Fig. 6.1.

The reader can experience active grouping processes by examining the illustrations in Fig. 6.2. Here, certain characters are inconsistent with others because of their size or orientation. Figure 6.2(a) presents differences in the size of a

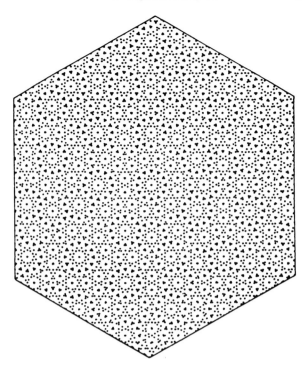

FIGURE 6.1. Evidence for the existence of active grouping processes. The pattern "seethes with activity" as rival organizations compete with one another. After Marroquin.[6.18]

bounded set of characters. These differences are said to be differences in first order statistics. Differences in second order statistics are depicted in Fig. 6.2(b), as differences in the orientation of a bounded set of characters. By glancing at these illustrations, the observer can use the preattentive phase of vision almost instantly to recognize the presence of an internal inconsistency[6.19] Identifying the nature of the inconsistency, however, requires a closer examination over a longer period of time. This examination is the attentive phase of vision. For most observers, the attentive phase takes longer for the illustration on the right than for the one on the left, because second order statistics require more time and attention for analysis compared with first order statistics.

Statistical order is a measure of the complexity in this relationship. In Fig. 6.2(a), the objects in the center of the image (the foveal field) differ in size and this difference is readily detectable during the preattentive phase of viewing. Differences in the size and density of objects in an image are examples of first order changes in image texture. In Fig. 6.2(b), the objects are all the same size and their differences in orientation are detectable by preattentive viewing. Greater effort is

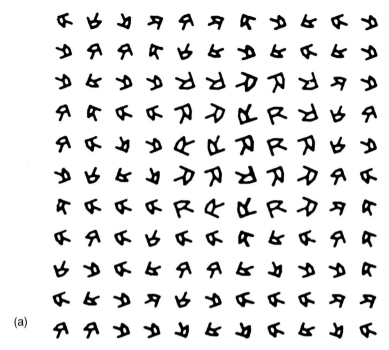

FIGURE 6.2. Preattentively distinguishable texture pairs. (a) With different first order statistics (difference in element size). (b) With different second order statistics (difference in element orientation). After Julesz.[6.19]

required during the attentive phase, however, to identify how the objects are different. The differences are detectable because the objects in the area of interest are all orientated in a similar direction that differs from the constant orientation of all remaining objects in the image. This type of pattern is an example of second order statistics.

Images with similar first and second order statistics, but with differences in the third or higher statistical orders, are usually not candidates for recognition of differences by preattentive viewing. An example of such images is shown in Fig. 6.3(a), in which the orientation of the letters appears random by global scanning and examining that the figures in the center of the image can be distinguished from their mirror image counterparts around the periphery.

Not all images that differ only in third or higher order statistics are distinguishable without careful attentive viewing. The image in Fig. 6.3(b) differs only in higher order statistics, yet the pattern of elongated blobs in the center is immediately distinguishable from that along the periphery. This distinction is probably not achieved through preattentive viewing. Instead it appears to result from rapid

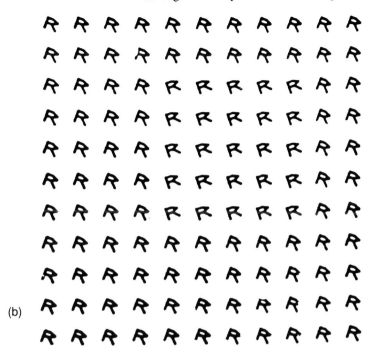

(b)

FIGURE 6.2. (*continued*)

visual processing of local information during the attentive phase. This process constitutes an intermediate phase of visual recognition that permits rapid extraction of local information by concentrating on the presence of local conspicuous features called "textons" by Julesz[6.23]: see also Sect. 2.7.5.

The elongated blobs in Fig. 6.3(b) are examples of textons. Only three classes of textons exist. They are color, elongated blobs, and termination number. Shown in Fig. 6.4 are examples of images with similar first and second order statistics, but which differ in their texture by being texton-deficient [Fig. 6.4(a)] and texton-rich [Fig. 6.4(b)]. It is the presence of textons in images that permits analysis of image texture. This analysis is becoming increasingly important in many imaging applications, including aerial photography and some types of medical imaging (such as ultrasonography). An example of an image that presents only limited statistical information is presented in Fig. 6.5. For those unfamiliar with the image, considerable study is usually required before the Dalmatian dog becomes visible.

Recently the concept of "fractal dimension" has been introduced to explain the interpretation of patterns of visual texture. As defined by Mandelbrot,[6.24] fractals are sets of numbers with the properties of similarity and randomness extended over a range of dimensional scales. The range is defined as the fractal dimension. Interest is developing in applying fractal analysis to many areas of imaging science, including applications to medical imaging such as ultrasonography, nuclear imag-

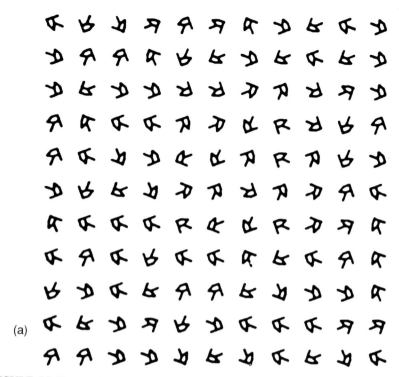

(a)

FIGURE 6.3. Preattentively undistinguishable texture pairs with identical first and second order statistics but different third and higher order statistics. (a) Composed of randomly thrown and not overlapping similar micropatterns and their mirror images. (b) Composed of elongated blobs that are quickly recognizable by rapid processing of information during the attentive phase. After Julesz.[6.19]

ing of the liver and lungs, mammographic interpretation of breast parenchymal patterns, and lung morphogenesis.[6.25–6.28]

6.4   Illusions

Knowledge of the external world, and our consequent behavior in it, depend strongly on detection and interpretation of visual clues through the process of forming and testing hypotheses about the origins of the clues. The fragile nature of this process is exemplified by how easily we are fooled by illusions. Illusions occur when visual signals lead to internal inconsistencies within an image, or when conclusions are reached that are clearly contradictory to our preconceived notion of reality. Illusions bring into question our ability to interpret and understand visual signals. Their universality is apparent when several observers arrive at similar conclusions when viewing an image, even though they all agree that the conclusions are nonsensical. These occasions present formidable challenges

(b)

FIGURE 6.3. (*continued*)

to those who believe that knowledge is derived only from experience, because illusions would not be recognizable by the observer if the empirical viewpoint were correct.[6.29]

Some optical illusions are physiological in origin. For example, afterimages that remain after viewing a particularly bright object (e.g., an intense light) are caused by a temporary desensitization of the retina following overstimulation. Most illusions, however, originate from cognitive rather than from physiological processes and are more difficult to explain. These illusions result from the interpretation of visual signals using assumptions that are incorrrect or inappropriate. Gregory[6.30] has classified cognitive illusions into four categories: ambiguities, distortions, paradoxes, and fictions.

6.4.1 Ambiguities

Some images offer two or more interpretations and the viewer can switch from one interpretation to another, usually at will once the different interpretations are known to be possible and some clues have been identified about the different

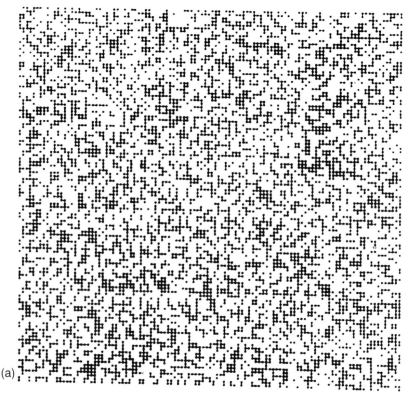

(a)

FIGURE 6.4. Images that present the observer with different opportunities for rapid extraction of texture information. (a) Texton-poor image. (b) Texton-rich image. After Julesz.[6.19]

interpretations. Only one interpretation is possible and the image does not favor any particular one over another. Ambiguities result when multiple interpretations are possible and the image does not favor any particular one over another. Ambiguities reveal the tremendous dexterity of perception, as the observer searches for a solution to a visual conundrum. Examples of ambiguities are shown in Fig. 6.6.

6.4.2   Distortions

Images often present distortions of size, length, and curvature that are confusing to the observer and challenging to interpret. The origin of many distortions is difficult to localize within the visual process; a few seem to be generated physiologically but most appear to be products of the cognitive processing of visual data. Some distortions are caused by misinterpretation of visual cues of depth extracted from images. For example, objects that are thought to be at a distance are expanded in size mentally to reflect the viewer's understanding of their true size. When the depth cues are misleading, this automatic cognitive process can lead to distortions. At other times, the depth cues may be straightforward but the

(b)

FIGURE 6.4. (*continued*)

viewer's preconceived notions of reality may result in a distorted image. Hence, distortions result from both "bottom-up" (visual cues) and "top-down" (assumed reality) misinterpretations of visual data.[6.31] Examples of distortions are shown in Fig. 6.7.

6.4.3 Paradoxes

Certain images present illusions that are known to be impossible even though they are clearly apparent in the image. The drawings of Escher are examples of paradoxes (Fig. 6.8), for instance where staircases appear to cycle without beginning or end, and waterfalls seem to have no origin.[6.32–6.33] Paradoxes provide a conflict between appearance and knowledge; we know that the image, although visually obvious, cannot be correct. Thus, visual cues from the image are in disagreement with our cognitive understanding of reality and the resulting paradox creates an unresolved dilemma.

FIGURE 6.5. Dalmatian dog. After Thurston and Carraher.[6.29]

6.4.4 Fictions

Often objects are assumed to be present in images when in fact they are figments of the imagination. Sometimes these assumptions are aided by postulating the presence of illusory edges and surfaces, such as those illustrated in Fig. 6.9, that are occluded by some nearer object or surface. The cognitive recognition of external reality depends in part on the projection of certain shapes onto visual images. Occasionally this projection leads to fictional representations that are not present in the images. Projection occurs very rapidly and may be essential to make sense of visual stimuli that impinge on the eyes at an extremely rapid rate. But at times, it can create visual fictions that are a form of optical illusion.[6.29] Fictions provides opportunities for the three-dimensional representation of objects in two-dimensional images. This technique is frequently employed in computer-generated "three-dimensional images" employed in medicine, engineering, and other disciples.

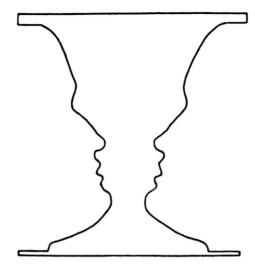

(a)

FIGURE 6.6. Ambiguous figures. (a) Vase or faces. (b) Duck or rabbit. (c) Black or white configuration. (d) Old or young woman. After Rock.[6.16]

6.4.5 Summary

The rule of "abhorrence of excessive coincidence" has been postulated by Rock to explain the projection of shapes onto images.[6.34] In the upper illustrations in Fig. 6.9, e.g., bright "phantom" triangles are seen in sharp outline even though the periphery of the triangle is not delineated. This type of image, known as a subjective or "illusory" figure, is the product of a visual hypothesis invoked to explain otherwise ambiguous visual data that appear to be the product of something more than coincidence. It illustrates the reliance of the observer on a top-down interpretation of visual data to make sense of perceived reality.

6.5   Color Vision

The perception of color has fascinated philosophers and physicists over the centuries. In spite of innumerable speculations and experiments, however, the ability of the human to perceive colors in images remains an enigma.

A prism separates a beam of white light into constituent wavelengths or spectral hues. Passing the constituents through a second prism increases the separation of the hues, but no new colors emerge as a consequence. In Western culture these hues have been identified as the seven colors violet, indigo, blue, green, yellow, orange, and red, in order of increasing wavelength. The colors are not separated by distinct boundaries. Instead, the gradations are gradual and poorly distinguished one from the other. More importantly, not all cultures have agreed on the categorization of seven colors. For example, the Greeks ordered colors in terms of variations in brightness or other characteristics rather than hue.[6.33] To the Greeks, green was

(b)

(c)

FIGURE 6.6. (*continued*)

(d)

FIGURE 6.6. (*continued*)

essentially "moist" and associated with life and youth. The terms "green wood" and "greenhorn" reflect this categorization.

Colors do not logically offer any particular sequence of linear order. Their categorization in terms of wavelength is a relatively modern development that reflects physical rather than perceptual properties. The difficulty of categorizing colors was addressed by Plato in *Timaeus*: "The law of proportion according to which the several colors are formed, even if a man knew he would be foolish in telling, for he could not give any necessary reason, nor indeed any tolerable or probable explanation of them."

Aristotle suggested that colors are variable mixtures of black and white, and that the rainbow contains the three colors red, green, and violet. His theory preceded by several centuries the physical definition of the three primary colors and suggested that the color of any object was strongly influenced by its surroundings.

Descartes[6.35] suggested that colors are properties of the ether, the invisible pervasive medium responsible for transmitting the forces of pressure and impact that explain abscopal effects (actions at a distance). He suggested that various colors are associated with different rotary velocities of ether globules, with faster rotations yielding the sensation of red, followed by slower rotations for yellow,

(a)

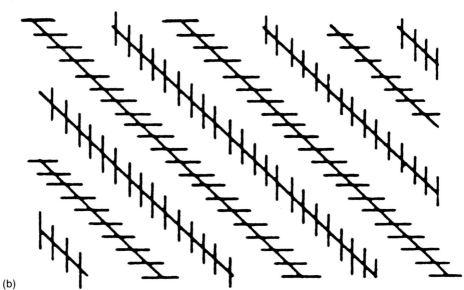

(b)

FIGURE 6.7. Distortions. (a) Concentric circles. (b) Parallel lines. (c) Equal-height cylinders. (d) Equal-length lines.

(c)

(d)

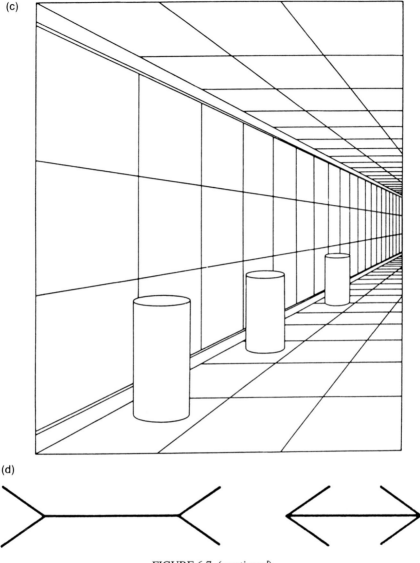

FIGURE 6.7. (*continued*)

green, and blue in that order. A light beam transmitted through the ether by static pressure is refracted to varying degrees by the rotating globules to provide the perception of color. Descartes' theory was challenged by Hooke,[6.36] who proposed that light is a "vibrative motion" propagated in waves that are deflected by reflections and refractions to yield the impression of color as they strike different parts of the retina.

In 1672, Newton showed that light is decomposed into spectral components as it passes through a prism and that subsequently it can be recombined into white light

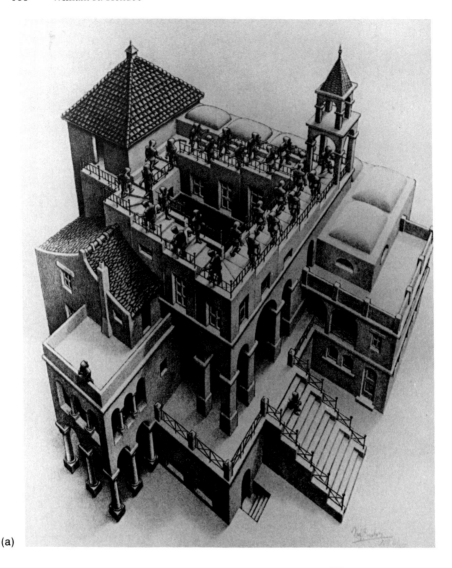

(a)

FIGURE 6.8. Paradoxes. (a) Nonending staircase. After Escher.[6.32] (b) Moving water without origin. After Escher.[6.32] (c) Impossible triangle. After Penrose.[6.33] (d) Impossible object. After Thurston and Carraher.[6.29]

of diminished intensity. Newton developed the color circle in which a perceived color depends on three properties: intensity, hue, and saturation. He demonstrated that a particular color can be obtained by multiple combinations of these three properties and that the impression of color is more than simply a manifestation of the physical properties of light. Newton suggested that the perceptual properties of the observer are also passively involved in the impression, perhaps by providing a

(b)

FIGURE 6.8. (*continued*)

direct correlation between the properties of the light and the physical construction of the eye that yields a sensory impression of the light.

Young was the first to propose that the observer is actively involved in the interpretation of color. In 1802, he wrote:[6.37] "As it is almost impossible to conceive each sensitive point of the retina to contain an infinite number of particles, each capable of vibrating in perfect unison with every possible undulation, it becomes necessary to suppose the number limited, for instance, to three principal

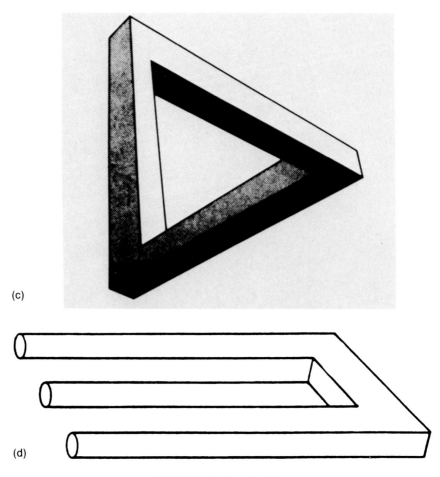

(c)

(d)

FIGURE 6.8. (*continued*)

colors...Each sensitive filament of the nerve may consist of three portions, one for each principal color."

A few decades later Young's theory was refined by Helmholtz and Maxwell to establish the modern theory of color vision. In analyzing the ability of a painter to reproduce the colors of a scene, Helmholtz[6.10] suggested that: "The painter can produce what appears an equal difference for the spectator of his picture, notwithstanding the varying strength of light in the gallery, provided he gives to his colors the same ratio of (intensity) as that which actually exists." That is, a painter can create the conditions of light stimuli to achieve an illusion that reproduces a "real" situation represented by receipt by the retina of light from an object.

The perception of color can be modeled by imagining a fictitious "color space" with points that correspond to distinct colored lights. Lights that appear different in any way correspond to different locations in color space, whereas lights that appear

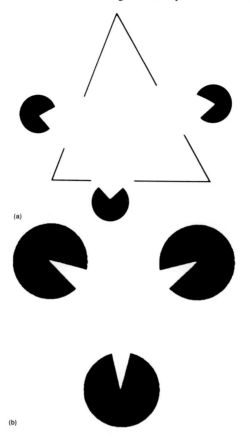

(a)

(b)

FIGURE 6.9. Fictions. (a) Subjective curvilinear triangle. (b) Concave equililateral triangle. (c) Cube. (d) Letters.

(c)

(d)

FIGURE 6.9. (*continued*)

identical correspond to the same location in color space irrespective of physical differences in intensity, hue, or saturation. The points in color space differ only according to perceived color and not according to the geometrical location of the origin of the color in the object. Any change in intensity, hue, or saturation of a point in the object. Any change in the location of the corresponding point in color space. A change in intensity or spectral composition of any object's luminance causes a shift of location in color space, e.g., because the composition of the light received by the eye changes in response to the luminance. If the intensity, hue, and saturation all change in synchrony, however, so that their ratios remain relatively constant, then no shift occurs in color space and the distribution of colors in the perceived object remains unchanged. For this reason, a painting may be viewed under lighting conditions different from those used when the painting is created, provided the difference affects the intensity, hue, and saturation about equally.

An explanation of color vision modeled on the concept of a fictitious color space presupposes that there is no such thing as absolute color. The perception of color is a property of the observer and is strongly influenced by the observer's perceptual characteristics. Color vision cannot be transcribed into absolute physical properties that are demonstrable through an "ideal color observer" who can be considered in isolation. In this manner, color vision resembes the physics of relativity in which no "ideal observer" exists who can quantify absolute motion.

6.6   References

[6.01] Barnes J. Perception: Early Greek theories. In: Gregory R.L., ed. *Oxford Companion to the Mind.* New York: Oxford University Press; 1987:603–604.

[6.02] Barnes J. Aristotle. In: Gregory R.L., ed. *Oxford Companion to the Mind.* New York: Oxford University Press: 1987;38–40.

[6.03] Scruton R. *Spinoza.* New York: Oxford University Press:1986.

[6.04] Locke J. *An Essay Concerning Human Understanding* (1690). Nidditch P.H., ed. New York: Oxford University Press; 1975.

[6.05] Newton I. *Opticks* (4th ed., 1704). London: Smith and Walford.

[6.06] Walker R.C.S. Leibniz's philosophy of mind. In: Gregory R.L., ed. *Oxford Companion to the Mind.* New York: Oxford University Press:1987; 433–434.

[6.07] Berkeley G. *An Essay Towards a New Theory of Vision* (1709). New York: C.P. Dutton;1910.

[6.08] Kant I. *Critique of Pure Reason* (1929, trans. Kemp Smith N.). London.

[6.09] Price H. *Hume's Theory of the External World.* Oxford: Clarendon; 1940.

[6.10] von Helmholtz H. *Handbook de Physiologischen Optik* (Part I, 1856; Part II, 1860; Part III, 1866. Leipzig: Voss. 1925, trans. Southall G.). New York: Dover; 1962.

[6.11] Gregory R.L. *Concepts and Mechanisms of Perception.* New York: Charles Scribner's Sons; 1974.

[6.12] Gibson J.J. *The Perception of the Visual World.* Boston: Houghton Mifflin; 1950.

[6.13] Kohler W. *Gestalt Psychology* (1929). New York: Liveright; 1947.

[6.14] Marr D. *Vision.* New York: Freeman;1982.

[6.15] Gregory R.L. *The Intelligent Eye.* London: Weidenfield and Nicholson; 1970.

[6.16] Rock I. *Perception.* New York: Scientific American Library; 1984.

[6.17] Zeki S. The visual image in mind and brain. Sci. Amer., 1992; 267:68–77.

[6.18] Marroquin J.L. *Human Visual Perception of Structure* (M.Sc. thesis; Department of Electrical Engineering and Computer Science). Boston: Massachusetts Institute of Technology; 1976.

[6.19] Julesz B. Vision: The early warning system. In: Gregory R.L., ed. *Oxford Companion to the Mind.* New York: Oxford University Press; 1987: 346.

[6.20] Mackworth N.H. Stimulus density limits the useful field of view. In: R.A. Monty and J.W. Sender, eds., Eye Movements and Psychological Processes. New York: Wiley; 1976:307–321.

[6.21] Bellemy L.J., Courtney A.J. Development of a search task for the measurement of peripheral visual acuity. Ergonomics; 1981, 24:597–599.

[6.22] Sanders A.F. Some aspects of the selective process in the functional visual field. Ergonomics; 1970, 13:101–117.

[6.23] Julesz B. Textons, the elements of texture perception and their interactions. *Nature* 1981; 290:91–97.

[6.24] Mandelbrot B.B. The fractal Geometry of Nature. New York: Freeman; 1982.

[6.25] Chen C.C., Daponte J.S., Fox AM.D. Fractal feature analysis and classification of medical imaging. IEEE Trans. Med. Imag., MI-8: 133–142; 1989.

[6.26] Cargill E.B., Donohoe K.J., Kolodny G., Parker A.J., Duane DP. Estimation of fractal dimension of parenchymal organs based on power spectral analysis in nuclear medicine scans. In: D.A. Ortendahl, J. Uacer (eds.), Proc. 11th Int. Conf. Inform. Proc. Med. Imag., 557–570; 1991.

[6.27] Caldwell C.B., Stapleton S.J., Holdsworth D.W., Jong R.A., Weiser W.J., Cooke G. Characterization of mammographic parenchymal pattern by fractal dimension. Phy. Med. Biol., 1990; 35:235–247.

[6.28] Nelson T.R., Manchester D.K. Modeling of lung morphogenesis using fractal geometrics. IEEE Trans. Med. Imag., MI-7:321–327; 1988.

[6.29] Thurston J.B., Carraher R.G. *Optical Illusions and the Visual Arts*. New York: Van Nostrand Reinhold; 1986.

[6.30] Gregory R.L. Illusions, In: Gregory R.L., ed. *Oxford Companion to the Mind*. New York: Oxford University Press; 1987:337–343.

[6.31] Hendee W.R. The perception of visual information. *Radiographics* 1987; 7(6): 1213–1219.

[6.32] Escher M.C. *The Graphic Work of MC Escher* (2nd ed.). New York: Hawthorne Books; 1973.

[6.33] Penrose R. Escher and the visual representation of mathematical idea. In: Coxeter H.S.M., Emner M., Penrose R., Teubre M.L., eds. *M.C. Escher: Art and Science* (2nd ed.). New York: North Holland; 1987: 143–157.

[6.34] Parks T.E. Illusory figures. In: Gregory R.L., ed. *Oxford Companion to the Mind*. New York: Oxford University Press; 1987:344–347.

[6.35] Descartes R. *Meteores*. Paris: 1638.

[6.36] Hooke R. *Micrographia* (reprint of 1665 ed.). New York: Dover; 1961.

[6.37] Young T. On the theory of light and colors. *Philos. Trans. R. Soc. London* 1802; 92.

7
Visual Data Formatting

Ulrich Raff

7.1 Introduction

A deeper understanding of the perception of visual information has puzzled researchers from a wide range of scientific disciplines including physiology, neurophysiology, neuroanatomy, mathematics, psychology, physics, and computer sciences. Although human vision is quite well described at a neuroanatomical level, the information processing tasks performed by the retina and the visual cortex of the brain remain largely unclear. The computational paradigms of biological vision are simply not understood.

What exactly is the human eye extracting from the contemplation of a scene? A theoretical approach to the vision problem should be able to explain views captured through a window, reflected from rippled water surfaces, or transmitted through surfaceless structures such as fog and smoke. Insight into the neural circuitry of the brain can be expected through an unambiguous description of visual information. Following this idea, neural network models have focused the attention of many researchers and have enjoyed spectacular results in pattern recognition tasks.

Image processing has emerged from signal processing theories and has become one of the multiple facets of modern information theory. An image stands, as its latin origin *"imago"* suggests, for the representation or description of an object, person, or scene. It carries information through light intensity distributions. Detection, identification, categorization, and cognition are the result of a series of physical phenomena including light absorption, refraction, and reflection from perceived objects of the real world.

Static scene description can be achieved by freezing external world information into two-dimensional light field distributions using external sensors. Most imaging information is analog in nature, although imaging data in digital form were introduced over two decades ago. At present, visual information is displayed, analyzed, and transmitted in digital form at an increasing rate in the industrial, scientific, and medical environments, in order to take advantage of the formidable flexibility of computer processing capabilities.

Temporal changes in visual information captured in a sequence of individual images allow data interpretation through the additional dimension of motion. Though all visual perceptions are based on temporal differences in light intensities, this adds considerably to our knowledge since we distinguish static from dynamic image sequences. Ciné presentation of visual data clarifies description tasks by virtue of an apparent continuous motion. Digital real-time applications which have considerably benefitted from very large scale integrated (VLSI) technology are currently adding new dimensions to the scope of visual information.

Information presented at a higher rate than 30 images per second is not integrated as a discrete sequence of images by the higher processing centers due to inherent limitations of the neural networks of the human brain. The slow propagation of neural signals, e.g., ranging from 0.1 to 0.01 seconds for eye-hand coordination, makes possible the illusion of continuous motion based on a discrete sequence of images. The high efficiency of assimilation of complex visual information is related to a powerful mechanism of selective omission of information by biological visual systems. Image information is to a large extent "integrated" as an entity, details being filled in dynamically as the eye captures three-dimensional scenes.

The mapping of a three-dimensional world into two-dimensional visual representations includes perturbations such as the fact that objects far away from the observer appear smaller than do similar objects nearer to the observer. This is easily "understood" in images so long as the identified objects are "known." An object unknown to the higher cognition centers can present an insurmountable interpretation task and scene understanding can fail. The extraction of "meaningful" content from visual information encoded in images is a crucial step in visual data formatting. How shall we coherently describe the contents of different classes of images such as weather modeling and medical imaging?

Descriptive routes to an understanding of visual information are highly subjective, leading to feelings and thoughts. The perception of art seems to confine itself into this subjective approach establishing its own *raison d'être*. Strong emotional components lead painters to express their perception of the external world. Throughout the history of artistic representations of visual data, surrounding objects have often been triggers to the creation of a perceived ambience rather than the recreation of the objects themselves. Contemplation and appreciation of art work seems not to rely on any kind of well-defined objective criterion. Paintings might well turn into "redefined" objects deeply modifying the inital perception of visual information. Nobody would disagree that Claude Monet, e.g., perceived the world "his way." The strong impact of his work has taught that he found techniques of presenting visual information in a unique way now appreciated as impressionism. His paintings seem to trigger, at least within a certain group of observers, some kind of "resonance" effect with the higher processing centers of the central nervous system. Nevertheless, we could try a scientific approach to "analyze" and understand Monet's perception of the interactions of light with objects, although the conspicuous presence of colors complicates this approach. This could keep the viewer away from possible illusions and false interpretations. Moreover, it might even give an objective insight into the relation of artwork formatted according to objective perceptions, their pictorial responses, and the higher processing centers of the brain.

A journey through the world of artistic visual perceptions dramatically illustrates the intricacies of the perception of visual information. Brightness, colors, contrast, details, edges, lines, boundaries, texture, and shapes all lead to delightful representations of the painter's mind. A comprehensive framework which includes these parameters could lead to a more profound understanding of the basic conditions linking vision and cognition.

7.2   Brightness, Contrast, and Details

The power of changes in light intensities in images is very well illustrated by painters such as Rembrandt, Monet, Goya, and so many others. Indeed biological vision is primarily sensitive to changes in light intensities rather than to the absolute light intensities themselves. Extremely small variations in light intensities can be visually registered even at very low light levels. The cat's so-called "night vision" is a good example of how well the eye is capable of adapting to light fields of low intensity with "minimal" changes to "see in the dark." This points out the fact that the overall light intensity is not nearly so important as might be thought. An exquisite example of subtle brightness changes carrying considerable information is given in the "Portrait of Seurat's mother" shown in Fig. 7.1. The extraction of basic attributes of light fields, i.e., characteristic features of an image, is the central issue in biological vision. In machine vision, this also has turned into one of the more intriguing and fascinating artificial intelligence (AI) problems.

In digital images, distributions of light intensities are encoded in pixel values through a quantitization process which attributes a reconstruction level in the form of a binary value to an analog signal amplitude. In practice, this can be accomplished in any spectral channel of the visible spectrum in which case spectral values of primary colors, tristimulus (amount of red, green, and blue primaries required to match a color) or luminance values can be used to describe light amplitudes in images. The information loss that has to be accepted in describing scenes with a set of numbers and colors is considerable. Indeed, a pixel described by its intensity and color should be able to characterize *inter alia* the luminance, location, and nature of light sources, light reflectance, transparency and opaqueness of objects, light transmission, refractance, absorption and scattering, and the optical and electrical properties of imaging sensors and other devices. This is clearly impossible with a collection of numbers and a set of primary color values. It also becomes immediately clear why machine vision is still in its infancy. A large number of channels in the nonvisible part of the electromagnetic spectrum, e.g., infrared, ultraviolet, and radar frequency range can conceivably be used as well to obtain an image with varying signal amplitude.

The overall intensity or brightness of an image is the first variable that strikes the viewer. The limited value of the overall brightness regarding the image content is evident with over- or underexposed images. We describe it with image amplitude values simply condensed in average intensity values which can be evaluated over the entire image as well as over some smaller pixel neighborhoods of a selected spatial location. Quite instantaneously, automatic visual attention is paid to possi-

FIGURE 7.1. Portrait of Seurat's mother. Drawing, about 1855. Slowly varying changes in light intensities across the image create a calm atmosphere and serenity emanates from the face of the painter's mother. Notice that virtually no edges are visible. Special effects are obtained through the texture of the paper.

ble variations in intensities to identify and try to understand the visual content of the image.

Changes in image intensities are spatial and temporal. In an experiment in which an observer is placed in an all-white environment in which not even the observer can be seen, no changes in light intensities are observed by the observer. Consequently, no information in the form of visual data is acknowledged, regardless of the absolute light amplitude value (which becomes a meaningless concept in this "trivial" case). Also, if an image window is fixed relative to the observer's retina, each individual photoreceptor is observing the same picture element and, if there is no intensity change in time at this location, no information is perceived by the brain's cognitive centers. Differences in spatial input intensities are converted to temporal input differences. Massive parallelism of neural activities converts this information back to a spatial information with varying intensities. Thus, these local image amplitude measurements are very important for the location and interpretation of perceived objects. Amplitude features of digital images can be used to estimate the distribution of these intensity values. They are translated into gray levels in monochrome (black and white) display devices.

The fundamental question of how to present data to the human eye for the "best" possible image analysis arises as a next logical step. These needs are highly goal-driven, though a bottom-up, i.e., data-driven, working hypothesis could have been adopted as well. How shall light intensity distributions and details be displayed to depict objects and their relationships in a scene? The need for a precise definition of image feature parameters is recognized. Enhancement techniques can be invoked to "improve" the image content. No mathematical theory of image enhancement is available, however, because we are unable to objectively describe the perception-cognition relationships. The simplest way to approach this complex task is to focus on basic concepts such as image contrast and detail. These concepts naturally lead to a meaningful description of pictorial scenes.

Contrast describes differences in light and dark regions of images without specification of absolute light intensities. Changes in contrast can be subtle throughout the entire image. Contrast might also vary precipitously over small and large regions. Details are perceived as a direct consequence of intensity changes over "small" image sections. The physical size of the details is inferred precisely from these changes. This is still ambiguous in a digital framework since we have not specified the spatial sampling rate at which the digital picture has been generated. We can clarify this situation in two ways: either we know the highest frequency present in the spatial light intensity distribution or we specify the spatial extension of the detail that it is desired to identify. The second option seems more logical, since, in practice, knowledge of a maximum frequency in signals is in conflict with the finite duration of signals recorded. Hence, we have to sample the signals "adequately." Theoretically, we need to fulfill the Nyquist criterion. In practice, signals are not band-limited and sampling is a finite duration operation, so that more samples are needed than theoretically predicted. The number of samples over a given region determines the number of pixels with varying intensities. Keeping this in mind, we have to live with aliasing and accept noise contributions ranging from quantization noise to all other electronic and physical sources contributing to image formation. These are, to a certain extent, "bad news" as we face the presentation of data for scene understanding.

Digital images are suitable for manipulation by computers to extract relevant parameters according to algorithms developed at varying levels of abstraction. Ideally, a presentation optimally suited to human vision is sought. This is well illustrated with radiological images whose diagnostic values are measured by their visual appearances. Machine vision can only succeed if the image properties relevant to the eye and higher cognition centers can be properly modeled and implemented. This requires the development of abstract computational algorithms to process images and the specification of their mapping onto implementation architectures. Abstract theories, algorithms, architectures, and implementations in terms of hardware and software have yielded interesting AI applications. Robotic vision, e.g., has gained increased attention and has enjoyed some success in industrial applications.

In a statistical representation of images, each pixel becomes a random variable. An image can then be interpreted as a two-dimensional sample function that describes an ensemble of images. The number of random variables can be quite large in discrete fields, e.g., 65,536 for a 256 × 256 image. A joint probability density function for all sample points would completely suffice to describe the stochastic process through which the image is obtained. Higher order joint probability distribution functions are difficult to measure. Models have to be invoked, which in turn might not be realistic. Usually, first and second order moments are used to describe the ensemble. First order probability density functions can be successfully estimated as well as modeled. The frequency of occurrence of image gray levels can be used as an estimate of the first order probability density distribution of image amplitudes.

The location and shape of an image histogram gives clues about the quality and character of the image. Each image has a unique histogram, while the reverse is obviously not true. Fig. 7.2(a) shows an aerial picture (512 × 512 pixels) with moderate contrast revealed by a narrow histogram. The dynamic range of the gray levels extends from 105 to 160 for this particular image [Fig. 7.2(b)]. Clipping of image intensities can be detected by inspecting the first order histogram. So-called "bimodal" histograms indicate two regions of clustered gray levels. Images with a "well balanced" gray level distribution show a broad distribution of gray level intensities. If only a small portion of the available dynamic range of image display devices is used, processing techniques can be invoked to modify the range for a data presentation better suited to visual tasks. The dynamic range of image display is generally between values of 0 and 255.

Digital contrast manipulations can be achieved through histogram specification, a procedure in which the histogram of the output image is specified in a desired form according to some objective or empirical criterion [Figs. 7.2(c) and 7.2(d)]. *Histogram modification* and *histogram equalization* are synonymous terms. An arbitrary transformation function (or mapping procedure) can be obtained, based on an empirical probability density function for the output image knowing the density distribution of the input image. Contrast manipulation based on histogram equalization, which specifies a uniform density function for the output picture, is available on many image display stations. This technique yields maximized zero order brightness entropy and maximum information for the observer. This is an optimal operation for data presentation, unless the useful information is known *a priori* to be concentrated within a known brightness range. An equalized gray level distribution, however, is not equal to a perceived equalized gray level distribution. Recognizing this, a histogram hyperbolization process has been introduced which produces a uniform perceived probability density function that includes the re-

(a)

FIGURE 7.2. Aerial photographs of highways. (a) This 512×512 image shows little dynamic range as depicted in (b); the first order histogram, resulting in moderate contrast. (c) and (d) A remapping of the highway picture after specifying a desired first order probability density of a Rayleigh type. Notice that the remapped highway picture has a histogram that exhibits the shape of a Gaussian function.

sponse function of the eye. The perception of brightness is not a linear function of light intensity. Images with well-balanced contrast that are pleasant to the human eye very often show a nonuniform distribution of gray levels, such as Rayleigh or exponential output probability density functions. The simplicity of first order histogram-based computer manipulations makes these techniques attractive for data display.

Common features of the first order histogram are its moments and entropy. The characteristics of first order histograms can be described with measures such as mean, variance, skewness, kurtosis, and energy. Sometimes, images are not suited to a histogram modification based on the gray level distribution of the entire image. Digital images can have a large dynamic range which can neither

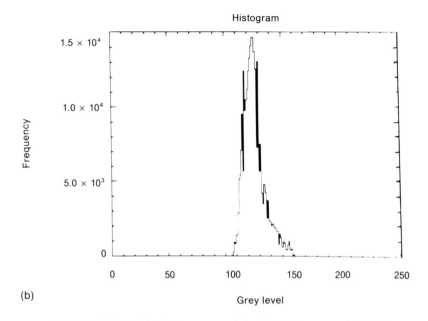

(b)

(c)

FIGURE 7.2. (*continued*)

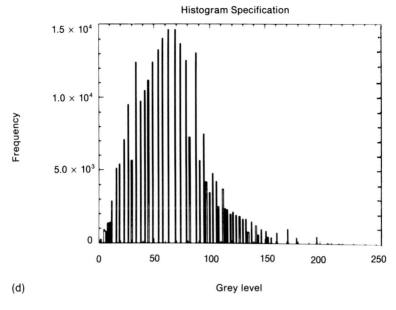

(d)

FIGURE 7.2. (*continued*)

be displayed nor perceived in its entirety, since most display stations afford only 256 gray levels. Even if the image memory uses 16 bits per pixel, the human eye is incapable of distinguishing more than approximately 60 gray levels in complex scenes.[7.1] Useful information can be spread over a wide range of values, e.g., computed tomography images usually cover a numerical interval of 4000. Windowing techniques are routinely used to select a range of gray levels to be displayed within the total interval. Data are viewed through windows of finite width at a selected base line.

Pizer, et al.[7.2] and Zimmerman, et al.[7.3] have recently modified the histogram specification technique to completely present data of wide dynamic range in one single image by using procedures known as *adaptive histogram equalization* (AHE) and *contrast limited adaptive histogram equalization* (CLAHE).[7.4, 7.5] In these techniques, small so-called "contextual" regions are used to cover the entire image and a histogram equalization process is performed locally within the contextual region. This technique allows the enhancement of all anatomical structures in the image, as can be seen in Fig. 7.3. The true relation to Hounsfield numbers is lost in this context; unprocessed windowed images are displayed with the enhanced images side-by-side for diagnostic purposes.

The extraction of useful features in images can also be approached in the transform domain. Many unitary transforms have been used in coding, filter design, and data compression. The most useful in digital signal processing is the discrete Fourier transform (DFT) which has been implemented in a variety of fast algorithms. In this representation, two-dimensional basis functions describe the brightness patterns in images. A weighted sum of these brightness patterns

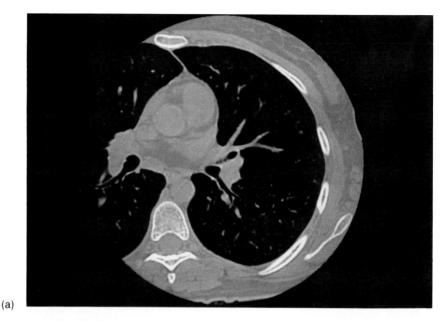

(a)

(b)

FIGURE 7.3. (a) Tomographic view through the chest of a patient with histiocytosis. Notice that little is visible in the lung regions which show up dark since their density is close to air. No window has been applied. (b) The same tomographic view after adaptive histogram equalization using an algorithm developed by Pizer, et al.[7.5] which performs local histogram equalization in so-called "contextual regions." Although the true density relation between various tissues is lost in terms of their Hounsfield numbers, all tissues involved are enhanced, especially the lung which displays honeycomb structures.

uniquely represents the input image. Detail can be characterized in the spatial frequency domain. In a Fourier representation, the zero frequency term is equal to the average brightness of the image. Transform features can be obtained through the application of zonal filters, apertures of simple structures are used to extract features that correspond to orientation, shape, and high frequency components in pictures. Other unitary tranforms, such as Hadamard, Haar, and Slant transforms, can be used as well to describe features, but have not yet been explored as thoroughly as the traditional Fourier representation.[7.6–7.7]

Second order probability distribution functions of image fields can be estimated by evaluating second-order spatial histograms. A second-order spatial histogram is an explicit measure of the joint occurrence of pixel pairs at a distance r and an angle θ with the horizontal axis. In digital images the distance r is measured in units of pixels and hence takes on discrete values as does the angle. Second-order histograms that estimate the second-order distributions can easily be obtained. If pixel pairs of amplitude values are highly correlated, the second order histogram exhibits clustered frequency values along the diagonal. Figures 7.4(a) and 7.4(b) illustrate this point. Figures 7.4(c) and 7.4(d) show the results obtained from the mandrill's face which displays regions of random gray level distributions where pixels are describing the primate's fur. Second-order histograms have been successfully applied in the description of texture. The mandrill's fur is shown in Fig. 7.4(a) and its second-order histogram features in Fig. 7.4(f).

7.3   Texture Discrimination and Edge Detection

Many natural scenes show a large variety of repetitive patterns consisting of entities of diverse shapes and sizes. The term *texture* (related to the word *textile*) has been introduced to describe these irregularities observed on surfaces. Despite its ubiquity in many images, no satisfactory description of texture has been established although a considerable amount of research towards its characterization has been carried out over the last decade. The concept of texture is not a new one. Forty years ago, Gibson stated that texture is a mathematical and psychological stimulus that describes surfaces in monocular vision.[7.8] Natural scenes exhibit mostly random texture (e.g., remote sensing from satellite imaging, or images of cell cultures), while artificial texture such as brick walls, cloths, and mosaics have a deterministic, often periodic texture. Texture is what Gibson would call the characteristic with respect to which a figure appears to be homogeneous.[7.8] Haralick[7.9] has reviewed the various statistical and structural approaches to texture and has discussed some of the models used by investigators to characterize it. According to Haralick, texture can be thought of as an "organized are phenomenon," which, when decomposable, has two basic conponents: one dimension is used to describe "primitives" out of which the image texture is made, and the second dimension is utilized to describe spatial distribution between these primitives in texture. Also described by Resnikoff,[7.10] texture is not itself a microscopic prop-

(a)

FIGURE 7.4. Probability of joint occurrence of gray levels in familiar scenes. This function is approximated by the second-order histogram. The distance between two points in the image is two pixels and the angle between them is 0 degrees. (a) Photograph of the model Cheryl Tiegs. (b) Joint gray level occurrence is clustered along the diagonal indicating a high correlation in the portrait of the model shown in (a). (c) Photograph of a mandrill. (d) High correlation in the bright regions of the image, the mandrill's nose, and an extended region with high probability of random occurrence of low and high gray levels corresponding to the fur of the mandrill's head shown in (c). (e) Fur, taken as an example of random texture, shows (f) A second-order histogram indicating uncorrelated structure where clustering of gray levels is lost.

erty of surfaces; rather it is a statistical property of surface features. The term *textons* has been coined to describe these characteristic features (see Chap. 6). Basic texture elements have also been labeled "textels," probably in recognition of the terms "pixels" and "voxels." The idea of statistical parameters to describe surfaces is a very attractive one, since macroscopic properties of matter are described statistically in physics. Somehow, texture conveys information and, within the framework of information processing systems, facilitates the task of machine vision. Preattentive exposure to the human visual system allows the discrimination of textures which has become a central information processing function.

The present state-of-the-art understanding of texture is mainly due to contributions from Julesz[7.11] who developed algorithms to construct synthetic textures with established properties and introduced parameters to describe the conspicuous features of texture. In 1981, Julesz[7.11] wrote "Research with texture pairs having identical second-order statistics has revealed that the preattentive texture discrimination system cannot globally process third and higher order statistics,

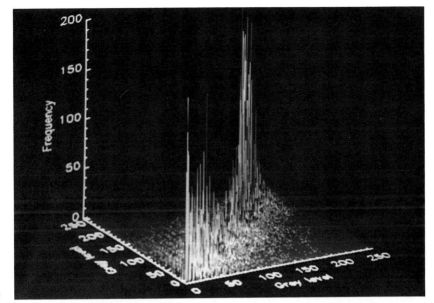

(b)

(c)

FIGURE 7.4. (*continued*)

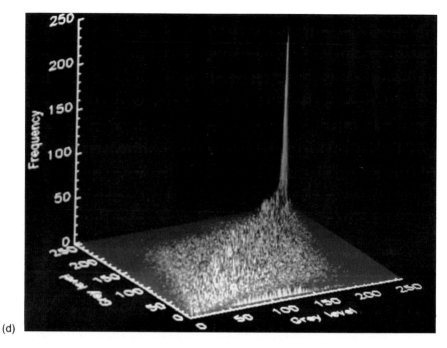

(d)

FIGURE 7.4. (*continued*)

and that discrimination is the result of a few local conspicuous features, called textons. It seems that only the first order statistics of these textons have perceptual significance, and the relative phase between textons cannot be perceived without detailed scrutiny by focal attention."

Before Julesz's contributions, the impact of Fourier analysis had naturally led to the inspection and characterization of texture with Fourier spectra. Second-order statistics are described by the probability that two randomly selected image pixels have the same intensity values. In binary (black and white) images, this reduces to computing the autocorrelation function of the images. The Fourier transform of the autocorrelation function, known as the *power spectrum,* does not convey any phase information (i.e., spatial location information). Hence, it has been suggested that textures with identical second-order statistics are visually indistinguishable. This does not always seem to be true.[7.12] On the other hand, Julesz, et al.[7.13] have found that textures with different second-order statistics may not be distinguishable. Moreover, regions of coarse texture should have spectral energy concentrated at low spatial frequencies, while fine texture is expected to have its energy concentrated at high spatial frequencies. Experiments have demonstrated, however, that these conclusions are not necessarily valid and that spectral overlap is observed for a variety of different natural textures.[7.14] Fourier spectral analysis has successfully been applied to detection and clasification of pulmonary disease.[7.15-7.16] Since Julesz's fundamental work, Fourier analysis is no longer predominant in describing the perception of texture features. Perception

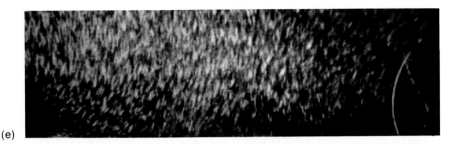

(e)

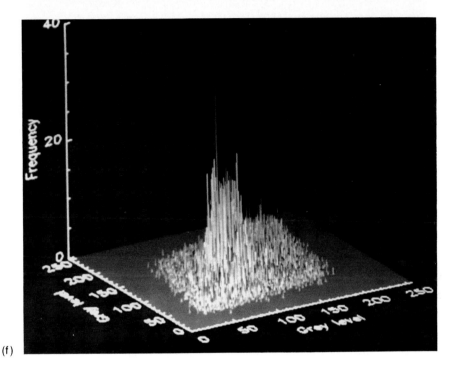

(f)

FIGURE 7.4. (*continued*)

and discrimination of texture seems to be dependent only on first-order statistics (i.e., texton densities).

Three texton parameters are used to describe texture features: these are color, elongated blobs which are quasirectangular patterns, and terminator number. Elongated blobs in turn are described by three parameters: orientation, length, and aspect ratio. If the aspect ratio of a blob, i.e., length divided by width, is close to unity, then the orientation parameter is no longer applicable. Terminators are the ends of elongated blobs. Ultrasound images of the human anatomy are examples where elongated blobs describe anatomical interfaces (see Fig. 7.5). The Dalmatian dog shown in Fig. 6.5 (in Chap. 6) is an exquisite example of a binary image consisting exclusively of elongated blobs as stimuli. Statistical properties allow us to discern

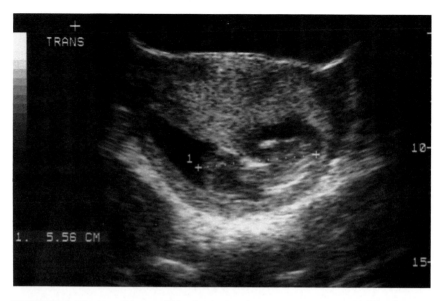

FIGURE 7.5. Ultrasound image. This picture shows an axial view through the lower pelvis of a pregnant woman. The fetus is 12 weeks old and the crown-rump length measures 55.6 mm. The head is on the left-hand side (left "+" sign). Notice that the entire view seems to be generated with black-and-white brushstrokes resembling those used by impressionists.

the image of the dog. Experiments support the conclusion that texton densities are computed over small areas.[7.10] Artists have been using painting techniques based on blobs for over a century. Figure 7.6 illustrates brushstroke-generated elongated blobs whose local random orientations yield a unique impression without the presence of any lines or well-defined edges. Signac, one of the central figures of neo-impressionism, used broadened brushstrokes to immobilize visions in place of the dot technique found in the "pointillism" school. One of his last canvases entitled "Le canal Saint-Martin" uses elongated blobs to create the special effect obtained by a mosaic-like texture (Fig. 7.7). Round dots distributed in varying densities and colors (not reproduced here) are used as a basic technique in Seurat's painting "La Seine á La Grande Jatte au printemps" emphasizing the presence of moving objects suddenly captured in time (Fig. 7.8).

More recently, new ideas have been introduced to help to understand textured patterns. The concept of "fractal dimension" has appeared as an attractive new analytic tool that might give some insight into the concept of texture. Fractals are sets of numbers that have the properties of self-similarity and randomness over many size scales at the same time. These sets can be described with an associated fractal dimension which is characteristic of the physical system that it is describing. Mandelbrot's work [7.17] has been extended to image-texture description.[7.18] Pentland has shown that the fractal dimension of image texture correlates highly with the subjective ranking of observers.[7.19] Growing interest is observed in applying fractal geometry to many different areas of science, engineering,[7.20–7.21]

FIGURE 7.6. "Le Jardin á Giverny," Monet 1902. This masterpiece shows a delightful garden in the French countryside, captured by the painter with his typical brushstrokes that resemble elongated blobs of varying sizes. Neither edges nor lines are visible in the painting. The distribution of blobs, however, displays the interaction of light and shade along an alley with its bordering vegetation in the garden at Giverny and leaves the viewer with the artist's "impression" of a peaceful environment.

and medical imaging.[7.22–7.23] Applications to ultrasonic imaging,[7.24] modeling of lung morphogenesis,[7.25] mammographic parenchymal patterns,[7.26] and nuclear medicine liver and lung scans[7.27] are only a few examples of the potential of these techniques applied to medical imaging. The computation of so-called attractors with deterministic iterated function systems (IFS) sheds a completely new light on

FIGURE 7.7. "Le Canal Saint-Martin." Signac 1931–33. This shows another exquisite example of artwork where elongated blobs are used to create a mosaic texture capturing in a classic composition this scene of metallic structures, sky, and water. Notice that many brushstrokes have the same orientation adding regularity to the texture, yielding the impression of immobility in the scene. This is one of Signac's last canvases and shows that the artist remained faithful to the technique he used during his entire life.

the concept of texture and patterns in images. Virtually all types of patterns have been generated using IFS codes. A theoretical overview of the problem of encoding and decoding patterns into their respective IFS has been given by Barnsley, et al.[7.28–7.29]

Many natural and computer-generated images do not exhibit texture and the perception of details has to be approached differently. Edges, lines, and contours shape the image and introduce information at a variety of levels. In general, abrupt changes in intensity define the boundaries. The classical experiment of two black-and-white half-planes demonstrates the basic principle underlying the identification of changes in intensity values. If the two half-planes are stabilized on the retina, all the photoreceptors of the eye receive a constant illumination and so no information is extracted at any level. We conclude that objects can be discerned either through motion of the scene with respect to the observer, or by the well-known saccadic motion of the eye. The receptors are perceiving temporal intensity variations in a real-time mode and information changes are computed based on these changes. In the saccadic eye motion, i.e., horizontal or vertical, the intensity changes at the dividing line of the two black-and-white half-planes can be identified. Edge detection indicates that biological vision has a built-in differentiator which continuously computes variations in perceived light

FIGURE 7.8. "The Seine at La Grande Jatte in the Spring," Seurat, oil on canvas about 1887. This painting shows a clear spring day on the banks of the Seine and illustrates particularly well the technique used by Seurat to freeze in space the present scene. Texture in the form of round blobs becomes a central tool to convey a unique impression in the artist's composition.

intensities from different regions comprising the image. Variations in texture can also be interpreted as boundaries and add significantly to the content description of scenes. Boundaries can also be described with an adequate use of colors: cartoons are a good example of images where edges are often perceived through simple color changes. In true color scenes and pseudocolored pictures, edge information, and hence details, can be extracted through variations of brightness, hue, and saturation of involved colors. Nevertheless, it seems that intensity changes in monochrome black-and-white scenes convey more edge information than can be discerned in color images. This is particularly striking in Goya's drawings.

Architectural object surrounding observers have simple geometric shapes and therefore are mapped into fairly simple image features: intersecting straight and curved lines define whole classes of objects. Natural scenes, such as an English garden, Monet's painting "Le Jardin á Giverny" (Fig. 7.6) or the famous mandrill known as Comtal's Trademark [Fig. 7.4(c)], are very complex due to the lack of order; edges constantly change direction over small regions of the picture. The information-processing task based on edge discrimination is further complicated if an imaging modality such as radiology generates pictures of objects with unknown shapes.

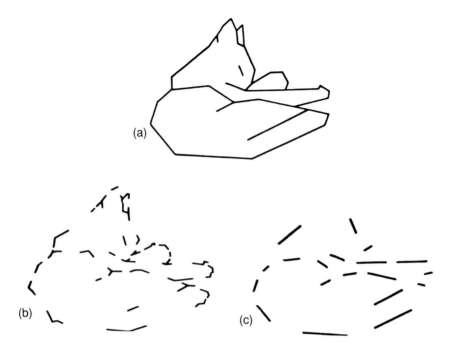

FIGURE 7.9. (a) Attneave's cat. (b) The "cheshired" cat is displayed on the left. The sleeping cat is easily identified after deletion of straight line segments, leaving in place line segments with high curvature radii. (c) Corners have been deleted leaving straight segments of the cat's outline. Without this information, the visualization of the cat becomes difficult. After Resnikoff.[7.10]

How are details perceived in images? In 1954 the psychologist Attneave[7.30] showed that information is concentrated at points of maximum curvature and that "redundant visual stimulation results from either (a) an area of homogeneous color ("color" is used in the broad sense here, and includes brightness), or (b) a contour of homogeneous direction or slope." Observers who had to locate 10 points around a closed curve with the intention of approximating it, accumulated the points on segments of highest curvature, i.e., smallest curvature radius. Attneave's famous cat, shown in Fig. 7.9(a), demonstrates that large sections of straight line segments can be left out without altering the image content. An abstraction of 38 points of maximum curvature were used and connected appropriately to represent a sleeping cat. Only the spatial location of points in space and the knowledge of how to connect them with straight line segments is required for the full information. The feline is easily identified in the picture of the "cheshired" cat as shown in Fig. 7.9(b). Considerable difficulty in extracting information arises, however, once corners and points of discontinuous derivatives are deleted in the cat's drawing, as can be observed in Fig. 7.9(c). This, as well as other similar experiments, indicates that boundaries defined by little or no curvature are not "needed" for the interpretation of scenes and that the higher cognitive centers "fill in" the missing information.

Edge detection in biological vision plays a central role in identification of information and scene interpretation. It is widely used as the driving force to implement machine vision in artificial intelligence. There is, nevertheless, striking evidence of cases where edge detection does not play any role: in Monet's "Le Jardin à Giverny" (1902), no boundaries are detected (Fig. 7.6), the garden being well visualized. The Dalmatian dog (see Fig. 6.5 of Chap. 6) falls into the same category. The correlation of blobs present in these pictures yields the information of a garden and a dog, apparently, through some autoassociative mechanisms which integrate blob densities without any outlines. In a "representational framework for vision," Marr[7.31] coined the term "primal sketch" as a representation which "makes explicit important information about the two-dimensional image, primarily the intensity changes there and their geometrical distribution and organization." Marr's image of two leaves (Fig. 7.10) shows a striking example of how higher cognitive centers identify the perception of both leaves in the absence of sufficient intensity differences in their overlapping parts based on numerical values alone. Numerical values within the outlined rectangular region of interest reveal clearly discontinuous surfaces. This seems to imply that biological vision extracts surface information directly without paying immediate attention to specific boundaries defining an object. While a whole series of boundary detection processes define the primal sketch, it remains unclear which technique should be preferred.

Edge detection is based on a differentiation process applied locally to gray level differences. This makes this process extremely sensitive to noise and to the accuracy of the numerical data. Keeping in mind that the understanding of biological edge extraction mechanisms yields a framework for machine vision, scene understanding can be approached through implementation of automated edge extraction in digital monochrome images. The goal is localization of high spatial resolution objects with sharp features. Its implementation, however, is very difficult. Even virtually noise-free images analyzed with algorithms based on gray level differences have a tendency to show spurious edges. The first breakthrough was achieved with the basic work of Hubel and Wiesel[7.32–7.33] showing that the visual system of mammals has the potential to identify points of tangency and changes of direction in tangent lines between high and low intensity regions.

The scientific literature offers a myriad of methods and techniques for detecting and labeling discontinuities in two-dimensional data with varying intensities. Step edges are important because they determine the end and the beginning of different surfaces. Creases are special types of step edges within the same object. Visual reconstruction, as introduced by Blake and Zisserman,[7.34] describes the computational task of reducing visual data to stable descriptions invoking filters to extract discontinuities. There are many ways of approaching the problem of edge identification in digital images. In high contrast scenes, simple thresholding techniques can yield some limited edge information. Since local variations in luminance are described with derivatives in continuous fields $f(x, y)$, operators based on differences of pixel gray levels or some combination of them can be used

FIGURE 7.10. Image of two leaves. There is not a sufficient intensity change everywhere along the edge inside the marked box to allow its complete recovery from the intensity values alone. The viewer has no difficulty in "seeing" the two leaves, however, demonstrating how the higher cognition levels fill in the missing information. After Marr.[7.31]

to characterize spatial variations in digital images $f(m, n)$. Maximum gradient values associated with the direction of the gradient operation yield edge information. Two kind of operators, *gradient operators* and *compass operators,* have been used extensively in computer applications[7.7, 7.35] in the form of 2 × 2 or 3 × 3 masks. Finite-difference approximations to orthogonal gradients and directional gradients are used in the form of these masks to extract edge content. Gradient images $g(m, n)$ are generated and then a suitable thresholding operation is applied to the digital gradient images $g(m, n)$, leaving edge pixels whenever the gradient value exceeds the threshold T. A binary edge map can then be used to trace the boundaries of objects. If the threshold level T is too low, noise is interpreted as edge feaures in the image. Spurious edges and hence "false" details make their appearance. This may be of considerable importance, depending on the type of information inspected. If the threshold is chosen to be too high, low amplitude structures are not detected in the image and corresponding edges are ignored. Details, as well as overall large structures, may get lost in the process. This shows that

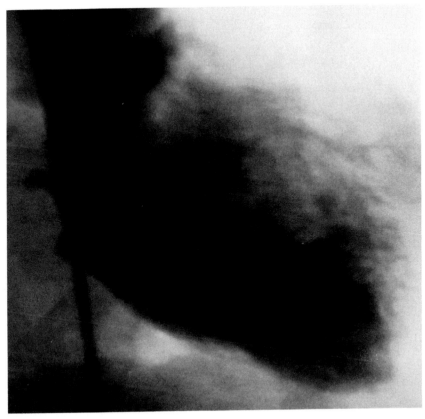

(a)

FIGURE 7.11. (a) A left ventriculogram obtained with digital cardiac imaging techniques in a 30° right anterior oblique (RAO) view after injection of contrast medium through a catheter which reaches inside the ventricle. The image is a 512 × 480 matrix and represents the left ventricular cavity at diastole. Notice the local variations in gray levels over the entire image. (b) A traditional edge detection operation based on adjacent gray level differences or combinations of them shows an image which cannot be interpreted due to excessive noise contributions.

in low contrast images the edge detection approach is likely to fail. An example is shown in Fig. 7.11 where a Sobel 3 × 3 mask[7.36] is applied to a digital cardiac left ventriculogram with a fair amount of statistical noise. The image reflects a scene lacking in complexity. Virtually no boundaries are detected since pixel values are varying mildly in crossing from intraventricular regions with contrast medium present to extraventricular regions with no contrast medium. In addition, noise is amplified in the process and interferes heavily with the identification of relevant information. Figure 7.12 shows the effect of some more sophisticated edge detectors applied to the photograph of the model shown in Fig. 7.4(a). Figure 7.12(b) shows the effect of applying Canny's technique[7.37] combined with thresholding. Figure 7.12(c) displays the effect of applying the Marr-Hildreth operator[7.38] at a

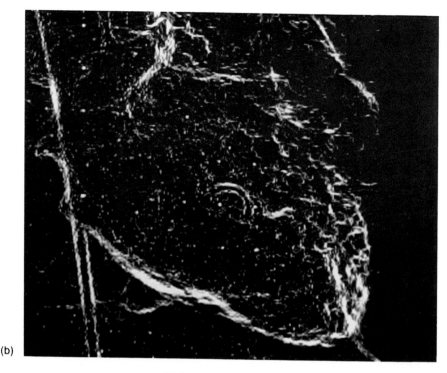

(b)

FIGURE 7.11. (*continued*)

scale that yields a meaningful edge outline of the model after adequate thresholding. This example illustrates the complexity of machine vision applied to familiar scenes.

Compass operators are finite windows, mostly 3 × 3, that measure gradients in some selected direction.[7.7] If 3 × 3 windows are chosen, eight compass gradients can be generated by rotating the boundary elements counterclockwise around the center pixel of the kernels. To illustrate the effect of such operations on digital images, the following North gradient operator g:

$$g = \begin{bmatrix} 1 & 1 & 1 \\ 0 & 0 & 0 \\ -1 & -1 & -1 \end{bmatrix}$$

and the additional seven compass operators generated from it have been applied to the 300 × 300 image shown in Fig. 7.13. Different North compass gradients can be generated by varying the magnitude of the 3 × 3 kernel elements. Edge maps obtained with the eight possible 3 × 3 masks corrresponding to this North compass gradient are shown in Fig. 7.13 without thresholding using a 300 × 300 section of a digital fruit basket image. Other nonlinear edge enhancement techniques have been introduced by Wallis,[7.35] Kirsch,[7.39] and Rosenfeld.[7.40] In noisy images, stochastic edge detection has been more successful.[7.41] In essence, the combination of edge

(a)

(b)

FIGURE 7.12. (a) A digital 300 × 300 image of Cheryl Tiegs is used to demonstrate the effect of two edge detectors based on some more sophisticated techniques. (b) and (c) The results of edge detection algorithms that use low pass filtering techniques prior to the actual edge-finding which is obtained through thresholding. Both techniques [Canny approach shown in (b) and the Marr-Hildreth method shown in (c)] yield similar results, although sufficiently different to make the identification step a considerable task for machine vision.

(c)

FIGURE 7.12. (*continued*)

enhancement techniques followed by thresholding operations has some value, although image detail is largely sacrificed when noise is present. This simply reflects that high spatial frequencies of arbitrary finite amplitudes that contribute to high resolution detail are visually lost to alleviate this impasse. Other techniques have to be invoked to alleviate this impasse.

More advanced techniques for locating discontinuities on various scales can mainly be divided into three categories. The first kind of edge extractor, based on the steepest gradient method for example, follows a linear low pass filtering of the image to increase the signal-to-noise ratio. Small objects corresponding to higher spatial frequencies start to disappear, fine detail is lost, and sharp discontinuities are blurred out. The major problems associated with these types of edge detectors are distortions which place discontinuities in wrong locations and round off corner information. Successful techniques based on the first approach can be found in the literature.[7.37, 7.38, 7.42]

A second edge detector category uses step-shaped templates which are fitted locally through small windows to the intensity values of the image. A "good fit," e.g., in a least square sense with the template, defines an edge in the window.[7.43]

The third approach to finding discontinuities is based on a global fit of a template over the entire data. Elastic membranes and plates under weak continuity constraints have been successfully applied to sparse and dense image data to "filter out" discontinuities of objects. Vision science has benefitted largely from the work of a series of authors who have developed algorithms that belong to this third category.[7.31] Blake and Zisserman[7.34, 7.44] have introduced algorithms that allow

FIGURE 7.13. Edge detection with simple 3 × 3 compass masks. A 300 × 300 fruit basket scene is submitted to eight different high pass kernels which extract maximum edge information whenever the intensity changes are oriented in one of the discrete eight directions (N, NE, E, SE, S, SW, W, NW) allowed by a 3 × 3 convolution operator. Observe that a combination of a variety of such operations can be used efficiently in AI tasks to extract information.

the extraction of discontinuities even in extremely poor signal-to-noise situations. In these algorithms, which fit weak strings (or rods) to one-dimensional data and weak membranes (or plates) to two-dimensional data, reconstruction of surfaces is achieved that fills in sparse and dense data, thereby minimizing the energy of these strings and membranes through the actual data. Blake and Zisserman solve these minimization problems by introducing the "graduated nonconvexity" (GNV) algorithm. In the GNC approach, the presence of local minima is avoided by defining a new function which is convex and hence has only one local minimum, which is at the same time the global minimum. Large amounts of computations such as in the "simulated annealing" method can be considerably reduced using the GNC approach.

The application of the Marr-Hildreth "theory of edge detection"[7.38, 7.45] has been of particular interest for the description of details, i.e., information at different scales. Following Marr and Hildreth's reasoning, two basic ideas underlie detection of intensity changes in images. These are that intensity changes occur at different scales in images in terms of their gray level values, and sudden changes in intensities result in peaks of first derivatives and hence "zero crossings" in the second derivatives. The filter that Marr and Hildreth introduced has the characteristics of a second derivative operator, with the capability to detect large blurry intensity changes as well as sharp details in images. The best operator found was the laplacean applied to a Gaussian distribution function with a variable standard deviation that can be used as a parameter. The operator can then be applied in its circularly symmetrical form; it takes on a radial function of the distance r in the shape of a Mexican hat (Fig. 7.14). The frequency spectrum shows clearly that this filter acts as a bandpass filter at different scales. The convolution of the Marr-Hildreth filter $\nabla^2 G(r)$ with an image reflects the detail corresponding to the choice of the standard deviation of $G(r)$. Binary images, where positive values are displayed with maximum intensity and negative values are set equal to zero, allow the extraction of zero crossings corresponding to a defined scale of image details [Fig. 7.15(a)]. The fact that details can be "captured" with a parameter and extracted at any scale makes the model very attractive. From the point of view of machine vision, several sizes of operators are required to cover the range of details in images. These steps define a model for information processing of human vision in its early stages. The "sombrero" (Mexican hat) filter introduced by Marr and Hildreth represents a viable model for the four psychophysical channels introduced by Wilson and Giese[7.46] and Wilson and Bergen.[7.47] It can approximate the difference of two Gaussians (DOG) as demonstrated by Marr and Hildreth in Fig. 7.16. These DOG functions describe the shape of the receptive fields of the retinal ganglion cells. The best approximation of Marr's filter by the DOG is obtained when the space constant ratios of the two Gaussians is 1 : 1.6. Wilson's estimate for the ratio of his channels was 1 : 1.75.

One basic question remains to be answered: How do the zero crossings obtained by different-sized filters relate to each other? In other words, how can we relate large scale intensity variations to small scale details in the same image? As pointed out by Marr, there is no reason why the zero crossings should relate *a priori* at different scales. Physical reasons can be invoked to claim a relationship: these physical assumptions were coined as "constraint of spatial localization" by Marr and Hildreth and reflect the physical nature of intensity changes in natural scenes. Figure 7.17 illustrates this situation very well, showing directional illumination, changes in surface orientations, and surface reflectance. Information is spatially localized at any scale. Hence we can conclude that large scale contours should be present at any smaller scale discontinuities by means of zero crossings obtained after application of the "sombrero" filter. At some scale this is, of course, no longer true. Interfering intensity changes in the large scale channel might prohibit the identification of details which require small scale intensity changes. A similar problem happens when different physical phenomena interfere to produce varying

FIGURE 7.14. The Marr-Hildreth "sombrero" filter is displayed as a shaded surface. The numbers on the z axis are arbitrary and point out the negative and positive portions of the filter. The filter, obtained by operating with the Laplacean on a Gaussian function, is circularly symmetrical. It is applied to generate an image where edges and details corresponding to different light intensities are determined through zero crossings which in turn define at a certain scale the objects to be identified in the scene.

intensity values in the same region at different scales. This is summarized in Marr's statement of spatial coincidence assumption: "If a zero-crossing segment is present in a set of independent channels over a contiguous range of sizes, and the segment has the same position and orientation in each channel, then the set of such zero-crossing segments indicates the presence of an intensity change in the image that is due to a single physical phenomenon (change in reflectance, illumination, depth, or surface orientation)." In practice this means that either discontinuities of adjacent scale sizes coincide, in which case the combined descriptions are corresponding [see Figs. 7.15(b)–7.15(d)], or detected discontinuities of adjacent scale sizes coincide, which implies that different regions or physical surfaces are in play. This led Marr to the fundamental idea of his "raw primal sketch" describing images with primitives such as edges, bars, blobs, and terminations. Figure 7.15(b) and 7.15(c) can be used to define the raw primal sketch of the scene shown in Fig. 7.15(a). A wonderful example to illustrate Marr's results has been demonstrated with a computer-generated block portrait of Lincoln from Harmon and Julesz.[7.48] The coarsely sampled and quantized picture of Lincoln is used to show the appearance of two sets of primitives to describe the picture at small scale and large scale (Fig. 7.18). Information from the larger channels can be obtained by viewing the drastically undersampled Lincoln picture from far away (approximately 6 feet). At a close distance, however, the picture conveys

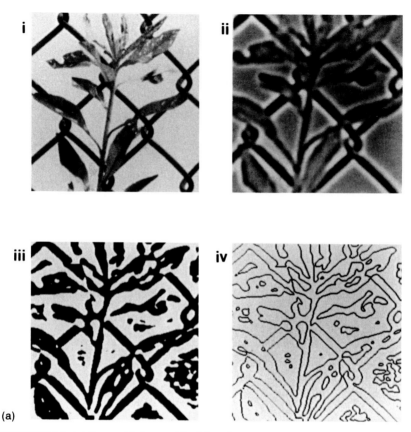

(a)

FIGURE 7.15. (a) Fencewith leaves. The Marr-Hildreth procedure to find boundaries in images; (i) original digital picture; (ii) filtered image (i); (iii) binary version separating positive from negative values; and (iv) edge extracted according to zero crossings in (iii). (b) and (c) The edges extracted at two different scales from image (a) after applying the Marr-Hildreth operator and extracting zero crossings in the image. (d) Superimposition of large scale and small scale edges show that any edge detected at the large scale is also present at the small scale, indicating that they truly belong to the identical physical phenomenon generating them. After Marr[7.31] and Hildreth.[7.45]

information obtained through the small scale channels without revealing the large scale contours. This indicates that there is information at large scale that is not used in the description of the information, although it is available to us. An intriguing artistically modified version of Lincoln's quantitized portrait by Dali is shown in Fig. 7.19. Rather than emphasizing the visual effect of coarse pixels in Lincoln's portrait, the artist has added details that absorb the viewer's mind at short distances. Only at fairly large distances does Lincoln's face become visible.

Knowledge-based inferences of three-dimensional world objects can be obtained from images. Varying contrast and boundaries are essential to describe shapes from objects in "real world" images. Though contours are intimately re-

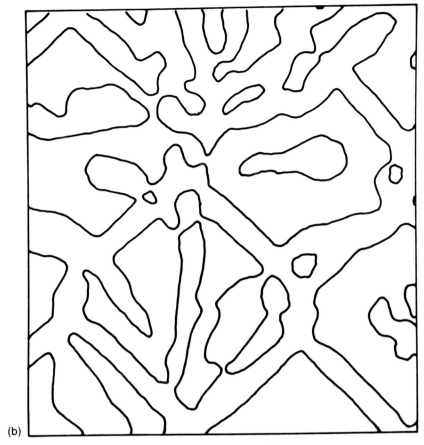

(b)

FIGURE 7.15. (*continued*)

lated to some form of detailed information, they might not give any clues regarding shapes as large as well as small scales. Contours, surface texture, and shading techniques have been used to generate a two-and-a-half-dimensional sketch of scenes. Illuminance and reflectance play a central role in visual data presentation and interpretation. The perceptual brightness of a surface can be computed as a function of the image irradiance independently of illuminance and reflectance. In general, illuminance is a component that is constant, or at the most varies slightly across an image, giving rise to small illuminance gradients. Reflectance gives rise to boundaries in scenes. Terzopoulos[7.49] has extended Grimson's[7.50] ideas to visual reconstruction approaching the shape-from-shading problem with multigrid methods[7.51–7.52] after reducing the problem to the solution of inhomogeneous partial differential equations with relaxation techniques. In the shape-from-shading problem, shapes of surfaces have to be recovered from image irradiance with constant illuminance, reflectance, and geometry.

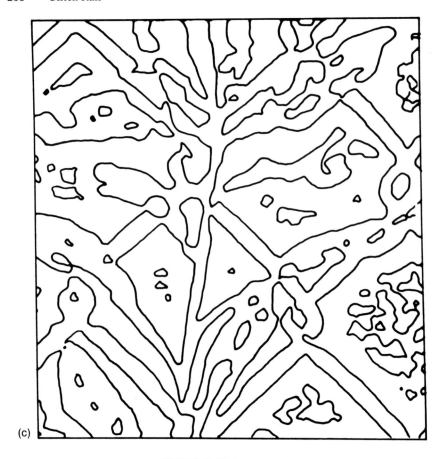

(c)

FIGURE 7.15. (*continued*)

7.4   Medical Imaging and Information

The ability to explore human anatomy without surgical intervention has enjoyed spectacular success over the last two decades. Medical imaging, especially radiological imaging, produces pictures that have little relation to natural scenes, where the complexity is determined by different physical phenomena. The "understanding" of natural scenes relies on image intensities, surface illumination, and surface reflectance. Thus, indoor scenes are more complicated to analyze and interpret because of secondary illumination effects, including light being reflected from multiple surfaces before reaching the human eye. The reflectance map depends largely on the physical nature of the surface, ranging in most cases between the Lambertian type (matt) and the perfect light reflector (specular). In addition, the geometry of surfaces and light sources plays a central role.

In traditional radiological images, image formation is based on the absorption of X rays in tissues of varying densities. In a second step, light is involved in the

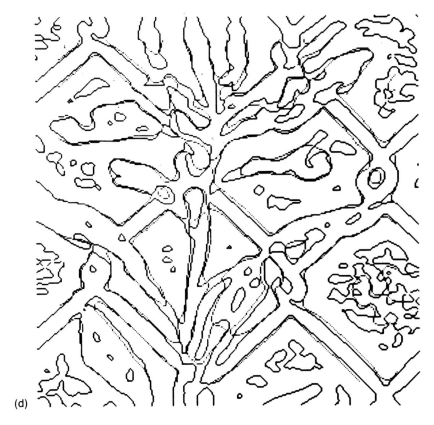

(d)

FIGURE 7.15. (*continued*)

image formation process through the use of intensifying screens. In traditional X ray imaging, X rays are used to provide a mapping of three-dimensional structures into a plane. In ultrasound imaging, ultrasound waves are reflected from various body interfaces to produce a map of internal structures. In nuclear medicine, radiolabeled pharmaceuticals are utilized to generate images according to their physiological distribution in organs. Tomographic imaging of the body with X rays in CT scan and radiowaves in magnetic resonance imaging (MRI) have helped to revolutionize radiological imaging. Images represent true slices through the human anatomy and details lost in planar imaging due to limited contrast are visualized in tomographic images.

Data presentation in radiological imaging is one of its greatest challenges. The discrimination of normal patterns and textures from abnormal ones relies on human abilities to "match," or "mismatch," visualized structures with a database which seems to be distributed over the higher cognitive centers for the trained individual. It seems that the "recognition" step following the detection of an object is much easier to achieve for real world objects than in radiological images. In medical imaging, contrast and detail are mostly dictated by physical processes, technology,

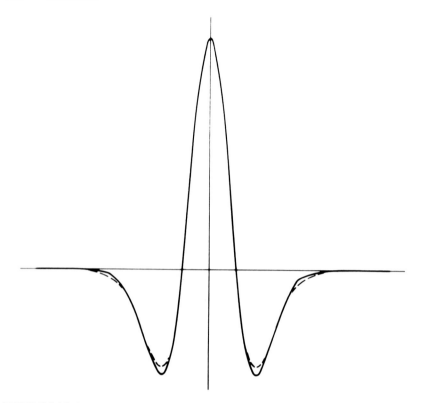

FIGURE 7.16. A cross section through the Marr-Hildreth filter (solid line) is closely approximated by the difference of two Gaussian functions (DOG). The DOG function has been proposed to model the shape of the spatial receptive fields of the visual channels.[7.46–7.47] This shows the similarity of the DOG model and the Marr-Hildreth filter. After Marr.[7.31]

and algorithms and, to a lesser extent, by image processing techniques and display media. The visual detection of minimal morphological changes in tissues is often impossible, due to a series of strongly correlated factors which prohibit their simultaneous optimization (e.g., noise reduction and resolution improvement). The strong limitation of the human visual system to discriminate more than approximately 60 gray shades in complex scenes hinders the perception of details in digital radiological images where the dynamic range can reach 16 bits of information per pixel. Simultaneous visualization of parenchymal details characterized by a wide range of numerical values obscures the true relationship among structures, but may lead to a better integral vision of their spatial relationships.

Small lesions can be difficult to detect in traditional radiological images due to poor contrast and noise. For example, pulmonary nodules involve difficult detection and identification tasks due to complex anatomy, limited dynamic range of the recording medium, scattered radiation, and low quantum efficiency. Tomographic imaging of the chest reveals far more detailed structure than does planar imaging, however, although not at the desired scale (or resolution). Small nodules

 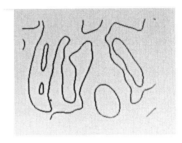

FIGURE 7.17. Artwork with light reflections. The result of the application of various scale Marr-Hildreth filters to the photograph of the sculpture (top left) is shown in the other three panels. Different information is picked up by the various filters. The difficulty lies in combining results into a single description. After Marr.[7.31]

down to millimeter size are most likely lost in noise, partial volume effects, and limited resolution. If individual details such as isolated nodules are not of primary concern, typical patterns and textures such as are found in emphysema and idiopathic pulmonary fibrosis can be used to identify and classify diseased pulmonary parenchyma. With an increasing use of digital imaging techniques and, inter alia, on-line enhancement techniques and real-time subtraction techniques using mask frames and dual energy images, detail and contrast have been improved to the benefit of medical diagnosis. Strangely enough, presentation of unprocessed images, where details are immersed in noise fields, remains important, indicating first a justified fear of artifacts, and second, that human vision is extremely powerful in sifting out relevant information. The human visual system can tolerate high amounts of noise which seems to be omitted by the higher cognitive centers during the identification process of objects at any scale. Small and large scale artifacts are not tolerated quite as well, due to their correlative characters. This partially explains why machine vision has enjoyed no breakthrough in the analysis of radiological images. Undoubtedly, neural processing models have to be invoked to understand these mechanisms.

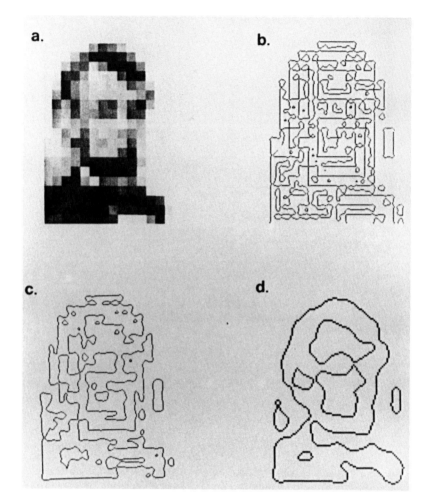

FIGURE 7.18. (a) Harmon and Julesz' quantitized portrait of Abraham Lincoln. At some distance it is easy to recognize Lincoln. (b), (c), and (d) The zero crossings obtained after use of three different sizes of Marr-Hildreth filter channels. Small channels emphasize the quantization of the picture which strikes the viewer at short distances. The large channel used in (c) provides the zero-crossings needed to recognize Lincoln. This demonstrates clearly the different types of information content of a particular scene that have to be unified for its complete description. After Marr.[7.31]

7.5   Visual Information and Communication

Images conveying two-dimensional or pseudo-three-dimensional information are presented to the visual cortex as still images or as continuous input that varies in real time, to which we respond *inter alia* with simple reactions to surrounding events using auto and hetero associations, or incorporating more complex neural activities involved in knowledge-based concrete and abstract decisions. With the

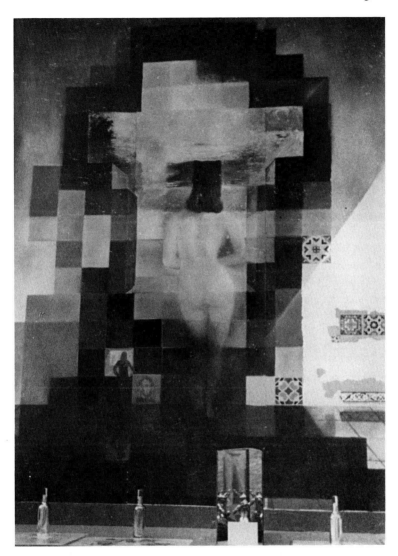

FIGURE 7.19. "Abraham Lincoln," Dali 1976. This preeminent artwork by Salvador Dali is shown as an artistic version of the undersampled Lincoln portrait displayed in Fig. 7.18(a). At close distance, true details are visible inserted according to the painter's imagination. Lincoln becomes visible at a fairly large distance corresponding to the effect of large channels in the Mar-Hildreth description of image content.

foundation of information theory by Claude Shannon roughly half a century ago, it became clear that communication of any kind of information would play a major role in today's society. Multimedia computer systems offering audio, graphics, animations, still images, and full-motion digital video are a good example of dissemination of information currently *en vogue.* The "world wide web (www)," which is available virtually to anybody operating a computer, shows how important and, at the same time, possibly ephemeral is the massive interchange of information of any kind.

Digital technology and networking are the tools of the present to handle and manage visual data ranging from simple static images, video telephones, animation sequences, and full-motion video to high-definition TV requirements, i.e. from 1 Hz frame rate to 60 Hz frame rate. While the lower range can be handled quite well, the real-time requirements have stimulated a series of new approaches to compress information while minimizing the loss of information by doing so. The compression of visual information has introduced the concept of *lossless* and *lossy* compression into our communication terminology. While lossy information may be acceptable in the multimedia world, the degree to which it is acceptable in the medical imaging world is not well understood. A quantitative measure is the only way out of the dilemma concerning the choice of lossy versus lossless medical images. The human eye is a poor integrator and in many situations cannot even tell qualitatively the difference of a lossy decoded image if the compression factor of the image is in the lower range, i.e. 3 to 6. Meanwhile the debate concerning possible medical image compression for diagnostic purposes continues. However, as archival media and networking capabilities improve, digital diagnostic data will increasingly be handled with lossless compression schemes in order to be recovered in their original form.

For many applications, ranging from the entertainment industry to high technology applications, transfer of images is accomplished with high degree of compression without suffering serious loss in quality. These needs have shed new light on the "understanding" of information in images. One of the most popular techniques has been coined the JPEG Standard for image compression (Joint Photographic Expert Group). The theoretical background of this method is found in the discrete cosine transform (DCT), one of many unitary transforms used since fast Fourier transforms became popular for processing images in dual space. The DCT of an image has the advantage of a "fast" computational algorithm for its numerical implementation while being nearly "optimal" in terms of its compression capabilities. In fact its performance is close to the Karhunen-Loeve transform (KLT) for signal representation in terms of the mean-square error criterion which applies when only a finite number of coefficients are retained to reconstruct the signal, i.e., the n largest ones, where n is some number less than the total number of transform coefficients. These theoretical characteristics have stimulated the implementation of efficient JPEG encoders and decoders. An excellent review of the JPEG standard used in multimedia compression techniques is the survey article by Borko Fuhrt.[7.53]

Probably the most intriguing lesson to be learned from compression algorithms is the shifting of relevant from visually "irrelevant" information. Another peculiar aspect of characterizing this "irrelevance" seems to be the natural redundancy of information in images, and even more so in images that are familiar. This is not true for information revealed in medical imaging of in vivo subjects. Natural scenes show a traditional repetition of straight lines defining edges, textured backgrounds, homogeneous surfaces, round and circular objects, and so on. Image features that are exactly self-similar remind us of fractals, such as Barnsley's *fern*[7.21] generated with a set of four simple affine transformations that require the storage of six numbers rather than individual pixels of the fern image itself (a decent resolution starts at 65536 pixels for an image pleasant to the human eye). It was Barnsley's idea that storing images as a collection of transformations might lead to image compression. As can be observed in the images of the Girl and Lenna in Fig. 7.20, there is some sort of self-similarity at different scales in the sense that some image regions resemble others at the same or different size, such as Lenna's shoulder, her hat, and some reflection in the mirror. This redundancy of information can be used to encode an image as transforms of parts of itself rather than as a copy of its entire self as in the case of the fern. The correct partitioning of images and affine mappings of finite-dimensional signals as well as other fractal compression techniques are described in detail in the state-of-the-art theory and application of fractal image compression in Fisher's monograph.[7.54] Traditionally, fractal image compression is comparable to JPEG and wavelet techniques in terms of performance.

Interestingly, image compression has also been looked at from the perspective of neural networks. The natural redundancy of information that can be observed in natural images can be exploited to compress imaged with artificial neural network (ANN) techniques. The idea is to encode the image with fewer bits than the original image. An application of a single-hidden-layer feed-forward neural network trained in an autoassociative mapping mode has an intriguing performance on images that are unknown to the ANN. The backpropagation technique has been used to "learn" the mapping of the image into itself. Cottrell and coworkers have proposed this technique for image compression.[7.55] A concise description of the technique has been discussed by M. Hassoun.[7.56] We have applied the method using 100 representative regions (each eight nodes on a side) of Lenna's image shown in the lower left of Fig. 7.20. This defines an input layer of 64 nodes for each input vector. The hidden layer was chosen with 32 nodes and the output layer had again 64 nodes to preserve the autoassociative mapping mode. The effect of the image compression on Lenna herself is shown in the lower right of Fig. 7.20 without quantization. The decoded image of Lenna is of high quality. The trained network was then applied to all consecutive 8×8 blocks of the Girl's image shown in the upper left of Fig. 7.20. Although there are, at least to some extent, true similarities in all human faces, it is quite surprising how well the Girl's face is reconstructed using the ANN trained with Lenna. As pointed out by Hassoun,[7.56] the hidden layer, which he calls the *representation layer,* attempts to extract some regularities from the input vectors used to train the ANN. Empirically it seems that the outputs of the representative layer span the subspace of the principal components of the image vectors used in the experiment. The results of this simple nonlinear autoassociative technique reveal that, from a small representative set of vectors describing Lenna, the net was capable of *generalization* when applied to another face. Figure 7.21 shows a close-up (zoom factor of 4) of Lenna and the Girl as generated by the trained ANN. Artifacts are

FIGURE 7.20. The Lenna image in a 256 × 256 resolution is shown in the lower left. The Girl image is displayed in the upper left with the same resolution. The lower right shows the result of recovering Lenna from the feed-forward ANN using 100 8 × 8 training regions. The upper right shows the result of applying the ANN trained with blocks from Lenna to all 8 × 8 blocks of the Girl's image. Notice the high quality of the "reconstructed'" image.

visible in both images; however, it is more than surprising that a single training image is sufficient to generate this quality in an image unknown to the ANN, which clearly shows the common feature content of natural scenes, i.e., inherent redundancy of information across images. Although the technique seems to be of no practical value for massive image compression and transmission, it provides some interesting theoretical insights into the features of visual information.

FIGURE 7.21. Zoom factors of 4 have been applied to the output images of Lenna and the Girl of the ANN trained with input vectors from the Lenna image. Notice the artifacts of incomplete information in the close-up of Lenna's face. The same artifacts can be observed in the Girl's image; however, it is surprising how well the simple ANN trained with blocks from one single image performs on the unknown image of the Girl.

7.6   Conclusions

The power of the human brain in omitting "irrelevant" information at any scale is unmatched. Gregory (see Chap. 6) has recognized the impact of omission of information on artificial intelligence:"The present state of Artificial (Machine) Intelligence has deep within it a strange paradox. It is just those aspects of control and the selection of relevant from irrelevant data which are the most difficult to mechanize—though they were the first problems to be solved by organisms." In medical imaging, machine vision faces a serious dilemma. Human anatomy is complex and no "normalized" or standard anatomy is yet available to train machine algorithms. The presence of noise, limited resolution, and low contrast, filtered out by the human observer, seriously hinders automated diagnosis. Many normal structures are ultimately irrelevant during the visual scanning process, yet essential as feedback to the detection of "relevant" abnormalaties. The training process used by biological systems is to a large extent based on accumulation of visual information. It is unknown how important subtle variations in normal structures at any scale could affect the machine detection of true abnormal structures.

Neural network models[7.57–7.59] of the human brain, together with growing VLSI technology, have opened the door to a better understanding of visual detection processes.[7.60–7.61] The concept of distributed memories[7.58] and the massive parallelism and lateral interconnections of biological neural networks are beginning to reveal the way to extract unknown from known features, i.e., abnormal from normal ones. Interesting models based on autoassociative memory paradigms have been developed by Kohonen and coworkers.[7.63–7.64] They allow, e.g., the identification of unknown patterns at any scale using sets of known "memorized patterns." Any image vector that can be written mathematically as a linear combination of known image vectors is not acknowledged as a new pattern. If normal patterns are stored, abnormal features can be extracted as "unknown" to the memory. This represents only one example of a new approach towards a global understanding of visual information. In the right context, this could shed some light on how to present visual data to the human observer by defining the complex relationships of imaging parameters whose improvements we have sought for so long.

7.7   References

[7.1]   Gonzalez R.C., Wintz P. *Digital Image Processing.* Reading: Addision-Wesley; 1977.

[7.2]   Pizer S.M., Zimmerman J.B., Staab E.V. Adaptive gray level assignment in CT scan display. *J. Comput. Asst. Tomog.* 1984; 8: 300–305.

[7.3]   Zimmerman J.B., Pizer S.M., Staab E.V., Perry J.R., McCartney W., Brenton B.C. An evaluation of the effectiveness of adaptive histogram equalization for contrast enhancement. *IEEE Trans. Med. Imag.* 1988; MI-7; 304–312.

[7.4]   Zimmerman J.B., Cousins S.B., Hartzell K.M., Frisse E., Kahn M.G. A psychophysical comparison of two methods for adaptive histogram equalization. *J. Digital Imag.* 1989; 2: 82–91.

[7.5]   Pizer S.M., Amburn E.P., Austin J.D., Cromartie R., Geselowitz A., Greer T. Adaptive histogram equalization and its variations. *Comput. Vis. Graph. Imag. Proc.* 1987; 39: 355–368.

[7.6]   Jain A.K. Advances in mathematical models for image processing. *Proc. IEEE* 1981; 69: 502–527

[7.7]   Jain A.K. *Fundamentals of Digital Image Processing.* Englewood Cliffs: Prentice Hall; 1989.

[7.8]   Gibson J.J. *The Perception of the Visual World.* Boston: Houghton Mifflin; 1952.

[7.9]   Haralick R.M. Statistical and structural approaches to texture. *Proc. IEEE* 1979; 67: 786–804.

[7.10] Resnikoff H.L. *The Illusion of Reality.* New York: Springer; 1989.

[7.11] Julesz B. Textons, the elements of texture perception, and their interactions. *Nature* 1981; 290: 91–97.

[7.12] Julesz B. Schumer R.A. Early visual perception. *Annu. Rev. Psychol.* 1981; 32: 575–627.

[7.13] Julesz B. Frisch H.L., Gilbert E.N., Shepp L.A. Inability of humans to discriminate between visual textures that agree in second order statistics—revisted. *Perception* 1973; 2: 391–405.

[7.14] Hawkins J.K. Textural properties for pattern recognition. In: Lipkin B.S., Rosenfeld A., eds. *Picture Processing and Psychopictorics.* New York: Academic; 1970: 347–370.

[7.15] Kruger R.P., Thompson W.B., Turner A.F. Computer diagnosis of pneumoconiosis. *IEEE Trans. Syst. Man. Cyberg* 1974; SMC-4: 40–49.

[7.16] Sutton R.N., Hall E.L. Texture measures for automatic classification of pulmonary disease. *IEEE Trans Comput.* 1972; C-2: 667–676.

[7.17] Mandelbrot B.B. *The Fractal Geometry of Nature.* New York: Freeman; 1982.

[7.18] Pentland A.P. Fractal surface models for communication about terrain. *Proc. SPIE* 1987; 845: 301–306.

[7.19] Pentland A.P. Fractal based description of natural scenes. *IEEE Trans. Patt. Anal. Mach. Intell.* 1984; PAMI-6: 661–674.

[7.20] Barnsley M.F., Deraney R.L., Mandelbrot B.B., Peitgen H.O., Saupe D., Voss R.F. *The Science of Fractal Images.* New York: Springer; 1988.

[7.21] Barnsley M. *Fractals Everywhere.* (second edition) New York: Academic; 1993.

[7.22] Lundahl T., Ohley W.J., Kay S.M., Siffert R. Fractional Brownian motion: A maximum likelihood estimator and its application to image texture. *IEEE Trans. Med. Imag.* 1986; MI-5: 152–161.

[7.23] Lundahl T., Ohley W.J., Kublinski W.S., Williams D.O., Gerwitz H., Most M.S. Analysis and interpretation of angiographic images by use of fractals. In: *Computers in Cardiology.* New York: IEEE; 1985: 355–358.

[7.24] Chen C.C., Daponte J.S., Fox M.D. Fractal feature analysis and classifiction in medical imaging. *IEEE Trans. Med. Imag.* 1989; MI-8: 133–142.

[7.25] Nelson T.R. Manchester D.K. Modeling of lung morphogenesis using fractal geometries. *IEEE Trans. Med. Imag.* 1988; MI-7: 321–327.

[7.26] Caldwell C.B., Stabelton S.J., Holdsworth D.W., Jong R.A., Weiser W.J., Cooke G. Characterization of mammographic parenchymal pattern by fractal dimension. *Phys. Med. Biol.* 1990; 35: 235–247.

[7.27] Cargill E.G., Donohoe K.J., Kolodny G., Parker A.J., Duane P. Estimation of fractal dimension of parenchymal organs based on power spectral analysis in nuclear medicine scans. In: Ortendahl D.A., Llacer J., eds. *Proc. 11th Int. Conf. Inform. Proc. Med. Imag.* 1991; 557–570.

[7.28] Barnsley M.Fm, Demko S. Iterated function systems and the global construction of fractals. *Proc. R. Soc. London A* 1985; 399: 243–275.

[7.29] Barnsley M.F., Ervin V., Hardin D., Lancaster J. Solution of an inverse problem for fractals and other sets. *Proc. Natl. Acad. Sci.* 1985; 85: 1975–1977.

[7.30] Attneave F. Some informational aspects of visual perception. *Psychol. Rev.* 1954; 61: 183–193.

[7.31] Marr D. *Vision.* New York: Freeman; 1982.

[7.32] Hubel D.H., Weisel T.N. Receptive fields, binocular interaction, and functional architecture in the cat's visual cortex. *J. Physiol. London* 1962; 160: 106–154.

[7.33] Hubel D.H., Weisel T.N. Receptive fields and functional architecture of monkey striate cortex. *J. Physiol. London* 1968; 195: 215–243.

[7.34] Blake A., Zisserman A. *Visual Reconstrcution.* Cambridge: MIT; 1987.

[7.35] Pratt W.K. *Digital Image Processing* (2nd ed.). New York: Wiley; 1991.

[7.36] Duda R.O., Hart P.E. *Pattern Classification and Scene Analyisi.* New York: Wiley; 1973.

[7.37] Canny J.F. *Finding Edges and Lines in Images* (S.M. thesis). Cambridge: MIT; 1983.

[7.38] Marr D., Hildreth E. Theory of edge detection. *Proc. R. Soc. London B* 1980; 207: 187–217.

[7.39] Kirsch R. Computer determination of the constituent structure of biological images. *Comput. Biomed. Res.* 1971; 4: 315–328.

[7.40] Rosenfeld A. A nonlinear edge detection technique. *Proc. IEEE Lett.* 1970; 58: 814–816.

[7.41] Jain A.K., Ranganath S. Image restoration and edge extraction based on 2D stochastic models. *Proc. Int. Conf. Acoust. Speech Sig. Process.* (Vol. 3). Paris; 1982; 1520–1523.

[7.42] Haralick R.M. Edge and region analysis for digital image data. *Comput. Graph. Imag. Process.* 1980; 12: 60–73.

[7.43] Hueckel M.G. An operator which locates edges in digitized pictures. *J. Assoc. Comput. Mach.* 1971; 18: 113–125.

[7.44] Blake A., Zisserman A. Some properties of weak continuity constraints and the GNC algorithm. *Proc. IEEE Conf. Comput. vision Patt. Recogn.* 1986; 656–661.

[7.45] Hildreth E.C. *Implementation of a Theory of Edge Detection* (M.S. thesis) Cambridge: MIT; 1980.

[7.46] Wilson H.R., Giese S.C. Threshold visibility of frequency gradient patterns. *Vision Res.* 1977; 17: 1177–1190.

[7.47] Wilson H.R., Bergen J.R. A four mechanism model for spatial vision. *Vision Res.* 1979; 19: 19–32.

[7.48] Harmon L.D., Julesz B. Masking in visual recognition: Effects of two-dimensional filtered noise. *Science* 1973; 180: 1194–1197.

[7.49] Terzopoulos D. Multilevel computational processes for visual surface reconstruction. *Comput. Vis. Graph. Imag. Proc.* 1983; 24: 52–96.

[7.50] Grimson W.E.L. *From Images to Surfaces,* Cambridge: MIT; 1981.

[7.51] Brandt A. Multilevel adaptive solutions to boundary value problems. *Math. Comput.* 1977; 31: 333–390.

[7.52] Briggs W.L. *A Multigrid Tutorial.* Philadelphia: Society for Industrial and Applied Mathematics; 1987.

[7.53] Furht B. A Survey of Multimedia Compression techniques and Standards. Part I: JPEG Standard. *Real-Time Imaging* 1995; 1: 49–67.

[7.54] Yval F., ed. *Fractal Image Compression, Theory and Application.* New York: Springer; 1995.

[7.55] Cottrell G.W., Munro P., Zipser D. Image compression by back propagation: An example of extensional programming. In: N. Sharkey, ed. *Models of Cognition: A Review of Cognitive Science.* Norwood: Ablex; 1989: 208–240.

[7.56] Hassoun M.H. *Fundamentals of Artificial Neural Networks.* Cambridge: The MIT Press; 1995.

[7.57] Grossberg S. Nonlinear neural networks; principles, mechanisms, and architectures. *Neural Nets* 1988; 1: 17–61.

[7.58] Kohonen T. An introduction to neural computing. *Neural Nets* 1988; 1: 3–16.

[7.59] Carpenter G.A. Neural network models for pattern recognition and associative memory. *Neural Nets* 1989; 2: 243–257.

[7.60] Fukishima K. A neural network for visual pattern recognition. *IEEE Trans. Comput.* 1988; C-21: 65–75.

[7.61] Fukishima K. Analysis of the process of visual pattern recognition by neocognition. *Neural Nets* 1989; 2: 413–420.

[7.62] Rumelhart D.E., McClelland J.L., PDP Research Group. *Parallel Distributed Processing* (Vol. 1). Cambridge: MIT; 1986.

[7.63] Kohonen T. *Associative Memory. A Systematic Theoretical Approach.* Berlin: Springer; 1977.

[7.64] Kohonen T. *Self-organization and Associative Memory* (2nd ed). Berlin: Springer; 1987.

8

Image Manipulation

Ronald R. Price

8.1 Introduction: The Digital Image

Much of what is now known about image manipulation and image-processing theory has been developed by those early researchers concerned with electrical signal processing. [8.1–8.7] Images, when converted to electrical signals, were soon recognized to be a potentially fertile area for the extension of signal processing techniques. Early on, image processing was carried out on "analog" signals as opposed to numbers in a digital computer. Analog image processing was often performed using optical techniques or through the use of specially built electronic circuits which could perform specified functions. The term *analog* refers to systems and processes in which signals with uniform and continuous variations are manipulated. The term *digital* refers to systems and processes which involve manipulation of discrete numbers. Even in the case of digital images, the image information has exited at some point in time as an analog signal. To create the digital image, the analog signal must then be converted to discrete numbers and stored in a computer memory. This conversion process is called analog-to-digital conversion (ADC).

An example of this process is illustrated in Figs. 8.1 and 8.2 in which a scene is viewed by an analog video camera whose output is converted into a digital image. In the video camera, each horizontal line of the image is represented as

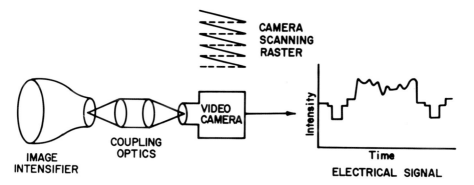

FIGURE 8.1. The video camera, through optical coupling, views the output phosphor of the image intensifier. The charge pattern on the light-sensitive faceplate pick-up tube is read out in a rectilinear scan pattern. A voltage signal corresponding to the light pattern is generated for each horizontal scan line. (After Price, et al.[8.10]).

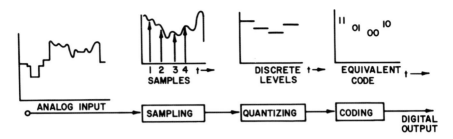

FIGURE 8.2. The analog voltage signal generated for each horizontal video line is the input for the analog-to-digital converter. The frequency at which the analog signal is sampled determines the spatial resolution of the digital image. The total number of samples across the horizontal line equals the horizontal dimension of the digital matrix. The vertical dimension equals the number of horizontal lines digitized. The quantizing phase of the ADC determines the number of shades of gray of the image. The coding phase assigns the binary number to each sample. (After Price, et al.[8,10]).

a voltage wave form, which varies in direct proportion to the brightness of the image at that point. The analog voltage signals are the input to the analog-to-digital converter (ADC) (Fig. 8.2). The ADC then samples the continuously varying (analog) signal at prescribed time intervals. It is the sampling frequency that determines the number of digital picture elements (pixels) that will be generated for each horizontal scan line. It is this feature of the ADC that determines the spatial resolution that can be achieved with a digital image. The range of magnitudes of the numbers generated by the ADC determines the contrast (or gray-scale) resolution of the digital images. The images in Fig. 8.3 illustrate how the number of samples into which the image is divided and the number of shades of gray used to display the image affect the visual perception of the digital image.

This chapter describes and illustrates some of the more common image manipulation algorithms and techniques used to process digital images in medicine.

8.2   Interpolation

Some image manipulation techniques have been developed which can affect the perception of an image without actually changing the information content of the image. One such technique is called interpolation (Fig. 8.4). The interpolation technique is usually applied to images which have small image matrices. Small matrices are often aesthetically displeasing because of the "blocky" appearance of the picture elements. By using a larger display matrix and "filling-in" between the actual picture elements, the image presentation can be improved without actually altering the original image data.

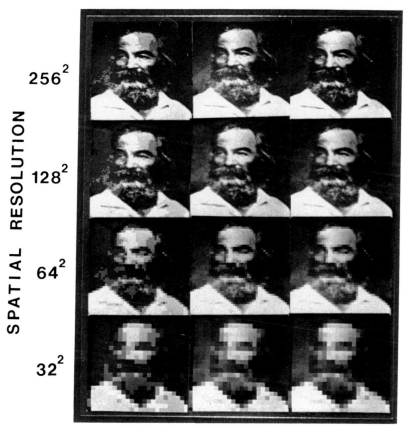

FIGURE 8.3. The ADC parameters determine both the spatial and contrast resolution (shades of gray) of the digital image. An easily recognizable subject (Walt Whitman) illustrates how image perception changes as the shades of gray and spatial sampling vary. (After Price, et al.[8.10]).

8.3   Gray-Level Manipulation

In a digital image, each pixel is represented by a number stored in a computer memory. In the presentation of the digital image, we have a choice of how the stored number relates to the brightness or to the color with which each pixel is displayed. In most cases, the choice is to use a linear gray-scale map. In a linear map, we attempt to display each pixel at a brightness directly proportional to the stored numerical value, i.e., pixels whose stored values are two times larger are displayed at twice the brightness. Nonlinear maps or look-up tables (LUT) can often be used effectively to enhance image features. It is worth noting again that

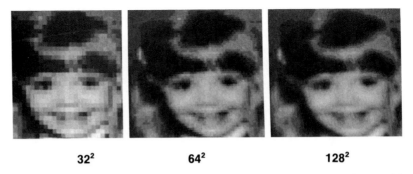

FIGURE 8.4. Interpolation can be used to improve the perception of a digital image without altering the image data. An image of the author's daughter was digitized originally into a 32 x 32 matrix with 16 shades of gray. By displaying the original data within larger matrices (64 x 64 and 128 x 128) and "filling-in" linearly between the original data, a more aesthetically pleasing image is obtained. (After Price, et al.[8.10]).

a display LUT does not alter the stored image data, but serves only to change the intensity of gray levels displayed on the screen.

Nonlinear gray-scale mapping can most easily be understood by observing the changes in the image histogram.[8.8] The image histogram is a plot of the number of pixels at a particular gray level versus the gray-level value. Frequently, the values of gray levels within an image do not span the range of available brightness levels that the system is capable of displaying. Figure 8.5(a) is an example of such an image, where most of the pixel values are contained within the lower half of the available display levels. Were it not for our ability to independently adjust the brightness of the image display, the image would appear dim relative to an image with a histogram where the pixel values were shifted to higher values. Lack of image contrast rather than image brightness, however, is the most important problem presented by the images which have histograms that are shifted to low brightness levels. An image with a histogram which spans a narrow range will be a low contrast image. The low contrast results because of small brightness difference between the various pixels. In general, a low contrast image has a histogram which is narrow, a high contrast image has a wide histogram. A dark image will have a histogram shifted toward low gray-level values and a bright image will have a histogram shifted to high pixel values.

One of the most useful gray-scale mapping algorithms used to improve low-contrast images is histogram-equalization or histogram stretching. In histogram equalization, we attempt to spread the image gray levels over the entire display brightness range without changing the shape of the histogram. The spreading of the histogram increases the relative difference between adjacent brightness levels, thus improving contrast. Figure 8.5(b) shows the result of a histogram equalization when applied to the image in Figure 8.5(a). The resultant image is somewhat more harsh in appearance; however, features which were previously undiscernible due

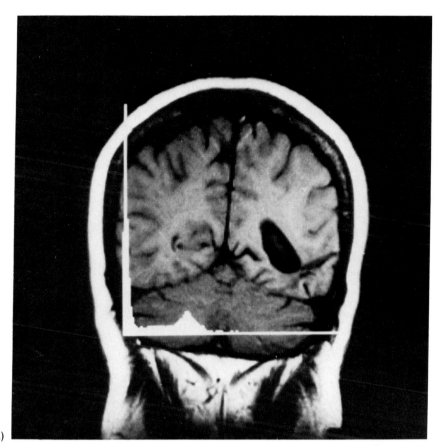

(a)

FIGURE 8.5. (a) An MRI scan in which the levels of pixel intensity are shifted to low values is a darker image as well as a low contrast image. The histogram of the pixel intensity is used to aid the understanding of what is seen. Since most display systems have independent control to allow adjustment of brightness, one may not readily recognize reduced brightness. (b) Contrast enhancement is achieved by histogram equalization. Background features which were not discernible are now visible. Other low contrast features are also now recognizable. The equalized histogram (shown as an overlay) maintains the same shape but expands to utilize the entire range of display. Larger gaps between gray levels increase contrast.

to low contrast are now visible. The corresponding resultant histogram shows that the desired goal of utilizing the entire range of display levels has been achieved.

Most modern digital display workstations provide for on-line image contrast adjustment which is capable of producing results similar to histogram equalization. This adjustment is generally achieved through two on-line controls called "window width" and "window level" or "center." Figure 8.6 illustrates the use of on-line window width and level adjustments to allow visualization of structures which would otherwise be invisible without gray scale manipulation.

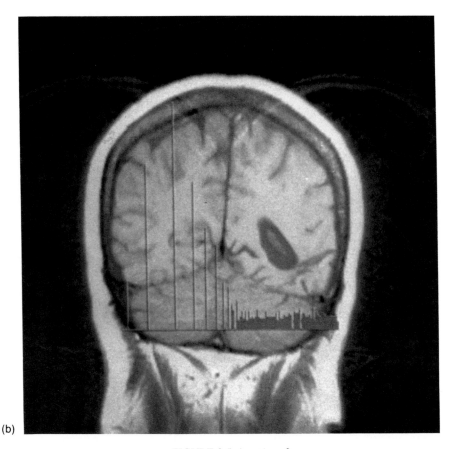

(b)

FIGURE 8.5. (*continued*)

8.4   Filtering

Filtering is one of the most important, as well as one of the most widely used, types of image processing. [8.5,8.8] Most of us can easily relate to the concept of filtering as it is used in everyday life with audio systems. In these systems, frequencies are enhanced or attenuated with a turn of the bass or treble controls or by using the similar, but more sophisticated, balance control. The frequencies contained in an audio signal can be used as an analogy to understand the "spatial" frequencies contained in an image and the corresponding filters used to enhance or attenuate the spatial frequencies in the image. By reducing the bass control, we filter out low frequencies and preferentially pass high frequencies; this is a "high-pass" filter. By reducing the treble control, we similarly generate a low-pass filter. An increase in the audio balance control at a given frequency range "boosts" frequencies; a decrease attenuates those same frequencies.

Through this analogy we can begin to understand the physical meaning of spatial frequencies by noting that the frequency of any wave is inversely proportional to

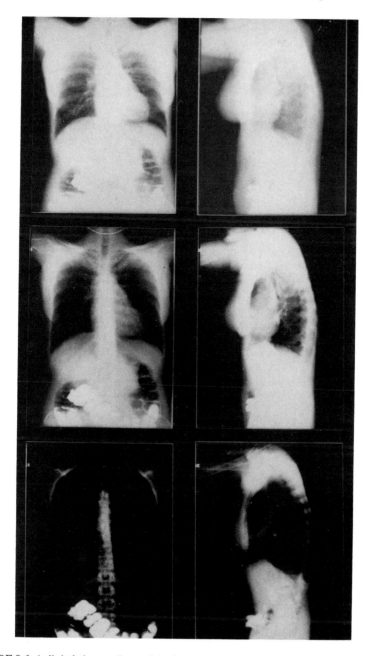

FIGURE 8.6. A digital chest radiograph is viewed with different window and level settings. The upper images illustrate a low contrast, high brightness setting. The center images are displayed to allow visualization of very bright objects such as the spine and the radio-opaque contrast material in the bowel. Further adjustment allows detail in the spine and bowel to be seen, but at the expense of the lung field. In this example, the image data actually span a greater range than the display system is capable of displaying. (After Price, et al.[8.10]).

its wavelength. Spatial frequency is defined as the number of times that a gray level in an image oscillates between black and white within a given distance. If the gray level oscillates from black to white and then back to black again returning to its original value in 1 mm, it will have a spatial wavelength of 1 mm and a spatial frequency of 1 cycle/mm. Often spatial frequencies are expressed as line pairs (black and white) rather than cycles; thus 1 spatial cycle/mm is the same as 1 line pair/mm. Two cycles/mm would be 2 line pair/mm, and so on. Image gray-scale oscillations which occur over short distances are said to have a high spatial frequency. Those which are slowly varying in distance are said to be low spatial frequencies. Thus, a small object in an image which spans a very short distance will contain high spatial frequencies and a large object spanning long distances will contain low spatial frequencies. The image background usually has slowly varying gray levels over large distances; thus, low spatial frequencies. An edge, on the other hand, presents a rapid change in gray level in a short distance and is thus characterized by high spatial frequencies. Random noise in an image presents itself as a "graininess" in the image and is usually superimposed over the true image structure. This graininess is usually of a low intensity yet very rapid variation in gray scale. As such, noise is often responsible for the highest spatial frequencies contained within an image.

In any imaging system, some structures will be too minute to be imaged. In general, a system's ability to image structures will be inversely proportional to the spatial frequency of the structure. A quantitative measurement of an imaging system's ability to respond to different spatial frequencies is called the modulation transfer function (MTF). The MTF is a plot of the percent of spatial frequencies that an image contains relative to the actual spatial frequencies present in the object. The inability of an X-ray system to accurately reproduce high spatial frequencies is illustrated in Fig. 8.7. In this image from a digital X-ray system, we can easily resolve lead strips which have wide separations (low frequencies), yet rapidly lose the ability to resolve the high spatial frequency strips. In more typical medical images, this same effect is manifest by indistinct or blurred edges and then the loss of detail in small structures.

Spatial frequency filtering can be carried out either in the frequency domain or in the spatial domain. [8.7] To understand this statement, we note that the image we see exists in the spatial domain. To convert the image to the frequency domain, the image must undergo a Fourier transformation in which the image structure is decomposed into its component spatial frequencies. This conversion is a mathematical manipulation of the data carried out by the computer. In this form, the design of filters which enhance or attenuate certain frequencies or frequency bands is relatively easy in that one simply multiplies the various frequency components by small numbers to attenuate them and by large numbers to achieve a spatial frequency boost. Once the frequency components are modified or eliminated in amplitude according to the desired filter function, an inverse Fourier transformation is used to restore the image back to the spatial domain for viewing. Even though the design of filters is often easier in the frequency domain, the actual application of the filter is usually performed in the spatial domain for reasons of

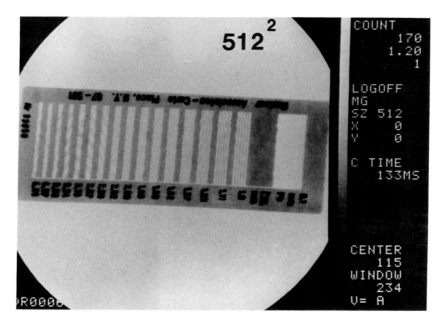

FIGURE 8.7. A radiograph of a resolution phantom consisting of lead strips of different spatial frequencies. The system is capable of resolving the lines up to about 1.8 lines/mm. Higher spatial frequencies are not seen because of system blurring.

convenience and speed. To filter an image in the spatial domain, we manipulate the pixel gray levels themselves directly without carrying out a mathematical conversion. The mathematical method for carrying out spatial domain filtering is called "convolution."(See Chapter 4, Section 4.2)

The actual convolution algorithm is performed by first defining a convolution kernel or filter mask. The filter mask is simply a square array of numbers. The numbers are in essence weighting factors which control the relative influence of surrounding or neighboring pixels on a particular pixel of interest. The filter masks can be designed to have the same effect as any frequency domain filter one may choose. The size of the square array of numbers (whether 3x3, 5x5, 9x9, or larger) will to some degree affect the accuracy with which the spatial domain filter can be made to match the frequency domain filter. The most common filter masks in current use are 3x3 and 5x5. Larger filter masks have the disadvantage of requiring longer computation times.

The convolution calculation may be visualized as a transparent overlay that is placed over the image matrix to be filtered. The center of the overlay is then moved in a systematic pattern over the original image. At each pixel position in the image, each value of the filter mask is multiplied by the value of the pixel over which it lays. The sum of all of these procedures (9 in the case of a 3x3 filter mask) is then placed at the position of the central element of the filter mask relative to the low-pass:

$\frac{1}{9} \times$

1	1	1
1	1	1
1	1	1

high-pass:

$\frac{1}{9} \times$

−1	−1	−1
−1	+8	−1
−1	−1	−1

high-boost:

$\frac{1}{18} \times$

−1	−1	−1
−1	17	−1
−1	−1	−1

gradient:

$\frac{1}{9} \times$

−1	0	+1
21	0	+2
−1	0	+1

original input image. A filtered image is created pixel-by-pixel in this manner and is of equal dimensions to the original image.

The more commonly used convolution filters are low-pass or smoothing filters, high-pass filters, and high frequency boost filters. The low-pass filter is designed to suppress high frequencies while leaving low frequency objects unaffected. Since random noise is high frequency in character, a low-pass filter suppresses random noise and is thus referred to as a noise reduction filter. If we can define the signal in an image as the intensity of the low frequency (large) structures and the noise as the variations (usually standard deviation of pixel values over some relatively large uniform area) in the pixel values. The signal-to-noise ratio in a low-pass filtered image is increased relative to the original image, because the signal is left largely unaffected while the high frequency variation (noise) is reduced.

High-pass filters pass high frequencies and suppress low frequencies, thus emphasizing edges and small structures. Unfortunately, high-pass filters also enhance random noise. A high frequency boost filter is used to actually amplify high frequency structures in the image.

Typical 3 × 3 filter masks are shown in the following matrices. [8.5,8.8] The factor in front of the filter is a normalizing factor which is multiplied by each element before summation to ensure that the output pixel value is not outside the valid range.

Blurring resulting from the imaging process itself can also be thought of as a convolution. That is, each point in the image results from the summation of some proportional amount of many points in the object. This smearing is called the point-spread function and is characteristic of the imaging instrumentation being used. When the effect of the point spread is added up for each point, a blurred image results. An imaging system in which the shape of the point-spread function does not change across the image field is called a stationary system. Imperfect imaging systems introduce blurring that limits image quality by eliminating high frequencies. The result is degraded sharp edges and blurred small objects. The imaging process also introduces high frequency random noise into the image. Thus, imaging systems both eliminate desired high frequency components, and introduce other undesired high frequency components. If we knew the blur function of the system exactly and the system were noise-free, we could in principle construct an exact filter mask which could recover the lost resolution. In fact, since the blur function is not generally known, and high-pass filtration enhances noise, there is a limit to the amount of resolution that can be recovered.

The most common mask filter currently used for the reduction of random noise is binomial smoothing. Smoothing can be applied in either one dimension, as in time-intensity curves, or two dimensions, as in images. Probably the most common two-dimensional smoothing mask is the nine-point spread function illustrated in the following matrix.

binomial smoothing: $\frac{1}{16} \times$

1	2	1
2	4	2
1	2	1

Greater or lesser degrees of smoothing can be achieved through variations of the weighting factors. The low-pass mask filter above essentially replaces the central value by the average of all its neighbors to produce a greater degree of smoothing than the binomial filter mask. The particular binomial weighting illustrated here attempts to weight the influence on a particular picture element from neighboring elements by a factor which is inversely proportional to the distance. That is, elements which are far away will produce the least influence. One should further note that the sum of the nine weight functions (multiplied by the normalization factor) is unity. The advantage of a weight function which is normalized to unity is that it conserves total intensity. That is, the total intensity of the raw image and the smoothed image will be the same. This fact is important both for quantitative assessments and for image display. The nine-point smoothing is also frequently used as a preprocessing technique prior to application of edge-detection algorithms. This preprocessing is important since many edge-detection algorithms rely on gradient or derivative calculations that are very susceptible to noise, and the nine-point smoothing operation is used to suppress noise.

Figure 8.8 illustrates application of several 3x3 convolution filters as applied to magnetic resonance images of the head. Edge enhancement is illustrated by both a gradient filter and an unsharp mask (described in Sec. 8.5). The gradient filter, using a simple difference filter, approximates a first derivative. At edges, the first derivative (slope) of the gray-scale is large and provides a mechanism for edge detection. A less common edge-detection algorithm is the second derivative filter (Laplacean) which produces two maxima at each edge. The double line is often confusing and frequently more noisy. The unsharp mask processing is a two-step convolution filter which can be used as a high frequency boost filter for edge detection. In unsharp masking, the image is first preprocessed using a smoothing filter. The new smoothed image is then subtracted from the original unprocessed image. The smoothed image may be weighted before subtraction

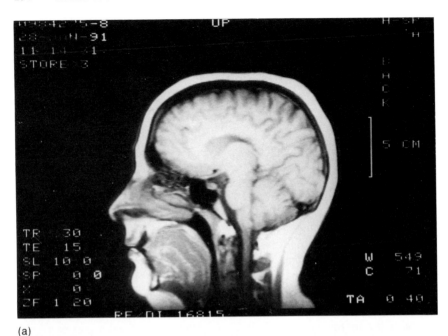

(a)

(b)

FIGURE 8.8. (a) Original unprocessed MR image of the head. (b) Gradient edge-detection filter producing a single edge. (c) High frequency boost filter enhancing small structures and background noise. (d) Noise reduction low-pass filter. (e) Unsharp mask image to enhance edges. (f) Laplacean edge-detection filter producing a double edge.

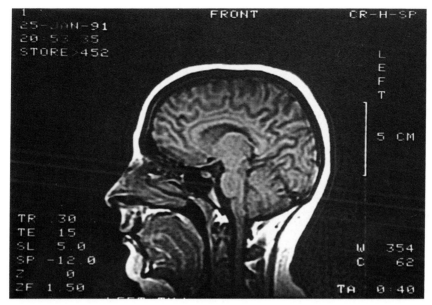

(c)

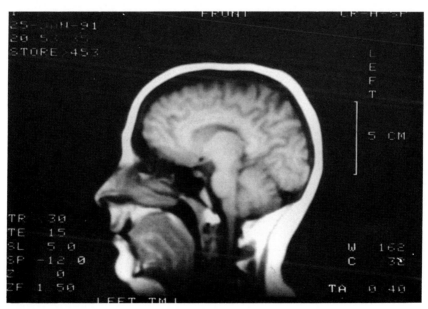

(d)

FIGURE 8.8. (*continued*)

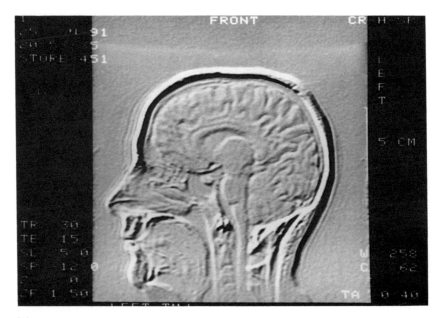

(e)

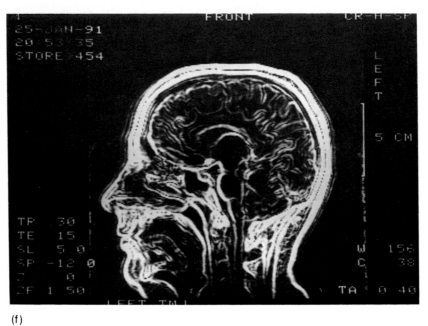

(f)

FIGURE 8.8. (*continued*)

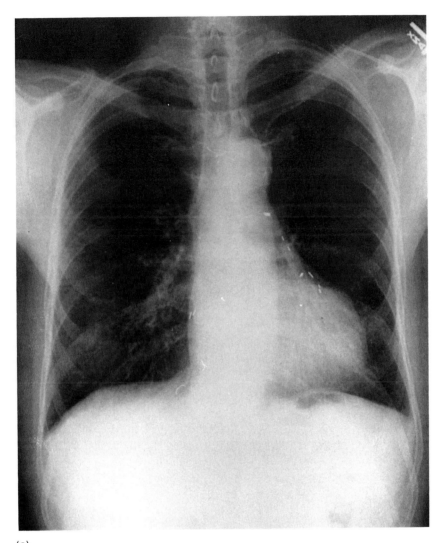

(a)

FIGURE 8.9. (a) Original digital chest film. (b) High frequency pass filtered imaging using a weighted unsharp mask technique to enhance edges and small structures. Rib edges for example are now more distinct, as well as lung parenchyma detail. (Courtesy Fuji Medical Systems, USA.)

to achieve different degrees of high frequency or edge enhancement. Figure 8.9 illustrates the effects of unsharp masking as applied to a digital chest radiograph.

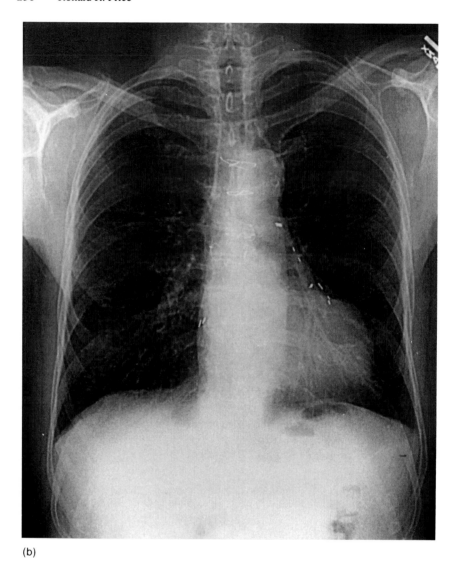

(b)

FIGURE 8.9. (*continued*)

8.5   Geometric Processing and Image Co-Registration

Most image manipulation algorithms are performed to alter the intensity values of the pixels in order to achieve a specified change in perception. Another type of image manipulation which is concerned with pixel location as well as pixel intensity falls into the general category of geometric manipulations. [8.9]

Geometric manipulations may consist of simple pixel shifts, translations or rotations to align images with "reference" images, or they may consist of more complex image distortion-correction algorithms. [8.10] The alignment of images of the same subject taken at different points in time, either with the same imaging modality or with different imaging devices, is referred to as image "co-registration."

One type of image distortion manipulation used in image co-registration is referred to as "rubber sheet" mapping or image "warping." The term rubber sheet mapping gives a good idea of how to visualize what is taking place during the mapping calculations. Images which undergo geometric manipulation behave as if the image were painted on the surface of a rubber sheet which is stretched and distorted to achieve some desired configuration.

The actual computational algorithm being used is a coordinate transformation. Each pixel in the original image with spatial coordinates (x,y) is sent to a new spatial coordinate (x', y') in a new "transformed" image. The new output image may or may not have the same number of lines or pixels per line as the input image. In addition, multiple input pixels may be mapped onto a single output image pixel. In order to avoid holes in the output image, the transformation is frequently carried out in reverse. That is, each pixel in the output image is scanned to determine the origin of the input pixel. In addition, when mapping from the input image to the output image, input pixels may not necessarily have integer coordinates and will not necessarily be mapped into the center of the output image pixel. These problems require decisions about how to divide the gray-level intensity into the neighboring pixels and what to do if a pixel is missed. The use of blur functions and interpolation are often used to yield aesthetically pleasing results. [8.11]

At this point, we have neglected to specify the exact form of the transformation. Most medical applications have found it adequate to use polynomial transformations with orders no higher than quadratic. Similar mappings employed for aerial reconnaissance may require much higher orders to achieve the necessary degrees of freedom in the warping calculation. In our laboratory, coordinate mappings of the following form have been found to yield adequate degrees of freedom in most mappings involving X-ray images. Each pixel coordinate (x,y) in the original image is mapped onto the new coordinates (x', y') using the following mapping function:

$$x' = a + bx + cy + dxy \quad y' = e + fx + gy + hxy$$

This particular mapping allows the geometric degrees of freedom of translation, rotation, magnification, and hyperbolic warping. Other polynomials can be chosen to allow other degrees of freedom.

In the implementation of the above transformation, the letters a through h are unknown constants which must be determined for each image transformation. A convenient way to determine the constants is to observe the location of clearly definable fiducial points in an undistorted reference image and then to use the coordinates or the reference points in the distorted image to solve the above equations for the constants. In the above transformation, to solve for the eight constants, we must identify at least four coordinate pairs for fiducial points. The four coordinate pairs produce eight simultaneous equations which can be solved for the eight unknown constants.

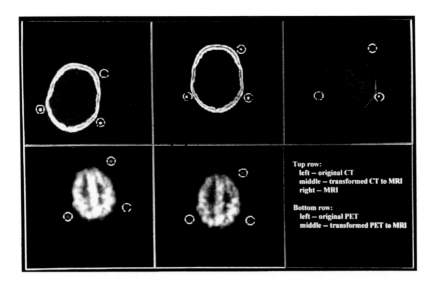

FIGURE 8.10. Corresponding point co-registration of CT, MRI, and PET images. Top row from left: original CT image; CT image transformed to MRI reference frame; original MRI image. Bottom row from left: original PET image; PET image transformed to MRI reference frame. (Courtesy of J. Michael Fitzpatrick, Ph.D., Vanderbilt University.)

The application of aerial reconnaissance uses the initial aerial photograph of a land area as the reference image and all subsequent photographs are mapped onto the same reference frame as the initial photograph. The purpose of this entire operation is to look for any changes that may have occurred between photographs. Since the photographs will likely be made from different angles and altitudes, there will be an apparent distortion in subsequent images. By locating several clearly identifiable points that are common between the initial and later photographs, the constants of the mapping can be determined.

A similar approach to the above is used to co-register medical images. At one time, it was thought that it would be possible to use a baseline chest radiograph as a reference, and by polynomial mapping we would be able to reregister subsequent images to produce difference images. It was hoped that these difference images would be more sensitive to the detection of disease. Even though the idea is still valid in theory, practical constraints have prevented the concept from being fully tested.

A number of successful applications of image co-registration in medicine, however, can be noted. Figure 8.10 illustrates the co-registration of images of a patient's head taken with three different medical imaging modalities: computed tomography (CT), magnetic resonance imaging (MRI), and positron emission tomography (PET). The patient had been fitted with a rigid frame that was visible by each modality and provided common fiducial markers allowing a unique transformation to be defined. As illustrated, the reference was taken to be the

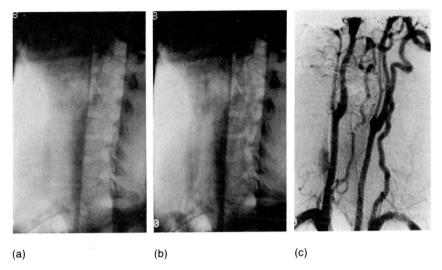

(a)   (b)   (c)

FIGURE 8.11. A digital subtraction angiogram of the carotid arteries. (a) Mask image acquired before the injection of contrast material. (b) Image acquired following contrast injection. Opacified arteries are present but due to overlying tissue and bone cannot be easily seen. (c) Subtraction image showing carotid arteries.

MRI image (top right); all other images were then transformed to the same reference frame. Figure 8.10 illustrates the use of the "corresponding points" method of co-registration.[8.12] An extension of the method to co-register curved surfaces composed of multiple discrete points is referred to as the "corresponding surfaces" method.[8.13–8.14] The corresponding surface method has been used to co-register three-dimensional image data sets from CT, PET, and MRI.

8.6   Image Subtraction

One of the simplest, yet most powerful, algorithms for image feature extraction is image subtraction. Image subtraction enhances features by increasing conspicuity. Image subtraction obviously cannot generate features that are not already present in the image, but subtraction can make features more conspicuous by eliminating structures that are unwanted or that tend to confuse the scene.

The use of subtraction is clearly limited to those applications where a "mask" image can be acquired; subtraction angiography is an example. The images shown in Fig. 8.11 illustrate the use of subtraction angiography for the carotid arteries. The mask radiograph is taken prior to the injection of a vascular contrast agent (Fig. 8.11(a)). The mask image is then subtracted from a radiograph taken at the time at which the contrast material arrives at the region (Fig. 8.11(b)). The resultant image (Fig. 8.11(c)) is an image of the vascular tree without the confusion of the overlying soft-tissue and bony structures.

8.7   Segmentation

A new emphasis on image segmentation has resulted directly from the development of new and powerful three-dimensional image display systems and techniques. Segmentation is the term most commonly used to describe feature detection and extraction in digital images.

Feature extraction in this context basically refers to the differentiation of an object from its background. The segmentation algorithm must obviously be designed to recognize whatever feature the object possesses that is not in common with its background. Thus, segmentation algorithms can be as varied as the objects of interest. Traditionally, segmentation has relied on geometric, textural, and intensity differences.

The simplest segmentation algorithm in common use is segmentation by thresholding. The thresholding algorithm reduces a conventional gray-scale image into a binary image. Image pixel values above a specified threshold are defined as white, those below the threshold as black. Following segmentation, the display or measurement algorithm can choose to display or measure only white features or only black features. In a similar manner, we can segment an object with a gray level l within a window of gray levels l_1 to l_2 such that

$$l_1 \leq l < l_2$$

can be segmented.

An example of thresholding segmentation used in routine clinical practice is the three-dimensional display of soft-tissue and bony structures from sequential multi-slice (volume) computed tomography (CT). If we have closely spaced (adjacent) thin-slice (1 mm) CT images throughout a specified volume, then segmentation can be carried out on a volume basis. Bone is coded as all pixels with CT values above +200, soft tissue as all CT values between -200 and +200 and air as all values below -200. From this volume-coded data, surface rendering can be carried out for either the skin surface (Fig. 8.12(b)) or the bone surface (Fig. 8.12(a),(c)). With the addition of a vascular contrast agent, blood vessels may also be surface-rendered (Fig. 8.12(d)).

Within the computer, lines of sight or rays (passing through the volume) are used to detect the pixels representing the first occurrence of the interface either between air and soft tissue or between soft tissue and bone. By shading the detected interface pixels in inverse proportion to their distance from a chosen origin, one can achieve a shaded surface rendering of either the skeletal structure or the skin surface. Three-dimensional images rendered in this manner have been particularly useful for planning and evaluation of reconstruction surgery and for disc and spinal cord assessments where planar images may remain questionable or ambiguous.

More sophisticated segmentation algorithms are now under investigation which have the potential of detecting and segmenting tissues and structures through the use of multiple interdependent parameters. One such method is artificial neural networks (ANNs). [8.12-8.14] ANNs are multiparameter mathematical models which can relate a series of input patterns to a specific output pattern. Through the use

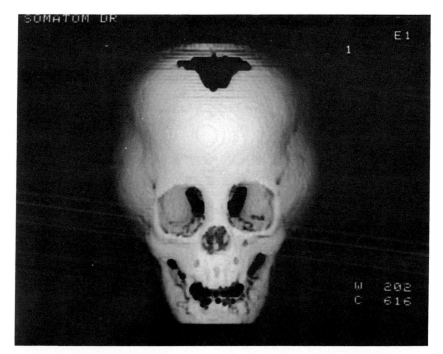

(a)

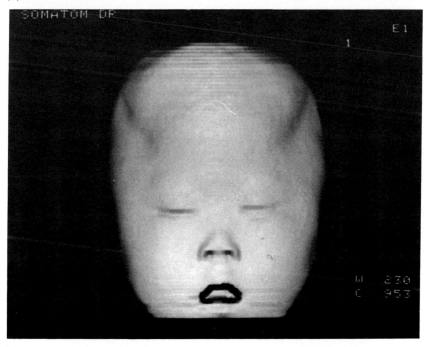

(b)

FIGURE 8.12. Three-dimensional surface-rendered images from a volume set of thin-slice CT scans. (a) Surface-rendered image of the skull. (Courtesy of Siemens Medical Systems, Iselin, NJ.) (b) Surface-rendered image of the skin (Courtesy of Siemens Medical Systems, Iselin, NJ.)

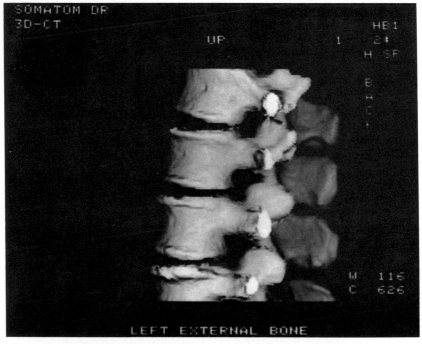

(c)

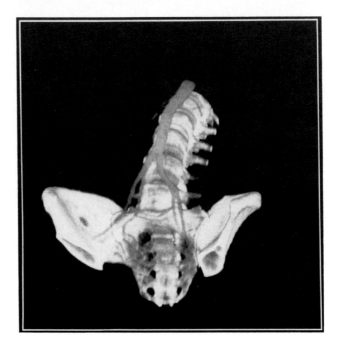

(d)

FIGURE 8.12. (*continued*). (c) Surface-rendered image of the lumbar spine (Courtesy of Siemens Medical Systems, Iselin, NJ.) (d) Surface-rendered image of the pelvis and aorta. (Courtesy of Picker International, Inc., Cleveland, Ohio.)

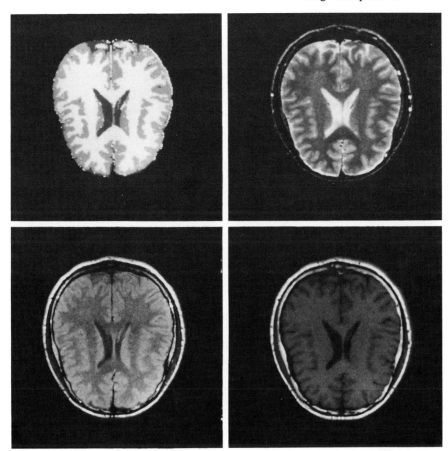

FIGURE 8.13. Image segmentation of MRI brain scans into gray-matter, white matter, and CSF. Top row; (left) MRI proton-density weighted image; (right) T_1 relaxation time weighted image. Bottom row (left) T_2 relaxation time weighted image; (right) ANN segmented image. Automatically segmented image of gray matter (gray tone), white matter (white tone), and CSF (black) resulted from manually defined small regions from each of the above MR images. (Courtesy of Benoit Dawant, Ph.D., Vanderbilt University.)

of a small, yet known set of input and output patterns, the ANN can be trained to apply the same associations to an unknown data set.

ANNs have been applied to MR imaging to detect and segment gray matter, white matter, white matter regions, cerebral spinal fluid (CSF), tumors, and tissue infarction. In the case of MR, each image pixel can be represented as a vector whose components are the values of the pixel's proton density and magnetic relaxation times T_1 and T_2. The model can easily be expanded to include other data. From a training set in which the vectors are measured for independently verified tissue types, the ANN can be trained to actually associate and identify new output patterns.

Figure 8.13 shows a series of MRI input or "basis" images for the ANN segmentation process, and the resulting segmented image. The input images are the proton-density weighted image (upper left), the T_1-weighted image (upper right) and the T_2-weighted image (lower left). From each of these three images, manually drawn training regions for gray matter, white matter and CSF are identified and then used by the ANN to characterize the entire image. Alternatively this training set may also be used to segment images from a different patient. The resulting segmented image is shown in the lower right of Fig. 8.13. The ANN-identified gray matter is displayed in the segmented image as gray, white matter as white, and CSF as black. The potential of ANNs and other similar methods is to offer the physician a tool for rapid and accurate screening of large volumes of image data now being generated by high speed imaging systems.

8.8   Maximum Intensity Projection

Many medical imaging modalities are now capable of producing isotropic three-dimensional image data of the human body. In the case of computed tomography and tomographic nuclear medicine (SPECT and PET), these volumes are assembled from spatially contiguous thin image slices. In the case of MRI, the data may be acquired either as slices or as an entire volume which is then mathematically sorted into image slices at any desired orientation. The most common method of viewing these volumes is through the creation of projection images. The most common projection image technique is the Maximum Intensity Projection (MIP). The basic concept of MIP is to produce two-dimensional projection images from a three-dimensional data set by constructing parallel rays through the entire three-dimensional data set at a desired viewing angle (Fig. 8.14). Each ray defines one pixel in the projected image. The gray level assigned to this pixel will be set equal to that of the maximum-intensity voxel encountered along the ray through the volume. MIP reconstructions of CT angiograms of the cerebral circulation (Fig. 8.15(a)) and of the pulmonary vasculature (Fig. 8.15(b)) are now used routinely. A number of variations on the basic MIP algorithm are under development. These include provisions for thresholds to be defined and for the sum of a specified number of pixels along the ray to be projected instead of the single maximum-intensity voxel value. In the sum projection, one may be able to partially recover the characteristic high intensity of overlapping vessels that is lost in the basic MIP algorithm.

Often the MIP pixels from an oblique viewing angle may project onto some intermediate location between pixels in the projection image and between pixels in the three-dimensional data set. In these situations, some interpolation or nearest-neighbor algorithm may be used to determine the intensity of the MIP pixel. Cine presentations of multiple MIP viewing angles give viewers a three-dimensional perception of the vascular structures.

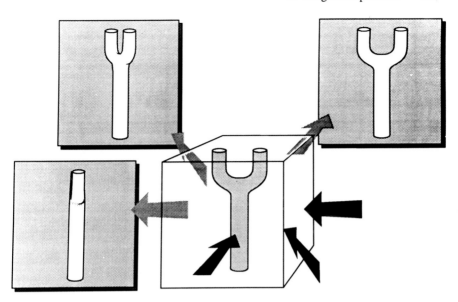

FIGURE 8.14. Maximum Intensity Projection (MIP) reconstructions of three-dimensional image data sets are used to produce projection images at various viewing angles. By viewing the MIP images sequentially in a cine or movie mode playback, the viewer is able to appreciate the three-dimensional nature of the structures.

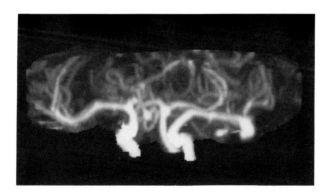

FIGURE 8.15(a). MIP image taken at one viewing angle of the cerebral vasculature from a three-dimensional data set acquired with thin-slice contrast enhanced CT (CT angiogram).

The contrast between a structure of interest (possibly a blood vessel) and its surroundings in a MIP image will generally be reduced as the thickness of the three-dimensional volume increases. This is a natural result of the MIP algorithm, which also selects the maximum background voxel value. Thus, it is desirable to reconstruct only the minimum volume that contains the vessel of interest. The technique of selecting a smaller volume for reconstruction from the full

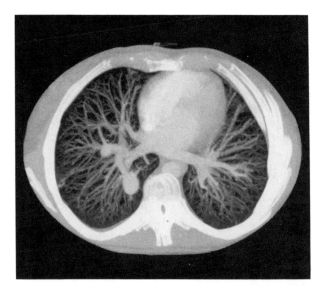

FIGURE 8.15(b). MIP of the pulmonary vasculature from a three-dimensional CT angiogram. (Courtesy Picker International, Inc., Cleveland, Ohio.)

three-dimensional data set is called targeted MIP. Targeted MIP reconstruction frequently also can help clarify uncertainties resulting from overlapping structures.

8.9   Conclusion

Evolution of medical imaging technology continues to move inexorably toward a completely digital discipline. The digital medical image has become a fertile area for the development and investigation of image manipulation techniques.

These techniques are subdivided into image intensity manipulations and geometric manipulations. Of these techniques, one includes algorithms which can enhance, eliminate, or extract a given feature for analysis. Other algorithms are designed to correct for imperfect image acquisition or image display systems.

At this time, many image manipulations remain a point of curiosity without having attained a definitive position in medical practice. Other techniques, however, have already achieved a degree of acceptance and are in routine use. These "accepted" techniques are, unfortunately, few at the present time. It is anticipated that as the development effort continues and the discipline matures, digital image manipulations in medicine will be commonplace.

8.10   References

[8.1]   Andrews H.C., Hunt B.R. *Digital Image Restoration.* Englewood Cliffs, NJ: Prentice-Hall; 1977.

[8.2]   Castleman K.R. *Digital Image Processing*. Englewood Cliffs, NJ: Prentice-Hall; 1979.

[8.3]   Williams L.E. *Nuclear Medicine Physics*, Boca Raton, FL, CRC Press, Inc.; 1987.

[8.4]   Hall E.L. *Computer Image Processing and Reconstruction*. New York: Academic Press; 1979.

[8.5]   Ballard D.H., Brown C.M. *Computer Vision*. Englewood Cliffs, NJ: Prentice-Hall; 1982.

[8.6]   Newman W.M., Sproull R.F. *Principles of Interactive Computer Graphics*. New York: McGraw-Hill Book Company; 1979

[8.7]   Pratt W.K. *Digital Image Processing*. New York: John-Wiley & Sons; 1978.

[8.8]   Hunter T.B. *The Computer in Radiology*. Rockville MD: Aspen Systems Corporation; 1986.

[8.9]   James A.E. Jr., Anderson J.H., Higgins C.B. *Digital Image Processing in Radiology*. Baltimore: Williams & Wilkens; 1985.

[8.10] Price R.R., Rollo F.D., Monahan W.G., James A.E. Jr. *Digital Radiography: A Focus on Clinical Utility*. New York: Grune & Stratton; 1982.

[8.11] Mackay R.S. Medical Images and Displays. New York: John Wiley & Sons; 1984.

[8.12] Hill D.L., Hawkes D.J., Crossman J.E. Registration of MR and CT images for skull base surgery using point-like anatomical features. *Brit. J. of Radiology*; 1991, 64(767): 1030–1035.

[8.13] Turkington T.G., Hoffman J.M. Accuracy of surface fit registration for PET and MR brain images using full and incomplete brain surfaces. *JCAT* 1995; 19(1): 117–124.

[8.14] Pelizzari C.A., Chen G.T., Spelbring D.R., Weichselbaum R.R. Accurate three-dimensional registration of CT, PET, and/or MR images for the brain. *JCAT* 1989; 13(1): 20–26.

9
Physical and Psychophysical Measurement of Images

Kevin S. Berbaum, Donald D. Dorfman, and Mark Madsen

9.1 Introduction

In this chapter, we consider how medical images may be evaluated in terms of the information they provide to human observers. Much of this book discusses what is known about the registration and interpretation of visual data within the human visual system. This knowledge, the product of a large psychophysical and neurophysiological literature, is fundamental to any attempt to characterize imagery: it specifies the dimensions, properties, and aspects of images that are informative. An understanding of visual perception should educate our attempts to characterize images by means of physical measurements. Beyond this, the psychophysical literature provides a family of methodologies for assessing diagnostic performance of imaging systems in which human observers serve as pattern recognizers. Psychophysical methods assess psychological response to variation in physical stimuli. These procedures can be applied even where little is known about the underlying recognition process itself or where the physics of the imaging process is not well understood. The best known and most widely used psychophysical method in medical imaging research generates receiver operating characteristic (ROC) curves. The second part of this chapter is devoted to an introduction to current use of these methods.

Much of the research that evaluates images based upon physical measurements tacitly assumes that human perception and pattern recognition are only affected in ways predictable from the substitution of one set of image parameters for another. For example, in considering the use of a cathode ray tube (CRT) for viewing medical images, a common assumption is that detectors designed to recognize some disease categories will register somewhat less contrast or fine detail information than if a conventional film image were inspected. Because of this assumption, studies addressing this issue often include only a few types of abnormality which are supposed to represent "biological criteria" for resolution (such as nodules for contrast resolution and pneumothoraces and interstitial lung disease for spatial resolution). Some studies go further along this line by substituting nonexpert for expert observers (thus ensuring that perceptual learning is not measured). So long as we are correct about the information being affected and its relevance for interpretation, the results from these few abnormalities can be applied to other types of abnormalities that share some of the same attributes. The fundamental assumption is that information carried from the individual through the image to the observer can be degraded within the channel but never improved. If the information is degraded in the image, then the facsimile of it registered in the observer is also degraded. Feature detectors responsible for classifying the patterns within the image then work less effectively because less information is available. Therefore, any change in the information represented in the physical image should have predictable effects downstream. The difficulty here is in identifying which aspects of images are relevant to the interpretive process.

Diagnostic systems require human observers for pattern recognition because humans are extraordinarily efficient classifiers of visual information and because the "effective procedures" allowing computerized recognition remain largely unknown. The latter reason seems to be based on a general lack of fixed criteria for determining a trivial change in a pattern versus a change sufficient to require a new classification. Analysis of images may remain all but impossible unless the analysis program embodies a logical model of the image so that such ambiguities can be resolved.[9.1] Almost any recognition problem can be simplified by a preliminary stage of analysis in which the image is "segmented" into meaningful units in terms of features of specific recognition categories which can serve as objects of focal attention (bringing all perceptual resources to bear on that part).[9.2] *A potential guiding principle for image processing research is that algorithms should attempt to achieve image segmentation that is consistent with morphological, physiological, or pathological segmentation.*

Examples may be useful. Commonly, colors have been used to represent ranges of continuous variables (e.g., temperature: white $= 101°$ to $110°$, blue $= 111°$ to $120°$, etc). These color boundaries suggest discontinuities where none actually exist. Therefore, this segmentation is not consistent with a conceptualization of temperature as continuous, and the discontinuities generated do not map onto structures. Consequently, such images tend to be ineffective representations for human observers. A better choice might be to use two colors to create a bipolar scale (e.g., red $=$ hot and blue $=$ cold) with the mix representing specific temperature, rather like a gray-scale image. Some other representation or device would then be needed to extract exact values for specific points in the image. Chan and Pizer,[9.3] and later Pizer and Zimmerman,[9.4] reviewed the literature on the optimal use of color in ultrasonographic display, and formulated pseudocolor scales that avoid this pitfall while improving contrast sensitivity. On the other hand, researchers developing algorithms for detection of interstitial lung disease based on measures of texture[9.5–9.6] needed a method for finding regions in rib interspaces from which to sample texture.[9.7] Biederman[9.8] has suggested that recognition in humans is based largely on distinctive three-dimensional pattern features. Given this hypothesis, an algorithm which projects two-dimensional segments into three dimensions may be very valuable. Some of the observer's models of anatomy may be three-dimensional, and even two-dimensional features distinctive of specific pathology may require reference to three-dimensional structure for their interpretation. In each of these examples, depictions that are consistent with human abilities to segment images into meaningful units of analysis allows classification to proceed more effectively.

Imaging in the future will rely more and more on currently underutilized capabilities (information channels) of observers, such as three-dimensional appreciation and motion perception. Images that provide the basic input to pattern-recognizing observers will function more effectively to the extent that they appropriately segment the data (form patterns) for the recognizer. To evaluate such images, we might choose the ROC approach (simple empiricism) rather than physical characterization because an adequate theory or description of the physical properties carrying information may not yet exist. The physics governing the transfer of information from an image to an observer sets fundamental limitations. From these limitations, we can deduce (at least in simple scenes) the conditions *necessary* for extracting information. Whether these conditions are *sufficient* also depends on the psychophysical state of the observer. To input a well-understood signal to a system (using a phantom) and evaluate the system response, we must know which dimensions of system response are informative to human observers. Otherwise, physical measurements fail to capture image qualities that are relevant for performance. There are no measures of image quality that are based on how well three-dimensional physiological structure or the motion of structures (or fluids moving through structures) is registered. Likewise, there are no measures of image quality based on how motion may be used to navigate the observer through a sequence of images or to communicate other diagnostically relevant information. No logical reason dictates that physical measurements cannot encompass such signals so long as the effective stimulus can be exactly specified. Therein lies the difficulty.

An example may prove helpful. In a recent evaluation of the usefulness of a picture archiving and communications system (PACS) workstation for interpreting body CT images,[9.9] cinématic presentation of a series of CT slices was frequently judged to be diagnostically helpful by providing new findings. The probable explanation for these results is that animation allows information distributed across several sections to be readily integrated.[9.10] For example, some abdominal structures can be difficult to identify clearly on some slices. The radiologist can search for those structures in adjacent slices where identification is easier and then extrapolate that information back to the slice containing the structure in question. The visual functions supporting this activity (in the inspection of images presented on hardcopy) are probably the same sort of comparative eye movements which have been described for inspection of CT images[9.11-9.13] and in the search of chest radiographs for nodules.[9.14] With animation of the series, motion perception of the temporally sequential images substitutes for the eye movements used to inspect spatially sequential images. Animation of slices allows the observer to follow an organ through the succession of slices. There are several ways that motion might enhance perception of medical images, as identified by studies of the effects of motion on primate visual systems.[9.15] In ciné presentation the image noise may also be integrated by the observer to effectively increase the signal-to-noise ratio. This effect may be small for CT images but more significant in gamma camera and ultrasound images.

Performance by trainee radiologists on a visual form reconstruction test, which attempts to measure aptitude for perceiving three-dimensional spatial relationships from two-dimensional images, has been found to correlate highly with diagnostic performance in subspecialities that use cross-sectional imaging methods, such as neuroradiology.[9.10] Deciding what two-dimensional structures should (or should not) be present in a particular cross-sectional image may depend on spatially visualizing how a section crosses various anatomic structures. Therefore, interpretation of images of sectional modalities might benefit from three-dimensional depiction, at least for novice interpretors. Also, radiologists who specialize in such modalities may tend to have above average three-dimensional perceptual ability.

These examples demonstrate that image quality cannot be measured solely in terms of information intrinsic to the image without considering the nature of the human observer who registers and interprets the information. The nature of the interpretative process defines the informative aspects of the image. The practical difficulty in benefiting from this knowledge is that, once we try to incorporate the perceiver, our measures of image quality may be affected by the state of the perceiver and by differences between observers. Furthermore, at present the interpretative process is not well understood: there are many competing models of various complex perceptual functions. On which model (or models) of the perceiver should we base the analysis of image quality? The task of discovering better ways to acquire and present images is already sufficiently difficult. Yet, little can be done to improve definitions of image quality without identifying the aspects of images that inform the perceiver.

9.2   Physical Measurements of Image Quality

The detection of useful information in an image is a complex process that depends on many physical factors as well as the state of the observer. What the observer perceives may be influenced as much by expectations and experience as by the quality of the displayed image. There are, however, fundamental physical limitations which are completely objective as to what information may be detected in images.[9.16–9.20] These limitations are characterized by the concept of the signal-to-noise ratio, which is determined by the quality of the imaging system, the size and contrast of the structures in the image, and the magnitude of the statistical fluctuations (noise). The signal-to-noise ratio scales the difference in magnitude between an object of interest and its surroundings in terms of the average fluctuation in the background. A mathematical model of these physical parameters can transform the problem of detection into an exercise in statistical optimization.[9.16-9.17] In practice, this technique is extremely difficult to apply to any real image because of the complex structures that are often present and because it requires accurate modeling of the viewer as well as the image. This technique can be applied to simple scenes, however, if we assume that the human visual system behaves in a mathematically optimal way and does not suffer from any additional degradation in resolution or noise. These criteria are embodied in the concept of the ideal observer. The ideal observer has the following information about the image: (a) the geometry of the object; (b) the distribution of count densities (or magnitudes) in the image; and (c) the probability distribution of the image noise.[9.17] Although the ideal observer is a poor approximation of the human observer, it is actually a useful approach. It establishes the relationships between the competing physical factors that govern the image quality and sets the bounds for how well a human observer is likely to perform.

9.2.1 *Image Formation*

To better understand the concept of signal-to-noise ratio and the factors that influence it, consider how medical images are generated. This task is summarized in Fig. 9.1. Within an individual, there exists a distribution of some physical parameter that we wish to image. This distribution is coupled to the modality that is chosen to make the image. Examples of the types of parameters measured by medical imaging devices are given in Table 9.1. While the choice of imaging modality depends on many factors, a major consideration is to decide which one yields the most information about a specific physical quantity. For example, a radiograph is highly specific and sensitive for imaging fractures of bone. Metastatic lesions to bone, however, are often more apparent on bone scintigraphic images even though the spatial resolution of the radiographic technique is superior. Because the radiograph is sensitive to changes in physical density, a lesion must be relatively large for its density to be altered with respect to normal bone. The scintigraphic study images differences in metabolic activity, which can be quite elevated in metastatic lesions as compared with normal bone. The important point is that detectability very strongly depends on the range of contrasts available to be imaged. All the other factors of resolution and noise are of little consequence if there is no distinguishing difference between the normal and abnormal features in the individual.

   All imaging modalities have a detection stage in which data from the object are collected. In this stage, information about the object is degraded by the finite spatial resolution of the detecting system and by the statistical fluctuations (noise) which are fundamental to all physical processes. The spatial resolution generally depends on the geometric configuration of the source and detector combination, and is also affected by scattered events. The noise at this stage is often spatially independent (uncorrelated or white) and is governed by Poisson statistics in photon detection systems such as radiography, CT, positron emission tomography, and gamma camera scintigraphy.[9.17]

   In digital imaging systems, the next stage is digitization. This stage may well be an integral part of the detection system, but it is useful to consider it separately. The data acquired by the detectors are sampled into a discrete array. Each element of

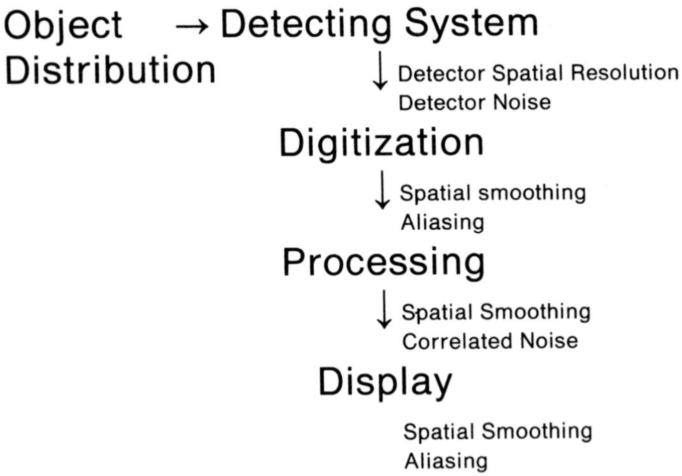

Object   → Detecting System
Distribution   ↓ Detector Spatial Resolution
                            Detector Noise

Digitization
                  ↓ Spatial smoothing
                  Aliasing

Processing
                 ↓ Spatial Smoothing
                 Correlated Noise

Display
              Spatial Smoothing
              Aliasing

FIGURE 9.1. Summary of Image Formation.

TABLE 9.1. Medical Imagin Modalities.

Modality	Measured Parameter
CT, radiography	Attenuation (physical density)
Ultrasound	Acoustic impedance
MR	Proton density, biophysical environment
Scintillation camera, PET	Physiology, tissue function

the array corresponds to a spatial or time interval (bin). Because the sampled data are averaged over this interval, an additional loss resolution occurs.[9.21] In addition, the finite sampling of the discrete array sets an upper limit on the frequency of information which can be accurately recorded. This limit is referred to as the Nyquist frequency and it is equal to (1/2 × bin width). Information in the sample which is modulated at a rate that exceeds the Nyquist frequency will create an artifact known as aliasing, which also degrades the image.[9.21]

In some imaging modalities, images are immediately generated at the detection stage (e.g., in chest X-rays) or at the digitization stage, (e.g., in digital radiographs). Computed tomography and other forms of digital image processing, however, are increasingly common. The algorithms used to reconstruct internal distributions from external measurements involve arithmetic operations that spatially correlate the data. This process not only results in a further loss of spatial resolution, but it also colors the noise. The correlated noise further obscures information in the images.[9.22]

Finally, the image information is displayed. For many modalities, a digital display is used. Potential losses exist here similar to those described in the digitization stage. If the display matrix is too coarse, aliasing and smoothing due to the pixel (or voxel) size will adversely affect the images. With these factors in mind, we now examine how the physical factors influence the detection of information in medical images.

9.2.2   Spatial Resolution

Spatial resolution is a very important but often misunderstood parameter. It does *not* refer to the minimum detectable object. Spatial resolution refers to the minimum image size of an ideal point source and therefore reflects an imaging system's ability to accurately reproduce the contrast available in the object. As shown in Fig. 9.2, spatial resolution is usually specified in terms of the point (PSF) or line spread function (LSF), or by the modulation transfer function (MTF), which is the frequency space representation of the PSF.[9.23–9.24] Both of these concepts assume that the imaging system is linear and stationary (i.e., that the point spread response remains constant over the entire field of view), which is not generally the case. For example, the spatial resolution of a positron emission tomography scanner is best at the center of the field of view and degrades rapidly at the edges. Furthermore, the spatial resolution is measured and specified at very low noise conditions which may never be reached in routine clinical imaging situations. Moreover, the overall system resolution is a function of the resolution of each of the stages, including any mathematical processing. The system resolution in terms of the MTF can be estimated from:

$$\text{MTF}_{\text{sys}} = \text{MTF}_1 \times \text{MTF}_2...\text{MTF}_i,$$

where MTF_i is the MTF of the ith stage.[9.24]

In the detection stage of many imaging systems, an inverse relationship exists between spatial resolution and sensitivity (detection efficiency). For example, the sensitivity of a gamma camera collimator decreases as the square of the full-width half-maximum (FWHM). Because the number of detected events limits the precision of the information contained in the image, improved spatial resolution is at least partially offset by increased noise levels due to reduced sensitivity. As a result, the optimal choice of the detection resolution and sensitivity depends on both the information flux and the power spectrum of the object under observation. As we see later, systems that image high contrast objects favor resolution over sensitivity, while the opposite is true for imaging large, low contrast objects.

A number of different measures have been used to evaluate image quality. Two that specifically refer to spatial resolution are the *noise equivalent passband* first introduced by Schade,[9.25] and its inverse, the *sample aperture,* introduced by Morgan.[9.26] The noise equivalent passband is equal to $\int \text{MTF}(f)^2 df$, where f is the (multidimensional) spatial frequency coordinate. Although both of these indices provide a scale for estimating the performance of imaging systems, they have limited usefulness. Imaging systems with very different MTFs can have identical noise equivalent passbands. Their performance, however, can vary markedly depending on the power spectrum of the object being imaged. Furthermore, as already noted, the MTF represents the spatial resolution capabilities of the imaging

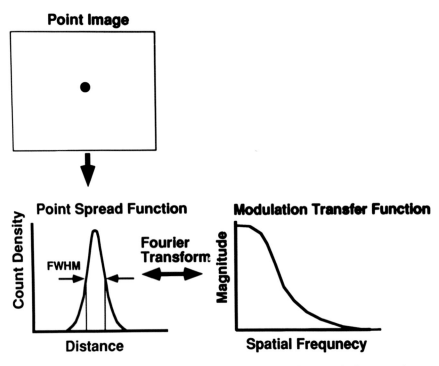

FIGURE 9.2. Point spread function and modulation transfer function. In linear, stationary systems, the point spread function and the modulation transfer function specify the spatial resolution performance.

system under low noise conditions. Since medical images are frequently noisy, measures which refer only to spatial resolution have limited application.

9.2.3 Noise

All real images have random, statistical fluctuations commonly referred to as image noise. Image noise is frequently related to the number of detected events. Therefore, the detection efficiency of the imaging system is an important performance parameter. In modalities that count photons, such as transmission radiography, CT, and positron emission tomography, the noise that enters the data in the detection stage is governed by Poisson statistics.[9.17] Poisson noise (for counts greater than 50) is well described by a Gaussian distribution with a mean equal to the number of detected counts and a standard deviation equal to the square root of the mean.[9.27] Thus, the average count density in a Poisson sample allows the straightforward calculation of confidence levels for the count levels in specific regions, and of the probability that deviations from the mean are only the result of random variations.

Image noise may be spatially correlated and have structure or texture, often best delineated by observing the image noise power spectrum,[9.28–9.30] denoted by $W(f)$. The noise power spectrum is obtained from the Fourier transform of

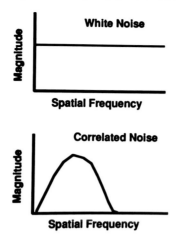

FIGURE 9.3. Noise power function. The frequency distribution of image noise is an important factor in detectability. If the noise magnitudes are independent of frequency, the noise is referred to as *white*.

the autocorrelation function.[9.31] It maps the fraction of the noise signal which is likely to occur in each spatial frequency band (see Fig. 9.3). Image noise which is independent of frequency is referred to as "white" noise. As already mentioned, the noise in the detection stage of most imaging systems is white, and those systems which produce images without a processing step yield images with (essentially) white noise, e.g., radiographs and scintigraphic images. Reconstructed images, such as those produced in any computed tomographic system, have noise which has been spatially correlated by the reconstruction algorithms. In fact, the noise power spectrum of early CT systems was used to deduce the reconstruction filter when this information was unavailable.[9.32] This noise is often referred to as *colored* or *correlated*. Although all image noise results in a loss of information, correlated noise interferes more because its signal falls into the same spatial frequency band as the image features.[9.22]

An index which has been used to specify an imaging system's ability to utilize the detected information is the *noise equivalent quanta* (NEQ). Noise equivalent quanta is a function of spatial frequency. It is an absolute measure of the system's performance as a photon detector[9.33]:

$$\text{NEQ}(f) \alpha \, \text{MTF}^2(f)/W(f).$$

As can be seen, NEQ includes both the effects of spatial resolution (through MTF) and noise texture [through the noise power spectrum, $W(f)$]. The NEQ(f) yields the apparent number of photons at each spatial frequency which contribute to the image.[9.17] If NEQ(f) is normalized by the total number of photons which were available to the detector (denoted by Q), we obtain another useful measure of system performance called *detective quantum efficiency* (DQE)[9.33]:

$$\text{DQE}(f) = \text{NEQ}(f)/Q.$$

The DQE gives the absolute efficiency of the imaging system as a function of spatial frequency. Both of these expressions are useful because they allow us to estimate the system's performance imaging specific objects.

9.2.4   Signal-to-Noise Ratio

The factors discussed above affect our attempts to glean useful information from an image. In a strictly physical sense, this task ultimately can be reduced to determining the probability that deviations from a mean count density in a particular area of the image are random. The problem of detection thus becomes one of statistical hypothesis testing.[9.16–9.20] The important parameter in this calculation is the magnitude of the deviation with respect to the standard deviation. This term is the signal-to-noise ratio, which has been shown to be the primary factor for image-based detection tasks.[9.34–9.36] The signal-to-noise ratio depends on the contrast and size of the area under consideration, the spatial resolution and sensitivity of the imaging system, the total number of detected events, and the probability distribution of the sampled counts. Each is now discussed in detail.

In this section we follow the example of Rose and other investigators in considering the detection of circular objects (hereafter referred to generically as *lesions*) located on a uniform background.[9.18] This model is illustrated in Fig. 9.4. The image on the left represents the object distribution, which is the actual physical distribution that a particular imaging system is attempting to record. Realize that the object distribution is itself an important factor in perception since it establishes the limiting size and contrast of the lesion. The figure on the right shows an image of this distribution obtained from a real imaging system with uncorrelated (white) noise. This figure can be used to establish some basic definitions. The average count density in the image is denoted by n. For photon detection devices, this number would correspond to the number of detected X or gamma rays per square centimeter. The standard deviation in a sample region of the image is denoted by σ. The size of the lesion on the image is an important parameter since the (ideal) observer integrates over this area while searching the image. The diameter of the lesion as observed on the image is denoted by d_i. If the spatial resolution of the imaging system is d_{res}, d_i is approximately equal to $(d_o^2 + d_{res}^2)^{1/2}$, where d_o is the actual physical diameter of the lesion. The contrast of the lesion with respect to its surrounding background is denoted by C. It is found from the ratio $\int \Delta s(x)dx/na_i$, where x is image coordinate (multidimensional), $\Delta s(x) = g(x) - n$, $g(x)$ is the count density over the lesion, and a_i is the image area of the lesion. The lesion signal is denoted by S and it corresponds to the recorded information resulting from the lesion. It is found from $\int_{a_i} \Delta s(x)dx$ using the same definitions as before. Thus, the magnitude of the signal depends on both the size and contrast of the lesion. From these quantities, we can define the signal-to-noise ratio, k. It is operationally defined as S/σ. More rigorously, it is found from

$$\int_{a_i} [\Delta s(x)]^2 dx / [\int_{a_i} \Delta s(x)C(x - x')\delta s(x')dxdx']^{1/2},$$

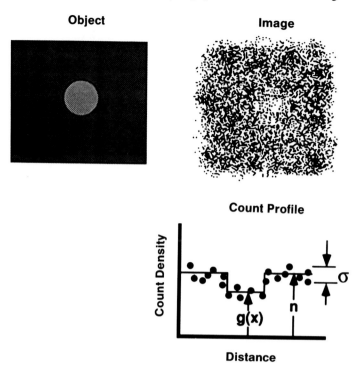

FIGURE 9.4. Contrast and signal-to-noise ratio. The contrast represents the signal size with respect to the background counts and the signal-to-noise ratio represents the signal size with respect to the average statistical fluctuation.

where $C(x)$ is the autocorrelation function.[9.17] For most real images, this integral is difficult to compute. For the example shown in Fig. 9.4, however, it can be easily calculated. This exercise is worthwhile because it establishes the relationship between the signal-to-noise ratio and the lesion's size and contrast. We shall assume in this example that the image counts are governed by the Poisson distribution. This assumption is not constraining, but is invoked because it simplifies the statistics.

The ideal observer scans the image searching for significant count differences in regions corresponding to the size of the lesion a_i. Using the previous definition, the average expected counts in the background found in a region with an area a_i is na_i with a standard deviation of $(na_i)^{1/2}$. If the actual counts within a specific area are denoted by x, the signal-to-noise ratio for this location is $k = (x - na_i)/(na_i)^{1/2}$. Since na_i is Gaussian distributed (in this example), the signal-to-noise ratio represents the number of standard deviations between x and na_i. Therefore, tables of normal probability distributions can be used to determine the probability of exceeding x from random fluctuations in the background. This provides objective criteria for deciding the likelihood that a particular region contains a lesion. The contrast between this area and the background is given by

$$C = (x - na_i)/na_i.$$

Expressing this relationship in terms of k yields

$$C = k/(na_i)^{1/2} \text{or} k = C(na_i)^{1/2} \simeq Cd_i n^{1/2}, \tag{9.1}$$

since the lesion area is proportional to the square of its diameter. Thus, in this example, the signal-to-noise ratio is directly proportional to the size and contrast of the lesion, and to the square root of the background count density.

This discussion describes the concept of the output signal-to-noise ratio, i.e., the signal-to-noise ratio calculated for a specific area of the image. Another similar concept is the threshold signal-to-noise ratio (k_{th}).[9.18–9.20] As the ideal observer scans the image, a judgment has to be made as to what minimum k-level corresponds to a true lesion. The choice of k_{th} is based on the objectives of the observer and the relative importance of correctly calling a true lesion (true positive) versus calling a lesion that is not there (false positive). The ideal observer can assign confidence levels to each of these possibilities. This concept is illustrated in Fig. 9.5, which shows the background $[pB(x)]$ and lesion $[pL(x)]$ probability distributions which have an output signal-to-noise ratio of 2.0. The false positive probability is the integral of $pB(x)$ from k_{th} to infinity, while the true positive probability is the integral of $pL(x)$ over the same limits. In general, a very low number of false positives is desirable. The number of false positives is determined by the product of the false positive probability and the number of decision samples which are made. If the area of the entire image is A_I, then this yields a minimum of $N = A_I/a_i$ samples. For example, if the image dimensions are 100×100 mm and $a_i = 5$ mm, then $N = 2000$. Thus, the false positive probability should be much less than 1/2000. For noise with a Gaussian probability distribution, this requires a signal-to-noise ratio on the order of 4.

Equation (9.1) can be arranged with contrast as the dependent variable to yield

$$C = k/(d_i n^{1/2}). \tag{9.2}$$

In this expression, C is interpreted as the lowest contrast at which a circular lesion of diameter d_i can be perceived. Importantly, both C and d_i refer to the size and contrast found in the image. A log-log plot of Eq. 9.2 with d_i as the independent and C as the dependent variable yields the contrast detail plot as shown in Fig. 9.6.[9.37–9.38] The contrast detail plot is a useful device for comparing the performance of imaging systems over a wide range of imaging tasks. For an imaging system with perfect resolution, the image size is exactly equal to that of the object. For these systems, which are referred to as noise-limited, the contrast detail plot is a straight line whose intercept is determined by the signal-to-noise ratio and the count density. All points to the right of the line represent objects which are perceptible, whereas points to the left of the line represent contrast-size combinations which are imperceptible. Increasing the image count density n moves the plot

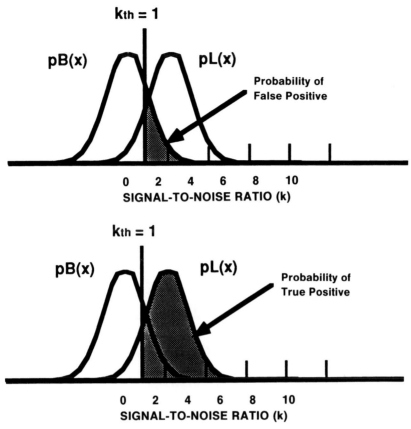

FIGURE 9.5. Statistical decision theory. The probability of correctly detecting the presence of a lesion depends on the threshold signal-to-noise ratio and the probability density function of the background and object. The true positive probability is found from integrating the object probability density function from k_{th} to infinity, while the false positive probability is found from integrating the background probability density function from k_{th} to infinity.

line to the left as shown in Fig. 9.7. Thus, as count density increases, the contrast necessary to perceive a lesion with a given diameter decreases. Imaging systems described by this model would place a very high premium on increasing the lesion contrast (e.g., using contrast agents) and increasing the system sensitivity.

Real imaging systems generally have finite spatial resolution, which, as already discussed, degrades the contrast in the image by blurring detail. This degradation is most evident for objects which are on the same order or smaller than the FWHM of the point spread function. If this is incorporated into Eq. 9.2, we have

$$C = kd_i/(n^{1/2}d_o^2)$$ (9.3)

where $d_i = (d_o^2 + d_{res}^2)^{1/2}$ and d_o and d_{res} are the actual lesion diameter and the system resolution, respectively. This change drastically alters the contrast detail

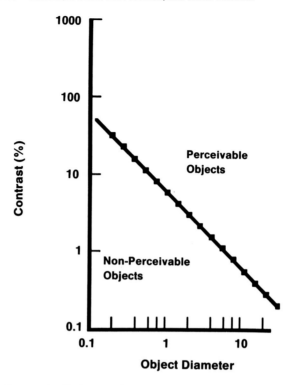

FIGURE 9.6. Contrast detail plot. The log of perceptible (image) contrast is plotted as a function of the log (diameter) for circular objects. Objects with contrasts that fall to the right of the plotted line are perceptible, while those to the left are not. The contrast detail plot is useful for comparing the performance of an imaging system under different operating conditions.

plot as shown in the right-hand portion of Fig. 9.6. In the region where $d_o \gg d_{res}$, the plot is essentially the same as that of the noise-limited system. As d_o approaches d_{res}, however, the minimum contrast required for detection increases rapidly. In this region, the detection of objects is limited by the system resolution much more than by sensitivity. It would seem, then, that systems designed to detect high contrast objects which are small compared with the spatial resolution should place a higher premium on spatial resolution than on sensitivity. This concept is illustrated in Fig. 9.8, which shows the contrast detail plot for two systems with different resolving power at the same count density. Note that there are limits to the degree that spatial resolution can be improved. Because of the inverse relationship that exists between spatial resolution and sensitivity, the increased noise that results from improvements in spatial resolution at the detection end requires more smoothing in the processing end, and this serves to set a lower limit on the size of d_{res}. Thus, optimal imaging depends on the objects being imaged as well as the sensitivity and resolution of the instrumentation. This idea is illustrated in Fig. 9.9 which shows the graphs of four contrast detail plots in

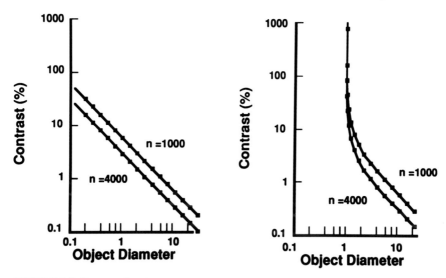

FIGURE 9.7. Contrast detail plots as a function of count density. For a truly count-limited imaging system, increasing the count density improves the detectability of all objects. In systems with finite resolving power, however, increasing the count density has relatively little effect on the detectability of small, high contrast objects.

which improvements in resolution are accompanied by sacrifices in sensitivity. The imaging of very small, high contrast objects is best performed by system A, but its performance rapidly degrades for larger, low contrast objects. Conversely, system D has the best performance for measuring large, low contrast objects, but performs very poorly in imaging the small, high contrast objects.

In this simple example, we have attempted to identify the physical features that limit the detectability of information in images. Ultimately, these features can be condensed into the concept of the signal-to-noise ratio. The signal-to-noise ratio determines the performance of the ideal observer.[9.16–9.20,9.34–9.36] It has also been shown to predict how well human observers perform in a variety of perception tasks. Humans fall below the ideal observer, however, by a factor of about 2.[9.34–9.36] There are many possible reasons for this. Human observers do not necessarily have access to the same information base as the ideal observer. For example, while the ideal observer knows what the noise power spectrum is and makes use of this information, the human observer generally cannot. As noted in Sec. 9.1, the ideal observer is invoked because of the extreme difficulty in mathematically modeling the observer (the eye-brain system). There are other methods which can be used to quantify the perception of image information, however, and these are discussed in the remainder of the chapter.

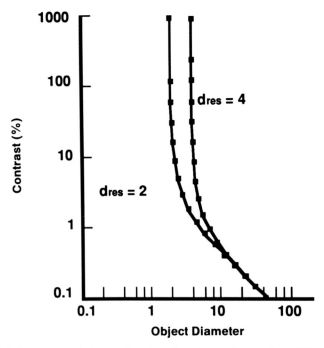

FIGURE 9.8. Contrast detail plot as a function of resolution. Changes in spatial resolution at a constant count density affect the performance of the system to image small, high contrast objects.

9.3   Limitations of Physical Analysis

With regard to human observers, physicists seem to tend toward philosophical reductionism whereas psychologists and computer scientists seem to tend toward philosophical holism. The ideal observer approach is based on the rationale that measuring the intrinsic information in the image for simple cases may be useful in characterizing the quality of more complex images. The quality of simple images and detection schemes represents a limiting case for complex images and perception. For this underlying assumption to apply, the same mechanisms of image interpretation must be involved in both simple and complex cases; which may not always be the case. There are image interpretation schemes (algorithms) that perform well with simple stimuli and tasks but cannot begin to cope with the kind of discriminations that support human intelligence. Of course, the goal of those researchers pursuing the ideal observer approach is different from those concerned with developing systems that mimic human pattern recognition. Any dissatisfaction of the latter may be based on the misconception that the ideal observer models attempt to characterize the same basic mechanisms as are involved in human and computer pattern recognition; in fact, ideal observer models cannot have anything in common with any realizable perceptual system. To quantify how much of a signal is conveyed by a medium such as an image, the ideal observer

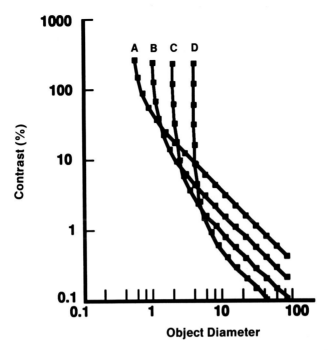

FIGURE 9.9. Contrast detail plots of Eq. (3) as a function of resolution and sensitivity. The four plots represent the system performance when improvements in spatial resolution are accompanied by a loss in sensitivity. In each of the plots, the count density is proportional to the d_{res^2}.

model must sacrifice abilities that the perceptual system of any viable organism must have, including the abilities to localize signals, differentiate among them, or classify them into groups. As soon as we attempt to build such requirements into an ideal observer, it can no longer perform optimally. Ideal observer models can be used to determine how much information is available in an image, and to gauge how much of the information in the image is acquired by a particular perceptual system. It can tell us nothing, however, about how to change an image to make whatever information that image contains more accessible to the human perceptual system. (An ideal observer model could be applied equally well to information channels to which humans are totally insensitive.)

In general, experimental psychologists and computer scientists do not embrace this theoretical approach because the "matched filters" used in ideal observers are too simple to account for complex perception and pattern recognition. The profound limitations of cross-correlation (template matching) to achieve pattern recognition have long been recognized in the perception and computer science literature. Psychologists and computer vision scientists need to challenge their models of perception with more difficult problems that would be encountered in image interpretation. Moreover, improved SNR may not result in improved detectability if a perceptual system is not attuned to the same information encoding as the ideal observer's matched filter. This mismatch is most often the case because human perception commonly succeeds with ill-defined stimuli. It should be clear from this discussion that models of perceptual systems and of ideal observers cannot be fruitfully combined either from the view of the acquisition of images from that of matching images to human observers.

One way to gauge the value of a theory is by asking whether it leads to experiments that generate useful information. Experiments derived from ideal observer theory tend to be limited to greatly simplified stimuli and tasks in which the theory may be expected to apply. To require an observer to perform a perceptual task in an unnatural way (i.e., in a way unrepresentative of the way the task would be performed in a clinical environment) is almost always a poor strategy. This lesson is the most basic teaching of experimental psychology. While adopting a wholly atheoretical approach would be unprofitable, researchers must be empirical enough to avoid simply finding out more about the experimental paradigm than about the perceptual capabilities of the observer. Signal-known-exactly and location-known- exactly methods seem to be required for ideal observer calculations (using a model based on cross-correlation). The concept of the ideal observer uses the total available image information to make a decision, and serves as a basis for scaling human performance relative to optimal performance (efficiency). These calculations are beneficial because they determine the best possible performance for specific stimulus conditions in order to scale human performance. The human performance between conditions can be compared directly, however, without such scaling. Inappropriate limitations on stimulus configuration and experimental task could result in inaccurate conclusions about the nature of perception.

The use of models of perception developed by experimental psychologists and computer scientists may be more valuable than ideal observer modeling in trying to understand how images and observers can be linked more effectively. Lessons from the study of visual perception and from machine vision can be brought to bear on problems in radiological image perception and in the task of improving image quality. Theories that describe the segmentation of images into meaningful units (e.g., perceptual grouping), pattern recognition, and perception of complex stimulus dimensions (e.g., motion) are needed to understand how humans interpret medical imagery. For example, the research literature in visual perception of random-dot cinematograms and relative motion suggests that signal information may be carried in the slice-by-slice correlations in the signal and in the noise. A theory of this perception must be based on temporally rather than spatially distributed information, or on differences in spatial information between images. Humans have information channels specialized for motion, and form (shape) information (e.g., signal) can be derived from motion. An idealized detector based on form-from-motion (interslice correlations) could be developed to scale human performance so long as the stimulus can be specified. How much would be gained from doing so may be open to question. To understand how images can better demonstrate to observers the information they contain, models of perception will be needed.

9.4   Measuring Observer Performance: Basic Principles of ROC analysis

This section briefly reviews the principles of receiver operating characteristic (ROC) methodology. The reader seeking a deeper understanding should consider several primary sources,[9.16, 9.39–9.41] the better and more complete introductions to the topic,[9.42–9.43] and some of the many excellent reviews of ROC methods applied to medical imaging found in the research literature.[9.44–9.57] The comments in Secs. 9.5 and 9.6 relate to our experiences and those of colleagues in attempting to apply such methods to specific research problems.

A traditional method of evaluating a diagnostic system involves analysis of a two-by-two decision matrix in which truth is crossed with the system's judgment of positive (abnormal) and negative (normal). With this characterization, difficulties arise in comparing two diagnostic systems. One system may yield not only more true-positives (disease-present, judged abnormal) but also more false-positives (disease-absent, judged abnormal). The fundamental problem with the decision matrix description of diagnostic systems is that it assumes a fixed threshold of abnormality for deciding whether to judge a case normal or abnormal. In fact, human decision-makers can change the decision threshold of abnormality upon which the normal-abnormal decision is based depending on costs of various kinds of errors. In other words, increase in true-positive rate can be traded for increase in false-positive rate. Thus, a family of decision matrices is needed to describe the performance of a diagnostic system.

Many of the assumptions of ROC methodology are similar to those underlying the ideal observer model as both are based on statistical decision theory. Instead of referring strictly to signal and noise in images, however, the theory of signal detectibility (TSD),[9.16] a model of how people detect signals, assumes an evidence variable by which the observer may infer states of the world. For medical imaging, this variable may be broadly defined as a dimension of "apparent abnormality" of the image (see Fig. 9.10). This dimension need not refer to the image itself but rather to the observer's internal representation of the image. Apparent abnormality is not restricted to simple variation in energy, but may involve an arbitrary degree of pattern complexity (e.g., pattern features, attributes, or properties). It is assumed that both normal and abnormal cases are normally distributed along this dimension, or more precisely that there exists a monotone rescaling of the decision axis that transforms the pair of latent distributions into Gaussian ones. The observer decides whether to report a case as normal or abnormal based on apparent abnormality.

While the overlap of the normal and abnormal cases is fixed by the image and the observer's level of perceptual learning, the level of apparent abnormality required to call a case abnormal is not defined by a fixed threshold.[9.58] The decision threshold for the normal-abnormal decision can be moved so as to divide the normal and abnormal distributions in different ways depending on the costs of various kinds of errors (see Fig. 9.11). The location of the decision threshold is under the observer's control and therefore is best thought of as an aspect

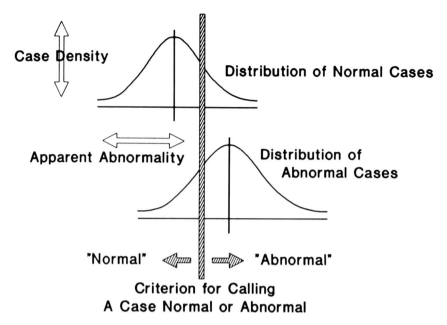

FIGURE 9.10. Basic assumptions of the theory of signal detectability (TSD). There is a dimension of "apparent abnormality," which may be defined in terms of observer perception, along which both normal (noise) and abnormal (signal plus noise) cases vary. The distribution of normal cases with regard to apparent abnormality is assumed to be Gaussian. The distribution of abnormal cases is assumed to be Gaussian but the mean of the abnormal cases is higher, or more abnormal, than that of the normal cases. The observer uses this dimension of apparent abnormality to decide whether to report cases as normal or abnormal.

of decision making. The decision threshold divides the decisions about normal cases into true negatives (TN) and false positives (FP) and divides the decisions about abnormal cases into false negatives (FN) and true positives (TP). Thus, the position of the decision threshold represents a tradeoff between specificity and sensitivity, where specificity is 1− the false positive rate and sensitivity is the true positive rate. Integrating areas under the distribution curves relative to the decision threshold produces conditional probabilities, $p(FP)$, $p(TN)$, $p(FN)$, and $p(TP)$, where $p(FP) = 1 − p(TN)$ and $p(TP) = 1 − p(FN)$. An optimal decision-maker decides whether a case is normal or abnormal by computing the likelihood ratio of the ordinate values of abnormal and normal distributions at that value of the evidence variable. *Decision rules* can then be generated of the form "if the likelihood ratio is less than or equal to 1, call the case normal, but if it is greater than 1, call the case abnormal." A real decision-maker, on the other hand, bases his decisions on the evidence variable, which may or may not be a monotone transform of likelihood ratio. When the two Gaussian distributions have equal variance, as in the current illustration, the evidence variable is a monotone transform of the likelihood ratio. Under this condition, decision thresholds on the

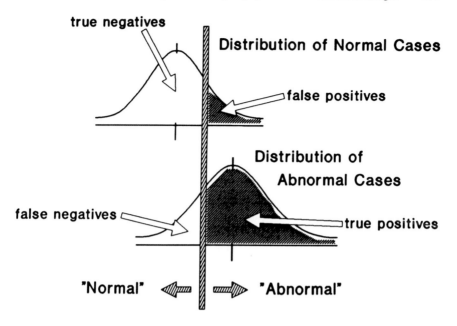

FIGURE 9.11. The decision threshold for the normal/abnormal decision can be moved so as to divide the normal and abnormal distributions in different ways depending on the costs of various kinds of errors. The decision threshold divides the decisions about normal cases into true negatives (TN) and false positives (FP) and the decisions about abnormal cases into false negatives (FN) and true positives (TP). Thus, the placement of the decision threshold represents the tradeoff between false positive and false negative rates. The position of the decision threshold is often referred to as in the equal-variance situation and Z_c in the unequal-variance situation.

evidence axis and corresponding decision thresholds on the likelihood ratio axis give identical ROC curves. When decision thresholds are placed on the likelihood ratio axis, they are often symbolized as β. The optimal value of β depends on the prior probabilities of normal and abnormal images (prevalence of disease), the costs of FPs and FNs, and the benefits of TPs and TNs. When decision variables are placed on the evidence axis, they are often symbolized as Z_c.

The fundamental relationship underlying ROC analysis is the relation of TP probability to FP probability (see Fig 9.12). In general, the ROC data are directly observed in experiments, whereas details of the overlap of normal and abnormal distributions in terms of apparent abnormality and the location of the decision threshold are not. ROC methodology permits the latter to be derived from the former. With Gaussian distributions, a nonlinear ROC curve results when y is the probability of TP, $p(\text{TP})$, and x is the probability of FP, $p(\text{FP})$. If the underlying distributions are Gaussian, then transformation of $p(\text{TP})$ and $p(\text{FP})$ to standard normal deviates transforms the ROC curve to a straight line. This underlying linearity allows ROC curves to be easily fitted by the method of maximum likelihood.[9.59] In the situation where normal and abnormal distributions have equal variance, d' is

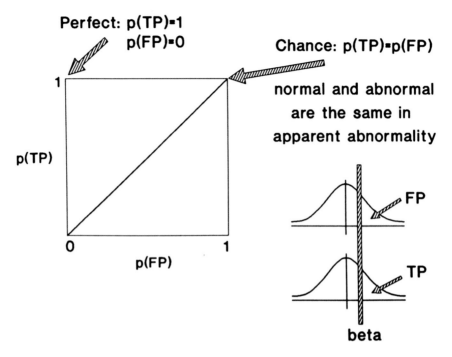

FIGURE 9.12. Basic attributes of receiver operating characteristic (ROC) space. Probability of TP response is plotted against the probability of FP response. Perfect performance is where TP response is certain (probability = 1) and chance of FP response is nil (probability = 0). If normal and abnormal cases had the same apparent abnormality (the same mean and degree of variation about the mean), then placing a decision threshold anywhere across the distributions results in equal probabilities of true and false positive response. Passing the decision threshold across such distributions traces out the chance line on the ROC graph (achievable by guessing).

defined as the difference between the distribution means divided by their common standard deviation. The parameter d' can be interpreted as the observer's internal or perceptual signal-to-noise ratio. When normal and abnormal distributions have unequal variance, Δm is defined as the difference between the distribution means divided by the standard deviation of the distribution of normal cases.

Figure 9.13 illustrates the relation of positions of the decision threshold across underlying distributions of normal and abnormal cases to the ROC curve for the equal-variance case. Each decision threshold leads to a decision matrix: an ROC point relating the probability of true positive to the probability of false positive response. The ROC curve can be interpolated and extrapolated from the observed points to describe the potential behavior of the system: every potential true positive rate is related to its corresponding false positive rate. This operation allows true positive rates of different systems to be compared at the same false positive rate. In Fig. 9.13, the distance between the distribution means is constant but the decision

Response Criteria

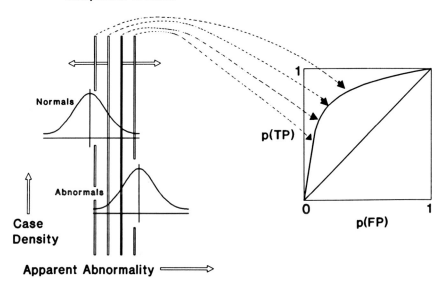

Case Density

Apparent Abnormality ⟹

FIGURE 9.13. The relationship between decision threshold placements across normal and abnormal case distributions and the ROC curve. The ROC curve that represents the performance of an observer or diagnostic system is plotted by forcing the decision maker to accept different criteria.

threshold is varied. As a result, the TP and FP rates vary together, generating a single ROC curve.

The alert reader probably wonders at this point how it is that several operating points or values of the decision threshold may be obtained. Several approaches exist. In a "yes-no" experiment, the observer states whether each case is normal or abnormal. Between sessions of an experiment, the observer is induced to change the decision threshold by manipulating the *a priori* probability of abnormals presented or by changing the costs of errors and benefits of correct responses. In medical imaging experiments, this approach is impractical because the costs and benefits are difficult to manipulate: costs and benefits are defined by the medical circumstances that ordinarily occasion the examination. Also, procedures that change costs and benefits are very inefficient (in that multiple sessions are needed). Another approach that is commonly used in evaluations of diagnostic systems uses confidence ratings. For example, the observer is asked to rate confidence in the abnormality of a case: 5 = definitely, or almost definitely, abnormal; 4 = probably abnormal; 3 = possibly abnormal; 2 = probably normal; 1 = definitely, or almost definitely, normal. By grouping the ratings in four ways (5 = abnormal and 1, 2, 3, 4 = normal; 4, 5 = abnormal and 1, 2, 3, = normal; 3, 4, 5 = abnormal and 1, 2 = normal; 2, 3, 4, 5 = abnormal and 1 = normal), four decision thresholds can be obtained in a single session. This method permits very efficient experiments because multiple ROC points are obtained in one session.

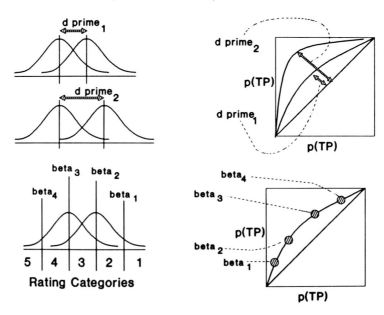

FIGURE 9.14. The fundamental value of ROC approach for characterizing diagnostic accuracy is that the ability to discriminate abnormal from normal cases is unconfounded with response bias. While response bias (β in this equal-variance example), expresses the differential placement of decision threshold upon underlying distributions without regard to mean separation, detection accuracy (denoted by d'), expresses the distance between the means of the underlying distributions divided by their common standard deviation. In ROC space, mean distance between distributions of signal and noise is reflected in the distance from the center of the ROC space [$p(\text{FP}) = 0.5$, $p(\text{TP}) = 0.5$] and the ROC curve. In ROC space, decision threshold is reflected in the position of ROC points along the curve.

The ROC approach to characterizing diagnostic accuracy offers a fundamental advantage because the ability to discriminate abnormal from normal cases is unconfounded with decision threshold. Figure 9.14 shows ROC curves generated by the equal-variance Gaussian model. Decision threshold, denoted here by parameter β, reflects the positions of ROC points along the curve and expresses the whereabouts of the decision thresholds on underlying distributions without regard to separation of the distributions. In contrast, detection accuracy, denoted here by d', equals the distance between the means of the underlying distributions divided by their common standard deviation. In binormal space, d' equals the y-intercept. The parameter d' does not depend on the placement of the decision threshold. In the top part of Fig. 9.14, d' increases, but β is constant. Consequently, the TP rate increases but the FP rate is unchanged.

Figures 9.15 and 9.16 contrast the independent variations in decision threshold and distribution overlap for the equal-variance Gaussian case. In Fig. 9.15, several choices of decision threshold are shown with fixed separation between normal and abnormal case distributions. The TP and FP rates vary together generating a single

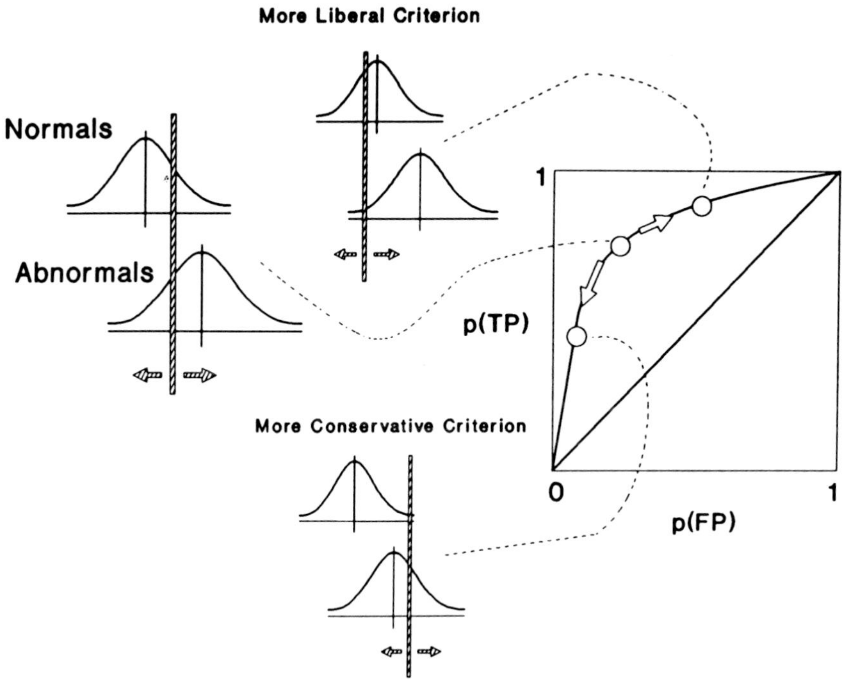

FIGURE 9.15. Variation in decision threshold with fixed separation between normal and abnormal distributions. If we have distributions of normal and abnormal cases that are unchanging with decision threshold free to vary, TP and FP rates vary together.

ROC curve. In Fig. 9.16, several systems are shown: several pairs of overlapping distributions with different distances between means are shown with the decision threshold placed equidistant from the distribution means. As the overlap between distributions decreases there are fewer FPs per TP and the resulting curve is elevated. As the overlap between distributions increases, there are more FPs per TP and the resulting curve is lowered. The theory of signal detectability provides methods by which values of d' and β can be computed for each ROC point. The inverse of the normal distribution function corresponding to the FP rate is subtracted from the inverse of the normal distribution function corresponding to the TP rate to obtain d'; β is the slope of the ROC curve at the selected point, or alternatively, the ordinate of the abnormal case distribution divided by the ordinate of the normal case distribution at the decision threshold. Points that lie on the same curve have the same value of d'.

Figure 9.17 illustrates how ROC analysis can be used to compare diagnostic systems. Figure 9.17(a) presents three ROC points for one hypothetical diagnostic system (black) and three ROC points for another diagnostic system (white). In clinical practice, each system (including the image interpreters) would tend to operate at a single point. That point, however, may lie anywhere on the curve. If we are given only that black operates at the highest point shown in Fig. 9.17(a)

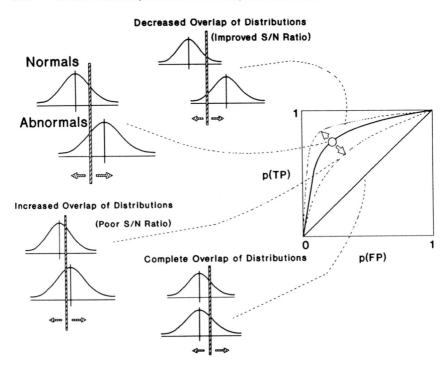

FIGURE 9.16. Variation in separation of normal and abnormal distributions. Several pairs of overlapping distributions with different distances between means are shown with the decision threshold placed equidistant from the distribution means. As the overlap between distributions decreases, there are fewer FPs per TP and the resulting curve is elevated. As the overlap between distributions increases, there are more FPs per TP and the resulting curve is lowered.

and white operates at the lowest point, then we have no adequate way to determine if the points for black and white lie on the same curve. By generating entire ROC curves, we obtain a meaningful comparison of the systems. A difference in true positive rates between systems can be attributed either to simple changes in willingness to call a case normal or abnormal, i.e., in the *decision threshold* [as in Fig. 9.17(b)], or to an actual difference in a system's ability to distinguish normal and abnormal cases, i.e., their relative accuracies [as in Fig. 9.17(a)]. Note that for both Figs. 9.17(a) and 9.17(b), the true positive rate for each black point is higher than for its corresponding white point. In Fig. 9.17(b), however, differences in true positive rates are completely compensated by differences in false positive rates. In the case of identical diagnostic accuracies and different decision thresholds, the ROC points for two diagnostic systems lie along the same fitted curve, with the points lying higher along the curve for one system [Fig. 9.17(b)]. In the case of varying relative accuracies, the ROC points for the two diagnostic systems lie along different curves [Fig. 9.17(a)].

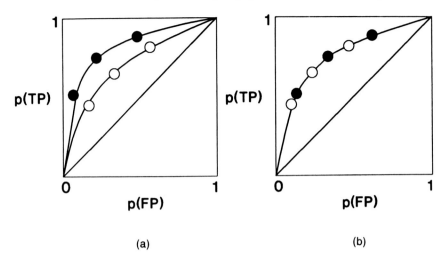

FIGURE 9.17. The performance of two hypothetical systems (black and white) in two experiments (a) and (b).

9.5   General Issues Regarding the Use of ROC Methods in Medical Imaging Research

As with most scientific methodologies, the practical application of ROC method in actual experiments reveals limitations unsuspected in a more theoretical consideration, requirements of effective use that are difficult to achieve in some kinds of inquiry, and details of procedure that must be made explicit and operational. Some of these considerations may seem obvious but they are nonetheless important.

9.5.1   Diagnostic Truth

"The validity of a diagnostic test cannot be established by use of the diagnostic test alone. It must depend on additional examinations, tests, observations and data provided by sources other than the specific diagnostic test under study." (Yerushalmy.[9.60])

Without some method of determining which patients have disease and which do not, there is no basis for determining which modality under evaluation is better (though it is possible to determine the similarity of their outcomes and which modality observers believe is better). ROC analysis requires an independent gold standard of diagnostic truth. The establishment of diagnostic truth in experimental studies has long been a matter of concern. In their seminal volume, *Evaluating Diagnostic Systems*, Swets and Pickett[9.56] assert that "the possibilities of introducing bias in favor of one modality and of introducing certain dependencies among the modalities are fundamental problems. . . Maximizing the credibility of truth data may also be in conflict with attempts to obtain a representative sample."

Metz[9.52] suggests that "the standards of truth are adequate for practical purposes if they are substantially more reliable than the diagnostic system undergoing study." Several additional issues, however, must be confronted. Exclusion from a study of patients who do not fall into one of the defined states of truth may lead to biased sampling. Also, a source of truth can introduce bias.

Revesz, et al.[9.61] have found that different methods of determining the correctness of observer decisions may lead to different conclusions. Methods in which information from the readers themselves is used to determine truth (such as a relative decision rate, majority rule decision, or consensus decision) may produce different outcomes than when nonreaders determine truth (such as in expert review or feedback review). These sources may also lead to results different from those obtained when clinical proof is available. In other words, radiological and clinical truth may differ.

Ker, et al.[9.62] have devised a formal protocol for establishing diagnostic truth in which there are six classes of verification certainty: (a) autopsy; (b) tissue-excision pathology report; (c) visual/palpation examination; (d) alternative imaging techniques; (e) clinician case history review; and (f) clinician case review with consultation. The purpose of this protocol is to specify the type and amount of information used as truth. This standardized verification procedure could increase replicability and even allow for the possibility of weighting cases depending upon verification certainty.

An independent standard of truth allows us to determine whether positive and negative responses are true or false relative to a decision threshold for each experimental condition. Often we are interested in a population of cases where independent truth is not available for some of the cases in the sample.

Because ROC studies virtually always involve a subset of cases with independent confirmation of disease status which may differ from cases without such confirmation, the representativeness of the case sample is often unknown. Non-ROC studies with no independent standard of diagnostic truth do not allow us to determine whether different rates of reporting abnormality represent better or worse accuracy. Therefore, it is impossible to determine which modality is better for the cases in which the modalities give different results. However, used in conjunction, both ROC and non-ROC studies provide useful information. Non-ROC studies with representative case samples are unaffected by verification bias. Economical non-ROC studies can serve as effective "pilot" experiments to prepare for ROC studies which consume more time and resources. The ROC studies can target those case types upon which modalities differ in the non-ROC studies. Studies of how similarly cases are classified by different modalities have value in that they provide an indication of the actual difference between modalities with real-world case samples. This objective is often difficult to accomplish with the ROC method.

9.5.2   Sampling

Ransoholf and Feinstein[9.63] show that many studies of diagnostic systems use case samples very different from any population to which one might want to generalize. Looking for relative difference rather than absolute level does not avoid this problem. Case sampling cannot be allowed to affect the ranking of conditions even though it can be allowed to affect absolute levels.[9.52, 9.57] Random sampling of examinations made within some interval is a powerful method to reduce bias.[9.56] Stratified samples can be as precise in representing the population with fewer cases, but a good description of the population composition from a previous sample is necessary.[9.52,9.57]

All other things being equal, a larger sample is more likely to represent a population, *but all other things must be equal*. If more than one trial per examination is collected or if trial images are generated from a few prototype images or simulated lesions, then the actual number of trials may not be related to the representativeness of a sample. Moreover, sources of truth can introduce bias in case selection and response scoring.

9.5.3   Statistical Precision and Power

Most of the fundamental questions regarding experimental design and conceptual approach to research objectives can be summarized with two words: *precision* and *power*. Statistical precision is the index of accuracy of an estimate where precision of an estimate is defined by the width of the confidence interval of the estimate. Statistical power is the index of experimental sensitivity, which is the ability of the planned study to detect differences or relationships when and if they exist. (Alternatively, power is the probability of rejecting the null hypothesis, which states that there is no difference between groups or, when the hypothesis is false, no relationship between variables.) Classical aspects of power estimation include sample size, significance level, strength of experimental manipulation, and control of error variance. Unfortunately, algorithms for estimating power require the user to provide assumed values for parameters which are hard to estimate with any precision without actually performing the experiment. When statisticians talk about power, they sometimes say that there is no substitute for experience in an area of research for determining the sample size needed. Experience gives us hunches about the kinds of differences that result for particular case samples for certain manipulations. Another approach to power analysis is to collect part of the data for an experiment and analyze it assuming various sample sizes before completing the entire experiment. Likewise, once the results of an experiment are known, the question of whether the case or reader sample size provided adequate precision and power can be addressed.[9.64] One way or another, power analysis requires pilot data. For ROC studies, attempting to estimate the amount of data needed for a given level of power is often not very helpful. In general, you cannot have too much statistical power, and we seldom have as much as we would like. Simply collecting as much data as possible is often best.

While it is difficult to estimate how many readers and cases will be needed before the experiment, methods to improve power are straightforward and well known. For case generalization, the number of cases is the main control for power. For reader generalization, the number of readers is the main control for power. Power can be improved by matching readers and cases and by using methods that take advantage of the resulting correlations.[9.56]

Efficiency of experimental design can be maximized where only relative (ordinal) differences between experimental conditions are of interest rather than absolute values of accuracy and magnitude of difference in the population. Instead of trying to represent proportionally all levels of case difficulty found in the population, only cases that are "subtle" (those for which there is a reasonable likelihood of detection in only one experimental condition) are used. Since absolute performance levels may vary without affecting rank order, the experimenter can determine the general positions of the curves by the difficulty of cases.

9.5.4   Absolute versus Relative Performance

It is more productive to document relative rather than absolute levels of performance of devices, individuals, or diagnostic systems. Experiments that measure relative differences provide information about diagnostic accuracy without a need to determine absolute level of differences in the total population of cases. Obvious abnormalities are readily detected by radiologists in most experimental conditions and, therefore, are not appropriate for most perceptual studies in radiology. Our approach to case selection involves choosing only target lesions that are "subtle," having a reasonable likelihood of detection in the experimental condition of interest but a lower likelihood of detection in the control condition (or vice versa). This approach to case selection maximizes the efficiency of experimental design when the observers are expert readers. If absolute values of accuracy are needed, the case sample must represent all cases interpreted by radiologists within the domain of interest. This necessity in turn may require stratified sampling of cases to ensure the appropriate range and spectrum of abnormalities.[9.52]

9.5.5   AFROC and FFE Analysis

The outcome of a properly designed ROC study are usually definitive, but the requirements of such studies may be awkward in the context of initial attempts to evaluate new forms of imaging with animal models. A lesion is produced within an animal, and histology provides proof of its presence and location. It is expensive and of secondary interest to generate images from completely normal (untreated) animals. Furthermore, the preparation may also generate multiple lesion sites offering the opportunity to test sensitivity [$p(\text{TP})$] with several experimental observations per animal. If there are numerous sites in which lesions might occur but do not, the experimenter may feel that such sites can justifiably be considered as "normals." A site-wise analysis is far more attractive than a case-wise or animal-wise analysis because many more data can be gathered per animal preparation.

The critical consideration for the validity of a site-wise ROC analysis is whether the occurrences of lesions and responses within an image are independent and whether the decision criterion is invariant over the number of targets reported within an image. In brief, it is assumed that the TPs and the FPs are independent and identically distributed within and across images. Satisfaction of search[9.65-9.69] occurs in interpretations of chest radiographs, skeletal radiographs, and abdominal contrast studies of actual patients: detection of one abnormality renders detection of another less likely (even though the one in no way physically masks the other). In some situations,[9.69] satisfaction of search effect involves changes in the decision threshold. Under these conditions, an analysis that treats multiple responses to an image as independent is not justified. With an animal model where the generation of lesions is under the experimenter's control, however, their occurrence can be made independent. Also, as Swensson has suggested (personal communication, 1990), a site-wise ROC analysis requires an appropriate distribution of the number of lesions per image so that the likelihood that another lesion is present remains the same regardless of how many lesions have been reported. Instructions to observers should explain this distribution and discourage limits on the number of responses per image that observers may ordinarily produce under more common reading conditions (with clinical images).

Chakraborty and Winter[9.70] developed very efficient methods for this situation that refine earlier free-response operating characteristic methodologies.[9.71] In alternative free-response operating characteristic (AFROC) analysis, the fraction of lesions with true-positive response is plotted as a function of the probability that an image will elicit a false positive response. The probability of an FP response per image (negative or positive images) is derived from the mean number of FP responses to the image. This method assumes a Poisson distribution of FP responses on images.[9.71] In the free-response forced error experiment (FFE), the observer is forced to respond to each image until a FP response is elicited. For this method, no Poisson assumption is required. AFROC and FFE methods allow every response to an image to contribute to the analysis, thereby providing more data which may allow more definitive outcomes. However, experimenters must be cautious in applying such methods. If AFROC or FFE methods are used when responses to the same image are not independent, the sample size used in the analysis can be greatly exaggerated.

Swensson[9.72] has proposed a general model for describing observer performance in detecting and localizing targets on images. A two-parameter binormal version of this model can be fitted to both ROC and localization-response LROC curves using the observer's image ratings and target localizations for a set of images. The additional information obtained from observer's localization performance may lead to improved methods of evaluating clinical imaging tests. If the assumptions of independent responses within an image and invariant decision criteria are met, the model can be extended to the free-response (FROC) curve and to the AFROC curve. This leads to a unified treatment of observer performance across various interpretation tasks.

9.5.6   Detection Methodology Applied to a Classification Problem

"'Generalized ROC curves' describing the relationships among the relative frequencies of the various types of decisions in almost any diagnostic decision-making experiment can be imagined and can, in theory, be determined experimentally. In some situations, however, the resulting relationships can be complex, difficult to measure to acceptable statistical accuracy without large numbers of trials, and difficult to interpret if an underlying theoretical structure is not available." (Metz, et al. [9.73])

The images of diagnostic radiology are inherently complex. In experiments using such images, a multitude of signal types may be present. Classification specificity is conventionally measured by plotting the joint probability of correct detection and identification against the probability of false-positive response. The ROC defined in this way is called a joint ROC curve.[9.56, 9.74–9.75] An accepted theoretical model that can be fitted to joint ROC data does not yet exist.[9.76] Another difficulty is the lack of a statistical method to test hypotheses of joint ROC parameters. If a difference in two joint detection and classification ROC curves were found using such an analysis, its meaning would still be ambiguous, however, unless no difference were found for the detection ROC curves. Joint ROC curves necessarily confound detection and classification.

Methodological obstacles in analyzing joint ROC data can be overcome. One promising approach might be to analyze detection and classification data separately. After a detection ROC analysis has been conducted, a classification ROC analysis can be conducted. Classification ROC data would be restricted to the subset of images in which an abnormality has been correctly detected. In fact, detection methods can be used successfully to investigate classification through pairwise rating ROC curves. A simple procedure would be to have the observer give separate confidence ratings to each possible disease type in a clearly defined set. This disease-specific scoring format has been used previously,[9.77] and Rockette[9.78] has provided a method for analyzing disease-specific format data within a detection framework. The method can be applied both to images with single diseases and to images with multiple diseases. In contrast to Rockette, we analyze detection and classification separately. For experiments concerned primarily with classification, one signal type can be treated as "noise" and the other as "signal." If many signal types are of interest, many such comparisons are required. If we try to analyze the overlap of distributions underlying each signal type, impossibly large numbers of cases may be required if more than a few signal types are included.

One way of implementing this method is to restrict the number of categories of disease to some small number. The response categories would be in one-to-one correspondence to those categories of disease. For each case, a confidence rating would be obtained for each classification category. All pairs of categories would be subjected to an ROC analysis. For each pair of disease classes, two ROC curves are computed: one in which the cases of the first disease class are treated as signal and the cases of the second disease class are treated as noise using the rating on the first disease class for analysis; and one in which the cases of the second disease class are treated as signal and the cases of the first disease class are treated as noise using the rating on the second disease class for analysis. For instance, if we have three disease classes and three response classes, there would be three pairs of ROC curves, two each for comparing the first with the second, the first with the third, and the second with the third disease categories. A confusion matrix can be defined as a table in which measures of accuracy such as ROC area can be entered in rows and columns that represent the disease categories treated as signal and noise. For example, the confusion matrix A can be represented as

$$\begin{bmatrix} * & A_{12} & A_{13} \\ A_{21} & * & A_{23} \\ A_{31} & A_{32} & * \end{bmatrix}.$$

The symbol A_{ij} is the area under the ROC curve when category j is treated as signal and category i as noise. A global measure of classification accuracy can be defined as

$$\bar{A}_{..} = \frac{1}{n(n-1)} \sum_{i=1}^{n} \sum_{j=1}^{n} A_{ij}(i \neq j).$$

The global measure of classification accuracy can be decomposed into two components, $\bar{A}_{i.}$, and $\bar{A}_{.j}$:

$$\bar{A}_{i.} = \frac{1}{n-1} \sum_{j=1}^{n} A_{ij}(i \neq j)$$
$$\bar{A}_{.j} = \frac{1}{n-1} \sum_{i=1}^{n} A_{ij}(i \neq j),$$

where $\bar{A}_{i.}$ denotes average diagnostic accuracy for category i averaged over the $n-1$ signal categories; and $\bar{A}_{.j}$ denotes average diagnostic accuracy for category j, averaged over the $n-1$ noise categories. In an idealized version of the model, $A_{ij} - 0.5$ is a true distance function, and it would have the following properties: (i) $A_{ii} - 0.5 = 0$; (ii) $A_{ij} = A_{ji}$ (symmetry); (iii) $(A_{ij} - 0.5) + (A_{jk} - 0.5) \geq (A_{ik} - 0.5)$ (triangle inequality).

Simonson, et al.[9.79] have recently used this classification methodology to assess the value of contrast enhancement in diagnosing major brain pathology.

The most fundamental lesson for experimental psychologists is that beings of a certain level of intellectual complexity[9.80–9.81] can simulate the behavior of simpler beings given a description of how to behave in particular circumstances. Human beings are of a level of self-organizing complexity[9.82] such that different recognition schemes may be used depending on task demands. If investigators think of human beings as detectors rather than classifiers, fundamentally different kinds of experiments will be conducted. Detection methodology often encourages us to ask rather simple questions about human perception, leading to results that are consistent with the human being performing as a rather simple detector. Medical imaging and image perception research need more and better methods. Detection ROC methodology has revolutionized our understanding of detection. We need new methodologies in order to enhance our understanding of classification.

9.6   Statistical Issues in ROC Analysis

"The main purpose in reporting an index of performance of a diagnostic system with a sample of cases, a sample of observers, and a sample of readings is not to tell the journal readership how well it performed in this particular sample that *was* studied, but to provide an estimate of how it would perform 'on the average' in those similar cases and observers and readings that were *not* studied." (Hanley,[9.83] p. 324).

There are two primary ways to improve diagnostic accuracy in diagnostic radiology. One method is to improve the image so that abnormalities are more evident. This approach involves the study of physical improvements in images. The other method is to improve the interpretive process, the most critical and least understood part of the diagnostic process. Through study of perceptual and cognitive aspects of the interpretive process, causes of interpretive failure may be understood and corrected. Study of psychological factors is necessary to understand interpretation of images. Recommendations from the first edition of this book emphasized that the choice of experimental design and statistical technique depends on whether statistical generalization to the population of observers or the population of cases is more fundamental, which in turn depends on the nature of the experimental question. Studies concerned with whether a physical manipulation influences image interpretability of a population of cases would use an error term based on variation among cases and allow for generalization to the population of cases. This design would treat observers as a fixed factor and images as a random factor. On the other hand, studies concerned with whether a psychological factor affects the observer's perception of a radiograph, but does not change the radiograph, would treat images as a fixed factor and observers as a random factor. This design would allow for generalization to the population of observers. Sometimes we wish to generalize to both the population of readers and the population of patients from which the corresponding test sample of readers and cases were drawn.

Dorfman, Berbaum and Metz (DBM)[9.84–9.86] have developed a new method for analyzing receiver operating characteristic (ROC) indices of accuracy from multiple readers. Their technique permits generalization to the population of readers, to the population of cases, or to both populations. In its general form, the DBM multi-reader, multi-case (MRMC) method assumes that both readers and cases are random effects, which permits generalization to both the population of readers and the population of cases from which the corresponding test samples of readers and cases were drawn. The method is applicable to all of the usual ROC indices of diagnostic accuracy, all of the standard bi-distributional models of the ROC curve, and to both discrete and continuous response formats. The key element in the methodology is the transformation of the raw ratings to pseudovalues which are treated as observations for purposes of statistical analysis. The pseudovalues are computed by the Quenouille-Tukey[9.87–9.89] version of the jackknife, and then analyzed by a mixed-model analysis of variance in which readers and patients are treated as random factors and treatments are treated as fixed factors. When the analysis of variance model does not apply, a general mixed linear model is fitted to the pseudovalues. This approach permits both hypothesis testing and estimation of confidence intervals. The new method develops ideas introduced by the Swets-Pickett methodology into a comprehensive and easy-to-use procedure. Studies designed to address only one dimension of generalization can also be accommodated within this framework: Treating observers as a fixed effect and cases as a random effect is equivalent to generalizing to the population of cases rather than readers; treating cases as a fixed effect and observers as a random effect is equivalent to generalizing to the population of observers.

In our first illustrative application of this methodology, we used a classical mixed-design analysis-of-variance model to analyze the data of Franken, et al.[9.90] on plain film versus CRT viewing of clinical neonatal radiographs. That study used a balanced factorial experimental design in which the same readers read the same cases in all modalities with no re-reading of cases. This design is often called a fully-crossed factorial design. When applicable, this is a very important experimental design because it is the most statistically powerful, and therefore the most manageable in terms of the number of readers, cases, and readings. At the single-reader level of analysis, the standard errors and the corresponding 95% confidence limits were quite similar for the CORROC[9.91] and jackknife methods. These findings agree with the conclusions of McNeil and Hanley, the first investigators to jackknife cases in a paired ROC design. The jackknife method gave standard errors and corresponding 95% confidence intervals in good agreement with the estimates of the CORROC model despite the fact that the jackknife method makes no assumptions regarding an underlying bivariate normal model for latent decision random variables. These results would seem to provide some support for the assumptions of the CORROC model as well as for the robustness and power of the jackknife method. Standard errors and 95% confidence intervals for multi-reader, multi-case analysis were smaller than for the single-reader multi-case analysis. Therefore, the DBM method was found to be more powerful than the statistical methods used by CORROC in our illustrative analysis. This additional power derives from the method's ability to reduce error by incorporating data from multiple readers whose readings are highly correlated between modalities and consistent from reader to reader.

Toledano and Gatsonis (TG),[9.92] using methods developed by Toledano,[9.93] confirmed conclusions from our method on the clinical neonatal radiographs of Franken, et al[9.90]: "our estimates of the areas under the ROC curves, their standard errors and the overall conclusions of the analysis agree with those reached by Dorfman, et al." Toledano and Gatsonis used an ordinal regression model in conjunction with generalized estimating equations in which readers were modeled as a fixed effect. Statistical conclusions derived from our method have also been confirmed by Efron's bootstrapping technique.[9.85] DBM and TG methodologies gave similar results for both individual readers and the averages over all readers. Thus, bootstrap provided some justification for DBM and TG methodologies. Moreover, bootstrap gave smaller standard errors for group than for individual reader means, thereby providing evidence for a tradeoff of readers and cases with regard to precision and power in this dataset. Finally, Metz[9.94] has recently provided strong support for the DBM MRMC method in computer simulations on fine-grained or quasi-continuous rating data.

9.6.1   Fitting Individual ROC Curves

RSCORE [9.95] and ROCFIT [9.57] are two rather well-known computer programs for fitting binormal ROC curves to discrete categorical rating data. Both programs use an iterative procedure called the method of scoring for maximum-likelihood estimation of ROC parameters. One major problem with method of scoring in ROC analysis has been that it sometimes mysteriously fails to converge. Metz made the very important discovery that this failure to converge is often caused by degenerate data sets.[9.57] Degenerate data sets result in exact-fit ROC binormal curves of inappropriate shape consisting of a series of horizontal and/or vertical line segments.[9.57, 9.96] Metz's ROCFIT program detects, classifies, and provides exact-fit solutions to degenerate data sets, but does not compute diagnostic accuracy in such situations. The classes of Metz-degeneracy identified by ROCFIT can be found within the ROCFIT code (subroutine DEGENE; variable ICLASS).

Such degenerate data sets have customarily been excluded from statistical analysis. Exclusion of data sets can be expensive and can introduce bias in data analysis because the excluded data sets are based on observer performance and may not represent a random sample from the population of observers. Whereas ROCFIT does not compute the area under the ROC curve, certain classes of degeneracy give unique exact-fit solutions and do, in fact, permit computation of the area. Although the shape of the exact-fit ROC curve is inappropriate, it does not follow that exclusion of degenerate data sets will yield more valid conclusions than the inclusion of those data sets.

Metz has pointed out that degenerate data sets tend to occur when sample sizes are small. When sample sizes are small, empty cells tend to appear in the ROC data matrix. Because multi-reader, multi-case ROC methods that use jackknife reduce each cell count by one unit, degeneracy will occur when there are single counts as well as zero counts in the cells. Therefore, degeneracy is more of a problem for these MRMC resampling methods. In a recent article, Rockette, et al.[9.97] have given an example in which the scoring method failed to converge in 30% of the jackknife subsamples for an observer in one condition. This demonstrates the need for a robust optimization procedure if one wishes to conduct this form of multi-reader multi-case ROC analysis.

The occurrence of empty cells is a common problem in the analysis of categorical data and has been addressed.[9.98] When a cell is empty due to small sample size rather than to theoretical impossibility, it is called a *sampling zero*. When a cell is empty because it is theoretically impossible to have counts in that cell, it is called a *structural zero*. If we assume that empty cells represent structural zeros, then infinite or zero values for the ROC slope and infinite values for the cutpoints are feasible values of the parameter space. If, on the other hand, we assume that empty cells represent sampling zeros caused by small samples, then such parameter values are not in the feasible parameter space. Inasmuch as empty cells seem to disappear with large samples when the number of rating categories is small, we believe it is reasonable to assume that empty cells appearing in such discrete ROC data matrices are sampling zeros, not structural zeros. We might expect, for example, that if a radiologist used a set of categories such as *definitely normal, probably normal, possibly abnormal, probably abnormal,* and *definitely abnormal,* and read cases for many, many years, there would be at least one normal and one abnormal case classified into each of the rating categories. To eliminate the problems introduced in computation by sampling zeros, Agresti[9.98] has suggested that researchers add a small constant to empty cells before conducting a categorical analysis. *Adding a small constant is a procedure for treating an empty cell as a sampling zero, not a structural zero.*

A robust new program, RSCORE4, was developed to estimate parameters in the face of degenerate data. To reduce the likelihood of convergence failure, we implemented Agresti's suggestion of adding small positive constants to empty cells, and supplemented the scoring method with a very robust pattern search algorithm.[9.99-9.101] We compared the two procedures, RSCORE4 and ROCFIT, in a series of computer simulations.[9.102] The population models for the computer simulations were derived from a classic population model published by Metz[9.57] that yields the most common form of degeneracy. Such population ROC curves occur when the observer tends to be very conservative with regard to confidence ratings.

We draw several major conclusions from this analysis. First, exclusion of degenerate data sets from analysis appears to be inappropriate in evaluating diagnostic systems because degenerate data sets are not a random sample from the population of data sets. Inclusion of Metz's exact-fit solutions to degenerate data sets gave more accurate measures of diagnostic performance than exclusion of degenerate data sets. Second, Metz's exact-fit solutions to degeneracy seem to be inferior, for the most part, to a dynamic Agresti procedure for dealing with zero counts when those counts are sampling zeros rather than structural zeros. Finally, RSCORE4 may provide a robust procedure for MRMC ROC analysis.

It should be pointed out that Metz[9.103] has very recently developed a robust method for proper binormal ROC analysis. Proper ROC curves are curves in which the decision dimension is strictly monotonic with likelihood ratio. Such ROC curves appear to eliminate the problem of degeneracy.

9.7 References

[9.1]   Hunt E.B. *Artificial Intelligence.* New York: Academic; 1985.

[9.2]   Minsky M. Steps toward artificial intelligence. *Proc. IRE* 1961; 49: 8–30.

[9.3]   Chan F.H., Pizer S.M. An ultrasonogram display system using a natural color scale. *J. Clin. Ultrasound* 1976; 4: 335–338.

[9.4]   Pizer S.M., Zimmerman J.B. Color display in ultrasonography. *Ultrasound Med. Biol.* 1983; 9: 331–345.

[9.5]   Katsuragawa S., Doi K., MacMahon H. Image feature analysis and computer-aided diagnosis in digital radiology: Detection and characterization of interstitial lung disease in digital chest radiographs. *Med. Phys.* 1988; 15: 311–319.

[9.6]   Katsuragawa S., Doi K., MacMahon H. Image feature analysis and computer-aided diagnosis in digital radiology: Classification of normal and abnormal lungs with interstitial disease in chest images. *Med. Phys.* 1989; 16: 38–44.

[9.7]   Powell G., Doi K., Katsuragawa S. Localization of inter-rib spaces for lung texture analysis and computer-aided diagnosis in digital images. *Med. Phys.* 1988; 15: 581–587.

[9.8]   Biederman I. Recognition-by-components: A theory of human image understanding. *Psychol. Rev.* 1987; 94: 115–147.

[9.9]   Berbaum K.S., Franken E.A. Jr., Honda H., McGuire C., Weis R.R., Barloon T. Evaluation of a PACS workstation for interpreting body CT studies. *J. Comput. Asst. Tomog.* 1990; 14: 853–858.

[9.10]   Berbaum K.S., Smoker W.R.K., Smith W.L. Measurement and prediction of diagnostic performance during radiology training. *Am. J. Roentg.* 1985; 145: 1305–1311.

[9.11]   Rogers D., Johnston R., Brenton B., Staab E., Thompson B., Perry J. Predicting PACS console requirements from radiologists' reading habits. *Proc SPIE* 1985; 536: 88–96.

[9.12]   Rogers D., Johnston R., Hemminger B., Pizer S. Development of and experience with a prototype medical image display. Presented at the Far West Image Perception Conference, Department of Radiology, University of New Mexico, 1986.

[9.13]   Pizer S.M., Johnston R.E., Rogers R.C., Beard D.V. Effective presentation of medical images on an electronic display station. *Radiographics* 1987; 7: 1267–1274.

[9.14]   Carmody D.P., Nodine C.F., Kundel H.L. Finding lung nodules with and without comparative visual scanning. *Percept. Psychophys.* 1981; 29: 594–598.

[9.15]   Franken E.A. Jr., Berbaum K.S. Perceptual aspects of cardiac imaging. In: Marcus M.L., Schelbert H.R., Skorton D.J., Wolf G., eds. *Cardiac Imaging–Principles and Practice: A Companion to Braunwald's Heart Disease.* Philadelphia: Sanders; 1991: 87–92.

[9.16]   Green D.M., Swets J.A. *Signal Detection Theory and Psychophysics.* New York: Wiley; 1966.

[9.17]   Wagner R.F., Brown D.G. Unified SNR analysis of medical imaging systems. *Phys. Med. Biol.* 1985; 30: 498–518.

[9.18]   Rose A. *Vision, Human and Electronic.* New York: Plenum; 1973.

[9.19]   Wagner R.F. Toward a unified view of radiological imaging systems. Part II: Noisy images. *Med. Phys.* 1977; 4: 279–296.

[9.20]   Schnitzler A.D. Analysis of noise required contrast and modulation in image detecting and display systems. In: Biberman L.C., ed. *Perception of Displayed Information.* New York: Plenum; 1973; 119–166.

[9.21]   Giger M.L., Doi K. Investigation of basic imaging properties in digital radiography: 3. Effect of pixel size on SNR and threshold contrast. *Med. Phys.* 1985; 12: 201–208.

[9.22]   Ohara K., Doi K., Metz C.E., Giger M.L. Investigation of basic imaging properties in digital radiography. 13. Effect of simple structured noise on the detectability of simulated stenotic lesions. *Med. Phys.* 1989; 16: 14–21.

[9.23]   Giger M.L., Doi K. Investigation of basic imaging properties in digital radiography. 1. Modulation transfer function. *Med. Phys.* 1984; 11: 287–293.

[9.24]   Dainty J.C., Shaw R. *Image Science.* New York: Academic; 1974.

[9.25]   Schade O. Optical and photoelectric analog of the eye. *J. Opt. Soc. Am.* 1956; 46: 721–739.

[9.26]   Morgan R.H. Threshold visual perception and its relationship to photon fluctuations and sine-wave response. *Am. J. Roentg.* 1965; 93: 982–997.

[9.27]   Sorenson J.A., Phelps M.E. *Physics in Nuclear Medicine* (2nd ed.). New York: Grune & Stratton; 1987; 115–142.

[9.28]   Kijewski M.F., Judy P.F. The noise-power spectrum of CT images. *Phys. Med. Biol.* 1987; 32: 565–575.

[9.29]   Moore S.C., Kijewski M.F., Mueller S.P., Holman B.L. SPECT image noise power: effects of nonstationary projection noise and attenuation compensation. *J. Nucl. Med.* 1988; 29: 1704–1709.

[9.30]   Giger M.L., Doi K., Metz C.E. Investigation of basic imaging properties in digital radiography. 2. Noise Wiener spectrum. *Med. Phys.* 1984; 11: 797–805.

[9.31]   Bracewell R.N. *The Fourier Transform and its Applications* (2nd ed.). New York: McGraw-Hill; 1978.

[9.32] Riederer S.J., Pelc N.J., Chesler D.A. The noise power spectrum in computed X-ray tomography. *Phys. Med. Biol.* 1978; 23: 446–454.

[9.33] Sandrik J.M., Wagner R.F. Absolute measures of physical image quality: Measurement and application to radiographic magnification. *Med. Phys.* 1982; 9: 540–549.

[9.34] Wagner R.F., Brown D.G., Pastel M.S. Application of information theory to the assessment of computed tomography. *Med. Phys.* 1979; 6: 83–94.

[9.35] Hanson K.M. Detectability in computed tomographic images. *Med. Phys.* 1979; 6: 441–451.

[9.36] Judy P., Swensson R.G., Szulc M. Lesion detection and signal-to-noise ratio in CT images. *Med. Phys.* 1981; 8: 13–23.

[9.37] Cohen G. Contrast detail analysis of imaging systems: Caveats and kudos. In: Doi K., Lanzl L. and Lin P.J., eds. *Recent Developments in Digital Imaging* (AAPM Medical Physics Monograph 12). New York: American Institute of Physics, 1985; 141–159.

[9.38] Cohen G., McDaniel D.L., Wagner L.K. Analysis of variations in contrast detail experiments. *Med. Phys.* 1984; 11: 469–473.

[9.39] Wald A. *Statistical Decision Functions.* New York: Wiley; 1950.

[9.40] Peterson W.W., Birdsall T.G., Fox W.C. The Theory of Signal Detectibility. *Trans. IRE Prof. Grp. Inform. Theory* 1954; PGIT-4: 171–212.

[9.41] Tanner W.P., Swets J.A. A decision-making theory of visual detection. *Psychol. Rev.* 1954; 61: 401–409.

[9.42] McNicol D. *A Primer of Signal Detection Theory.* London: Allen & Unwin; 1972.

[9.43] Gescheider G.A. *Psychophysics: Method, Theory, and Application* (2nd ed.). Hillsdale: Erlbaum; 1985; 135–166.

[9.44] Hanley J.A., McNeil B.J. The meaning and use of the area under a receiver operating characteristic (ROC) curve. *Radiology* 1982; 143: 29–36.

[9.45] Hanley J.A., McNeil B.J. A method of comparing receiver operating characteristic curves derived from the same cases. *Radiology* 1983; 148: 839–843.

[9.46] Kundel H.L., Revesz G. The evaluation of radiographic techniques by observer tests: Problems, pitfalls and procedures. *Invest. Radiol.* 1974; 9: 166–173.

[9.47]   Lusted L.B. General problems in medical decision making, with comments on ROC analysis. *Semin. Nucl. Med.* 1978; 8: 299–306.

[9.48]   McNeil B.J., Keeler E., Adelstein S.J. Primer on certain elements of medical decision making. *N. Engl. J. Med.* 1975; 293: 211–215.

[9.49]   McNeil B.J., Hanley J.A. Statistical approaches to the analysis of receiver operating characteristics (ROC) curves. *Med. Decis. Making* 1984; 4: 137–150.

[9.50]   McNeil B.J., Hanley J.A., Funkenstein H.H., Wallman J. Paired receiver operating characteristic curves and the effect of history on radiographic interpretation: CT of the head as a case study. *Radiology* 1983; 149: 75–77.

[9.51]   Metz C.E. Basic principles of ROC analysis. *Semin. Nucl. Med.* 1978; 8: 283–298.

[9.52]   Metz C.E. ROC methodology in radiographic imaging. *Invest. Radiol.* 1986; 21: 720–733.

[9.53]   Swets J.A. ROC analysis applied to the evaluation of medical imaging techniques. *Invest. Radiol.* 1979; 14: 109–121.

[9.54]   Swets J.A. Indices of discrimination or diagnostic accuracy: Their ROCs and implied models. *Psychol. Bull.* 1986; 99: 100–117.

[9.55]   Swets J.A. Form of empirical ROCs in discrimination and diagnostic tasks: Implications for theory and measurement of performance. *Psychol. Bull.* 1986; 99: 181–198.

[9.56]   Swets J.A., Pickett R.M. *Evaluation of Diagnostic Systems: Methods from Signal Detection Theory.* New York: Academic; 1982.

[9.57]   Metz C.E. Some practical issues of experimental design and data analysis in radiological ROC studies. *Invest. Radiol.* 1989; 24: 234–245.

[9.58]   Swets J.A. Is there a sensory threshold. *Science* 1961; 134: 168–177.

[9.59]   Dorfman D.D., Alf E. Jr. Maximum likelihood estimation of parameters of signal detection theory and determination of confidence intervals—rating method data. *J. Math. Psychol.* 1969; 6: 487–496.

[9.60]   Yerushalmy J. The statistical assessment of the variability in observer perception. *Rad. Clin. N. Am.* 1969; 7: 381–392.

[9.61]   Revesz G., Kundel H.L., Bonitatibus M. The effect of verification on the assessment of imaging techniques. *Invest. Radiol.* 1983; 18: 194–198.

[9.62]   Ker M., Seeley G.W., Stempski M.O., Patton D. A protocol for verifying truth of diagnosis. *Invest. Radiol.* 1988; 23: 485–487.

[9.63]   Ransoholf D.F., Feinstein A.R. Problems of spectrum and bias in evaluating the efficacy of diagnostic tests. *N. Engl. J. Med.* 1978; 299: 926–930.

[9.64]   Kundel H.L., Revesz G. The evaluation of radiographic techniques by observer tests: Problems, pitfalls, and procedures. *Invest. Radiol.* 1974; 9: 166–172.

[9.65]   Berbaum K.S., Franken E.A. Jr., Dorfman D.D., Rooholamini S.A., Kathol M.C., Barloon T.J., Behlke F.M., Sato Y., Lu C.H., El-Khoury G.Y., Flicking er F.W., Montgomery W.J. Satisfaction of search in diagnostic radiology. *Invest. Radiol.* 1990; 25: 133–140.

[9.66]   Berbaum K.S., Franken E.A., Rooholamini S., Coffman C.E., Cornell S.H., Cragg A.H., Galvin J.R., Honda H.H., Kao S.C.S., Kimball D.A., Ryals T.J., Sickels W.J., Smith A.P. Time-course of satisfaction of search in diagnostic radiology. *Invest. Radiol.* 1991; 26: 640–648.

[9.67]   Berbaum K.S., Franken E.A. Jr., Anderson K.L., Dorfman D.D., Erkonen W.E., Farrar G.P., Geraghty J.J., Gleason T.J., MacNaughton M.E., Phillips M.E., Renfrew D.L., Walker C.W., Whitten C.G., Young D.C. The influence of clinical history on visual search with single and multiple abnormalities. *Invest. Radiol.* 1993; 28: 191–201.

[9.68]   Berbaum K.S., El-Khoury G.Y., Franken E.A. Jr., Kuehn D.M., Meis D.M., Dorfman D.D., Warnock N.G., Thompson B.H., Kao S.C.S., Kathol M.C. Missed fractures resulting from satisfaction of search effect. *Emerg. Radiol.* 1994; 1: 242–249.

[9.69]   Franken E.A. Jr., Berbaum K.S., Lu C.H., Kannam S., Dorfman D.D., Warnock N.G., Simonson T.M., Pelsang R.E. Satisfaction of search in detection of plain film abnormalities in abdominal contrast examinations. *Invest. Radiol.* 1994; 29: 403–409.

[9.70]   Chakraborty D.P., Winter L.H.L. Free-response methodology: Alternative analysis and a new observer-performance experiment. *Radiology* 1990; 174: 873–881.

[9.71]   Bunch P.C., Hamilton J.F., Sanderson G.K., Simmons A.H. A free response approach to the measurement and characterization of radiographic observer performance. *Proc. SPIE* 1977; 127: 124–135.

[9.72]   Swensson R.G. Measuring detection and localization performance. In: Barrett H.H., Gimitro A.F., eds. *Information Processing in Medical Imaging.* New York: Springer-Verlag; 1993; 525–554.

[9.73]   Metz C.E., Starr S.J., Lusted L.B. Observer performance in detecting multiple radiographic signals. *Radiology* 1976; 121: 337–347.

[9.74]   Starr S.J., Metz C.E., Lusted L.B., Goodenough D.J. Visual detection and localization of radiographic images. *Radiology* 1975; 116: 533–538.

[9.75]   Swets J.A., Pickett R.M., Whitehead S.F., Getty D.J., Schnur J.A., Swets J.B., Freeman B.A. Assessment of diagnostic technologies. *Science* 1979; 205: 753–759.

[9.76]   Berbaum K.S., Dorfman D.D., Franken E.A. Jr. Measuring observer performance by ROC analysis: Implications and complications. *Invest. Radiol.* 1989; 24: 228–233.

[9.77]   Cooperstein L.A., Good B.C., Eelkema E.A., et al. The effect of clinical history on chest radiograph interpretations in a PACS environment. *Invest Radiol* 1990; 25: 670–674.

[9.78]   Rockette H.E. An index for diagnostic accuracy in the multiple disease setting. *Acad. Radiol.* 1994; 1: 283–286.

[9.79]   Simonson T.M., Yuh W.T.C., Crosby D.L., Michalson L.S., Dorfman D.D., Wiechert R.J., Lee H.J., Berbaum K.S. *The value of contrast enhancement in making the correct diagnosis of major brain pathology.* Scientific Presentation, American Society of Neuroradiology, 32nd Annual Meeting, Nashville, Tennessee, May 3–7, 1994.

[9.80]   Turing A.M. On computable numbers, with an application to the Entscheidungs problem. *Proc. London Math. Soc.* (Ser. 2) 1990; 42: 230–265.

[9.81]   Turing A.M. Computing machinery and intelligence. *Mind* 1950; 59: 433–460.

[9.82]   Von Neumann J. The general and logical theory of automata. In: Newman J.R., ed. *The World of Mathematics* (Vol 4). New York: Simon & Schuster; 1956: 2070–2098.

[9.83]   Hanley J.A. Receiver operating characteristic (ROC) methodology: The state of the art. *Crit Rev. Diagn. Imaging* 1989; 29: 307–335.

[9.84]   Dorfman D.D., Berbaum K.S., Metz C.E. Receiver operating characteristic analysis: generalization to the population of readers and patients with the jackknife method. *Invest. Radiol.* 1992; 27: 723–731.

[9.85]   Dorfman D.D., Berbaum K.S., Lenth R.V. Multi-Reader Multi-case ROC methodology: A bootstrap analysis. *Acad. Radiol.* 1995; 2: 626–633.

[9.86]   Dorfman D.D., Metz C.E. Rejoinder. *Acad. Radiol.* 1995; 2: S75–S77.

[9.87]   Quenouille M.H. Approximate tests of correlation in time series. *J. R. Stat. Soc.* (Ser. B) 1949; 11: 68–84.

[9.88]   Quenouille M.H. Notes on bias in estimation. *Biometrika* 1956; 43: 353–360.

[9.89]   Tukey J.W. Bias and confidence in not quite large samples. *Ann. Math. Stat.* 1958; 29: 614. Abstract.

[9.90]   Franken E.A. Jr., Berbaum K.S., Marley S.M., Smith W.L., Sato Y. Evaluation of a PACS workstation for interpreting of neonatal examinations: An ROC study. *Invest. Radiol.* 1992; 27: 732–737.

[9.91]   Metz C.E., Wang P.L., Kronman H.B. A new approach for testing the significance of differences between ROC curves measured from correlated data. In: Deconink F., ed. *Information processing in medical imaging.* The Hague: Nijhoff; 1984; 432–445.

[9.92]   Toledano A., Gatsonis C. Regression analysis of correlated receiver operating characteristic data. *Acad. Radiol.* 1995; 2: S30–S36.

[9.93]   Toledano A.T. Generalized estimating equations for repeated ordinal categorical data, with applications to diagnostic medicine. Unpublished doctoral dissertation, Harvard School of Public Health, 1993.

[9.94]   Metz C.E. The Dorfman/Berbaum/Metz method for testing the statistical significance of ROC differences: Validation studies with continuously-distributed data. Presented at the Sixth FarWest Image Perception Conference, Philadelphia, October 13–15, 1995.

[9.95]   Dorfman D.D. RSCORE II. In: Swets J.A., Pickett R.M. *Evaluation of Diagnostic Systems: Methods from Signal Detection Theory.* New York: Academic; 1982; 208–232.

[9.96]   Rockette H.E., Obuchowski N.A., Gur D. Nonparametric estimation of degenerate ROC data sets used for comparison of imaging systems. *Invest. Radiol.* 1990; 25: 835–837.

[9.97]   Rockette H.E., Gur D., Kurs-Lasky M., King J.L. On the generalization of the receiver operating characteristic analysis to the population of readers and cases with the jackknife method: An assessment. *Acad. Radiol.* 1995; 2: 66–69.

[9.98]   Agresti A. *Categorical data analysis.* New York: Wiley; 1990: 244–245, 249–250.

[9.99]   Chandler J.P. Subroutine STEPIT: An algorithm that finds the values of the parameters which minimize a given continuous function. A copyrighted program. J.P. Chandler, Copyright, 1965.

[9.100] Hooke R., Jeeves T.A. Direct search solution of numerical and statistical problems. *J. Assoc. Comp. Mach.* 1961; 8: 212–229.

[9.101] Dorfman D.D., Beavers L.L., Saslow C. Estimation of signal detection theory parameters from rating-method data: A comparison of the method of scoring and direct search. *Bull. Psychon. Soc.* 1973; 1: 207–208.

[9.102] Dorfman D.D., Berbaum K.S. Degeneracy and discrete ROC rating data. *Acad. Radiol.* 1995; 2: 907–915.

[9.103] Metz C.E. "Proper" binormal ROC curves: Theory and maximum likelihood estimation. Presented at the Sixth FarWest Image Perception Conference, Philadelphia, October 13–15, 1995.

10
Computer Vision and Decision Support

Henry A. Swett, Maryellen L. Giger, and Kunio Doi

10.1   Introduction

The preceding chapters describe basic mechanisms by which visual information is detected by the eye and brain, as well as fundamental principles of perception including texture and object recogniton and cognitive processing. In addition, Chaps. 7 and 8 discuss computer processing techniques that may make specific image features easier to perceive. In this chapter, we consider ways that computers can directly extract features from images (computer vision). and understand their meaning and support human cognition (decision support). For example, in medical imaging, computer-vision techniques may be used to detect and characterize a possible abnormality (such as a tumor mass) in an image of the breast and then artificial intelligence (AI) techniques may be employed to merge the extracted features into a diagnostic decision regarding the possibility of malignancy. Computer processing techniques include image processing, image segmentation, and feature extraction. Decision support tools include rule-based expert systems, discriminant analysis,, Bayesian methods, and neural networks among others.

Figure 10.1 presents an overview of some of the major components of image understanding and shows how computers and humans perform these functions separately and how they complement each other. When humans examine an image, a primary goal is the recognition and understanding of features and objects contained within the image. In medicine, the observer endeavors to classify observations as being characteristic of a group of diseases (a differential diagnosis) and when possible to recognize that the image features are unique to a single diagnosis. Often, careful consideration of available data may make it possible to reduce an initially large cluster of diagnostic possibilities to a small set or even a single diagnosis. Many image features are recognized almost instantly during a "preattentive" phase of perception (Chap. 6). Once the observer has rcognized that an unexpected feature is present, a more deliberate and conscious consideration of the nature of the detected abnormalities is pursued during an "attentive" phase of perception. In medicine, these two processes may occur very rapidly resulting in almost instantaneous recognition of a specific disease entity. Often, however, the observer must extend the attentive phase in order to first reach a differential diagnosis and, later, a more definitive diagnosis.

When an observer has failed to recognize and classify objects present in an image based on inherent knowledge and experience, additional knowledge may be sought to strengthen the attentive phase of perception. This knowledge may

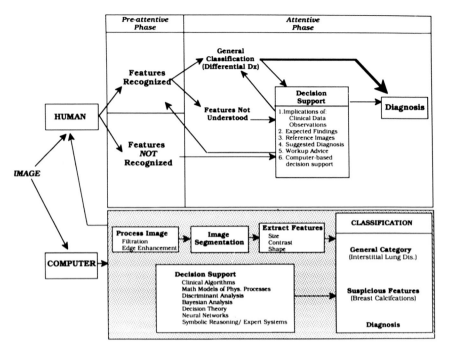

FIGURE 10.1. Image understanding by humans and computers.

"intelligent" computer systems. These sources may serve as a memory jog, or as a source of new knowledge which can be used to re-examine the original image and to reconsider the implications of new or previously rcognized features. In Sec. 10.4, we consider how computers can strengthen human cognition in this fashion.

Computers may participate in this decision process either independently, or in concert with the human observer. The image may be processed in order to lower the perception threshold of different classes of objects (e.g., pneumothorax) or in some cases to present visual or quantitative information that was not evident on unprocessed images. In medical imaging, techniques of computer vision may allow a general classification of findings (e.g., interstitial lung disease). Alternatively, given unique features of potentially significant abnormalities, the computer may flag findings for detailed inspection by the human observer (e.g., suspected malignant microcalcifications in carcinoma of the breast). The ultimate final decision about diagnosis and management remain with the clinician.

10.2   Computer Vision

Humans, through their visual system, perceive features (i.e., information) present in an image. Computer vision involves having a computer extract and determine meaningful descriptions of such features in an image. This image understanding is different from the image processing (Chaps. 7 and 8) of images to enhance features to be perceived by the human eye.

Computer-vision schemes can be used to extract features in an image that may or may not normally be perceived by the human, e.g., subtle thin lines or a specific texture. The development of computer-vision schemes requires *a priori* information about the medical image in question, and knowledge of various computer processing techniques and decision analysis methods. The required *a priori* knowledge about the specific image to be analyzed includes the physical imaging properties of the image acquisition system and the morphological information of the abnormality in question along with its associated anatomical background (i.e., a database). Computer processing techniques include image processing, image segmentation, and feature analysis.

In a clinical setting, a computer-vision scheme might be used as a computer-aided diagnosis (CAD) test invoked by the radiologist upon reviewing a radiographic case or as a routine screening procedure performed on all examinations, e.g., in breast cancer screening programs. Output from an "intelligent" workstation might be used as a "second opinion" in detecting or characterizing abnormalities in the image. A workstation could be configured to allow the radiologist to control the sensitivity and specificity of the computer output. Obviously, a choice of fewer false positive lesions, and vice versa. This tradeoff could be adjusted by a radiologist, depending on the nature of the case material and personal preference. For example, a radiologist might choose an output with high sensitivity for examining high risk patients being screened for cancer, whereas a lower sensitivity and correspondingly lower false positive rate might be desired for patients at low risk for cancer.

The remainder of this section briefly reviews some practical techniques used in radiographic computer-vision schemes. Various books on the general subjects of image processing and computer vision are readily available.[10.1-10.7]

10.2.1   Image Processing

Image processing techniques can be employed to enhance features for ease of extraction later or to de-emphasize unwanted features such as background noise. These techniques are applied to a digital radiographic image, which is basically a spatial matrix of gray levels. Each picture element (pixel) is composed of a gray level number that can be processed mathematically resulting in a transformed image in which the general format of the image (the spatial matrix of gray values) is retained. Both linear and nonlinear transformations may be employed.

The general linear spatial filter involves a transformation that modifies the frequency of gray level variations (the spatial frequency content) in the image. This type of manipulation can result, e.g., in an image that has sharpened edges, smoother transitions between gray levels, or reduced background noise depending on the filter employed in the processing. General linear filtering is done by centering a mask or (or kernel) over each pixel or group of pixels in the image and calculating, at each location, the weighted sum of the covered kernel. This process, known as convolution,[10.8] produces a new filtered image. In order to keep the average intensity level of the image similar before and after processing, the kernel function needs to be normalized, i.e., the coefficients of the kernel must add to one. Common linear filters include low pass filters that reduce high spatial frequency components in the image, high pass filters that have the opposite effect, high boost filters that amplify the high frequency elements without changing the low frequency components of the image, and bandpass filters that retain or enhance a selective range of spatial frequencies.

Another example of a common linear filter is the matched filter. This process is referred to as template matching[10.1] and is used to detect a particular feature in an image. The result of template matching is an image in which the locatiion of the particular feature produces the largest correlation with the template kernel.

Morphological filters and median filters are examples of nonlinear filters that can be used for signal enhancement or suppression. Mathematical morphology[10.2] involves a nonlinear filtering method that calculates the logical AND (erosion function) or OR (dilation function) of pixels within a kernel of some given size and shape. When extended to gray scale images the logical AND and OR operations can be replaced by minimum and maximum operations. By appropriately choosing the size and shape of the kernels, as well as the sequences of the AND operation and the OR operation, the filters can eliminate groups of pixels of limited size or merge neighboring pixels, and so on. The sequence of dilation followed by erosion is referred to as a closing operation and the sequence of erosion followed by dilation is referred to as an opening operation. Morphological filters have been used to process biomedical images in the study of Alzheimer's disease[10.9] and cytology.[10.10]

Processing the image to enhance edges or boundaries of a particular object in an image can be performed by a variety of kernels. A linear kernel that is used to calculate the gradient at each pixel location in the image has coefficients that add to zero so that areas of constant intensities yield a zero value in the processed image. Areas of high intensity in the new filtered image correspond to areas of large change with the respect to gray level in the original image. Morphological filters can also be used in edge detection. The difference of the original image and one processed with an erosion operation yields an image of the internal edges of objects. There are various examples of computer-vision techniques that have employed edge detection. Mathematical morphology was used in the development of a sifting theory for edge enhancement of X-ray images.[10.11] Edges of the heart and lungs in chest images have been located with the use of first derivatives.[10.12-10.14] In some such cases the gradient filter needs only to have one dimension, since *a priori* knowledge of the direction of the edge is incorporated in the computer scheme.

Histogram transformations are also used as initial processing steps.[10.4] The gray level histogram of an image indicates the frequency of occurrence of each gray level in the image in question. An example of such a transformation is histogram equalization in which the distribution of gray levels in the image is modified so that each gray value in the image occurs in an equal number of pixels. Another example is histogram specification, in which the histograms of two related images are matched to each other. Use of such a technique may be necessary in the comparison of radiographic images acquired at different times, or under different exposure conditions, or both.

Background correction can also be considered to be an initial processing step for use before further analysis. The process involves estimating the background and then subtracting this estimate from the original to yield a "flattened" background. Methods of estimating the background include using a low pass filtered version of the original image, or fitting the background with a low degree polynomial or with splines to approximate the low frequency background trend. Katsuragawa, et al.[10.15] used a two-dimensional polynomial surface to estimate the background trend within a ROI prior to calculation of texture measures in a computerized scheme for the characterization of interstitial disease in chest radiographs.

10.2.2   Image Segmentation

Segmentation of an image involves the separation of the image into regions of similar attributes. Basically, a segmentor only subdivides an image and does not try to recognize the individual segments.

Once the image has been processed to enhance particular features, the image can be segmented into areas of high and low enhancement by thresholding. For example, after the template matching of an image, gray-level thresholding can be performed in order to yield areas of correlation. In such a thresholding process, pixel values above some gray level cutoff are retained or set to some particular gray level, while all others are set to another gray level.

In edge detection, spatial filtering with gradient or Laplacean kernels followed by gray level thresholding can be performed to yield pixel locations corresponding to high gradient values, i.e., edges. Both Toriwaki, et al.[10.16] and Hashimoto, et al.[10.17] used edge detection approaches and thresholding in order to locate suspicious regions of the chest in a nodule search method.

Portions of an image with different "textures" can also be segmented due to the different structural or statistical patterns in two or more regions of the image. Parameters used to segment an image may also be useful in distinguishing patterns as normal or abnormal with respect to medical decision making.

10.2.3   Feature Analysis

Feature analysis entails the recognition of specific characteristics of the individual segments. Once features have been segmented from the rest of the image data, parameters relating to their particular characteristics can be calculated. Three common characteristics are size, contrast, and shape. The size of a feature can be calculated by counting the number of pixels within a particular feature. This process first requires the computer to find the pixels that actually belong to the feature. Region growing[10.1] is a technique in which the (x, y) location of pixels common to a feature are grouped together. Region growing entails some sort of connectivity requirement, such as eight-point connectivity that utilizes the relationship the eight neighboring pixels have with respect to a given pixel. Basically, the computer searches the image for a pixel that has a gray level above some cutoff value. Once found, region growing begins; any pixel that is connected to a pixel within the group and has an acceptable gray level becomes a member of the group. When no more connected pixels have acceptable gray-levels, the region growing stops.

Contrast can be measured by determining the difference between the gray levels in the features and those in the surrounding area. Features of high contrast can be extracted from image data using local gray-level thresholding techniques.

An example of a shape descriptor is a measure of circularity. The degree of circularity can be defined as the ratio of the area of the object within an equivalent circle to the total area of the object. A measure of circularity as a shape parameter was used in the detection of nodules in digital chest radiographs.[10.18] Another measure of circularity involves the ratio of the perimeter (in terms of the number of pixels along the border of the feature) to the area (in terms of the number of pixels within the border of the feature). This measure is sometimes referred to as a measure of compactness of the feature in question.

An example of curve detection involves preprocessing with a gradient operator, thresholding, and then use of the Hough transform[10.19] to locate a curve with a particular parametric description. Approaches based on the Hough transform to locate circular features have been used in the detection of tumors in chest images.[10.20–10.21] A combination of a gradient histogram analysis technique and the Hough transform was developed by Sanada, et al.[10.22] in the detection of rib edges and pneomothorax (a fine, slightly curved feature) in chest images.

In order to describe the pattern of an area of interest segmented from the image by some previous technique, texture[10.23-10.26] can be performed. Examples of texture measures include calculations of the energy, entropy, correlation, and moment of inertia, which characterize the spatial gray-level dependence of the region of interest. In the Fourier domain (spatial frequency domain), measures such as the zero and first moments of the power spectrum can be used to characterize the magnitude and coarseness of the pattern in the region of interest. In addition, by limiting the calculations to specific angular orientations, information on the directionality of the texture can be obtained. Various investigators[10.15,10.27-10.32] have developed computerized texture analysis schemes for characterizing lung patterns in digital chest images. Others have also employed texture analysis in the computerized detection of suspicious abnormalities in mammograms.[10.33-10.35] Magnin, et al.[10.36] and Caldwell, et al.[10.37] tried to evaluate the risk of developing breast cancer by computing texture measures from the mammographic appearance of breast parenchyma.

10.3   Computer Vision Examples

10.3.1   Mammography

Breast cancer is the most common malignancy of women and its incidence is rising.[10.38] Mammography is the most effective method for the early detection of breast cancer and studies have shown that regular mammographic screening can reduce the mortality from breast cancer in women.[10.39–10.42] The American Cancer Society and other medical organizations have recommended these screening guidelines: women between 35 and 39 should obtain a base line mammogram, those between 40 and 50 should obtain biannual mammograms, and those over 50 years of age should obtain annual mammograms. In other countries, including the U.K., the screening program commences later and involves fewer mammograms at less frequent intervals. Mammography is becoming one of the largest volume of X-ray procedures routinely interpreted by radiologists. In addition, radiologists do not detect all cancers that are visible on the images in retrospect.[10.43–10.49] The missed detections may be due to the very subtle nature of the radiographic findings or to oversight by the radiologist.[10.43–10.44] It is apparent that the efficiency and effectiveness of this screening could be increased substantially by use of a computer system that would aid the radiologist in detecting lesions and in making diagnostic decisions.

a.   Computerized Detection and Classification of Masses in Mammograms

A computer-vision scheme is being developed for the detection of masses in mammograms that is based on the deviation from the architectural symmetry of normal right and left breasts, with asymmetrics indicating potential masses.[10.50–10.51] Figures 10.2(a) and 10.2(b) illustrate a pair of conventional mammograms of the left and right breasts; note the mass as indicated by an arrow. A nonlinear subtraaction technique is employed in which gray level thresholding is performed on the individual mammograms prior to subtraction. Ten images threshold with different cutoff gray levels are obtained from the right breast image, and ten are obtained from the left breast image. Next, subtraction of the corresponding right and left breast images is performed to generate ten bilateral-subtraction images. This linking process accumulates the information from a set of ten subtraction images into two images that contain locations of suspected masses for the left and right breasts. Figures 10.2(c) and 10.2(d) demonstrate a pair of run-length images containing suspicious locations, prior to the application of feature-analysis techniques.

Feature-analysis techniques are used to reduce the number of false positive detections. These feature-analysis techniques include morphological filtering, and size and border testing. The sequential application of a morphological closing operation followed by a morphological opening operation is used to eliminate isolated pixels and to merge small neighboring features. A size test is used to eliminate features that are smaller than a predetermined cutoff size. A border test is employed to eliminate artifacts arising from any breast border misalignment that is present at subtraction. After the feature-analysis techniques, the results can

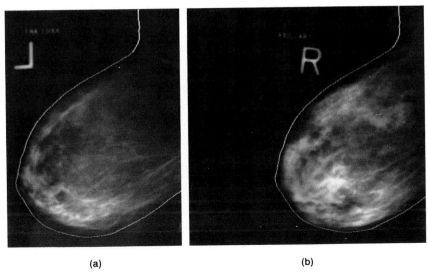

(a)   (b)

FIGURE 10.2. (a) Left and (b) right mammograms demonstrating the symmetry of normal breast architecture between the corresponding breasts. (c) and (d) Processed images were obtained using the bilateral-subtraction method. Suspicious areas can be seen in the left and right breasts, prior to the application of feature-analysis techniques. (e) and (f) Computer-reported results indicating locations of possible masses in the left and right breasts can be presented to the radiologist as an aid in making the final diagnosis.

be presented to a radiologist, as shown in Figs. 10.2(e) and 10.2(f), as an aid in making the final diagnosis.

The input to the computerized scheme, for a given patient, is the four conventional mammograms obtained in a routine screening examination: the right cranio-caudal and the left (CC) views, and the right and left mediolateral-oblique (MLO) views. Using 46 pairs of clinical mammograms, the approach achieved a true positive rate of 95% with three false positive detections per image.

Malignant masses can often be distinguished from benign masses due to their more spiculated appearance in the mammographic image. Thus, a computerized scheme is being developed[10.50–10.51] to distinguish between malignant and benign masses, based on the degree of speculation exhibited by the mass in question. Figure 10.3(a) shows an example of a malignant mass; note the spicules on the mass. In the computer-vision scheme, the mass is first segmented from the anatomical background of the mammogram using growing techniques [Fig. 10.3(b)]. After a mass has been isolated from the background, its margin information is extracted. This margin is then smoothed in order to substantially reduce possible spicules. The difference before and after the smoothing operation provides an indicator of the degree of margin spiculation, with a large difference corresponding to a high degree of spiculation and therefore, to a greatly increased likelihood of malignancy.

One measure of spiculation is the standard deviation of the fluctuations about the margin of the mass. After the margin of the mass is tracked [Fig. 10.3(c)], the x, y

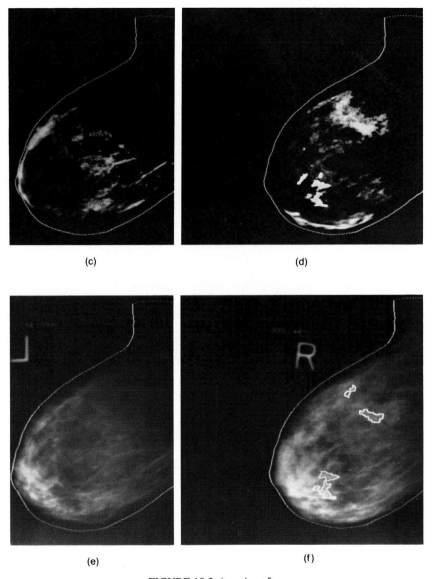

(c)   (d)

(e)   (f)

FIGURE 10.2. (*continued*)

positions are smoothed using a running mean (averaging) filter, yielding a smooth outline of the mass, as demonstrated in Fig. 10.3(d). The difference between the corresponding boarder points [Fig. 10.3(e)] is then used in determining the standard deviation. Another measure of spiculation is the normalized difference in areas of the mass with and without the presence of spicules. The mass, devoid of spicules, is obtained by a morphological opening operation. Currently, the scheme, using these two measures of spiculation, achieves a classification accuracy for

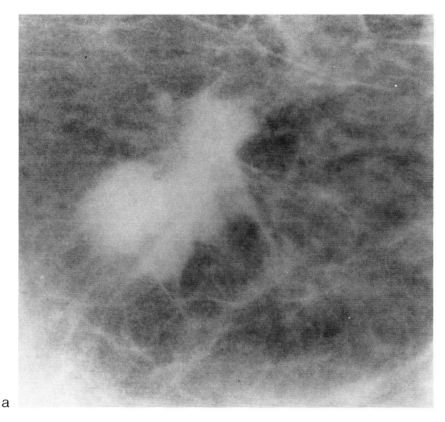

a

FIGURE 10.3. Examples of (a) a malignant mass and (b) the mass after extraction from the breast architecture using region growing techniques. Intermediate steps in the computerized measurement of spiculation are illustrated by (c) the original margin of the mass, (d) the smoothed margin, and (e) fluctuations about the margin corresponding to the spiculation.

malignancy of 97% with a false positive rate of 79%, for a clinical database of 50 masses having a sensitivity of 100% and a false positive classification rate of 95%.

b.   Computerized Detection of Microcalcifications in Mammograms

A computerized method for the detection of microcalcifications in mammograms is being developed by Chan, et al.[10.52–10.54] and Nishikawa, et al.[10.55] It includes a preprocessing step referred to as a difference image approach. Basically, the original digital mammogram is filtered once to enhance the signal of the microcalcifications. The difference of the two resulting processed images yields an image (a difference image) in which the variations in background density are largely removed. The spatial filter used in enhancing the microcalcifications is a filter matched to the size and contrast variations of a typical breast microcalcification. Filters that have been used in signal suppression include a median filter, a contrast

b

c

d

FIGURE 10.3. (*continued*)

reversal filter, and a box rim filter.[10.53] These are examples of spatial filters that attempt to remove the signals of microcalcifications from the image. Figure 10.4 demonstrates an original mammogram and its corresponding difference image. A cluster of microcalcifications is indicated by an arrow in Fig. 10.4(a).

Microcalcifications are segmented from the image using global gray-level thresholding and local thresholding techniques. With global thresholding, a pre-selected percentage of pixels with values at the high end of the histogram of the difference image are retained while all others are set to a constant value. With local thresholding, the mean and standard deviation are determined within a square kernel centered at the pixel of interest, in order to estimate the local background noise fluctuations. If the pixel in question has a value that is larger than the mean

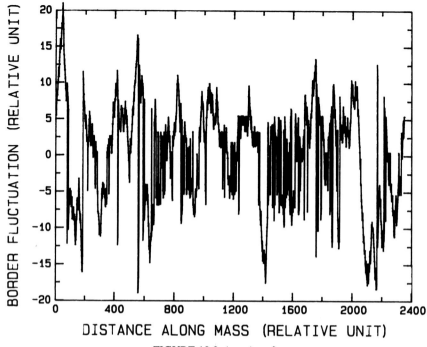

FIGURE 10.3. (*continued*)

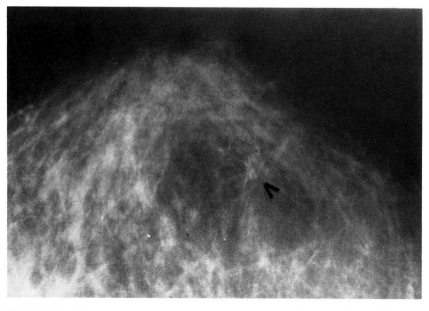

(a)

FIGURE 10.4. (a) Example of a mammogram with microcalcifications and (b) the difference image obtained using a 3 by 3 matched filter and a box-rim filter.

(b)

FIGURE 10.4. (*continued*)

pixel value by a preselected multiple of the standard deviation, then it is retained as a potential site for microcalcifications.

The segmented image is next subjected to feature-extraction techniques in order to remove features arising from structures other than microcalcifications. During the computer search of the segmented image, each feature is localized using region growing techniques and tested for size and contrast. The region of interest about remaining features is then subjected to low frequency background correction and is characterized by the first moment of its power spectrum, defined as the weighted average of radial spatial frequency over the two dimensional power spectrum. The final signal-extraction technique used is a clustering test in which clusters containing more than a preselected number of signals within a region of preselected diameter are identified by the computer. The preset values are determined using *a priori* knowledge from the clinical experience of radiologists.

The computerized scheme of Chan, et al.[10.54] was tested on a database of 60 clinical mammograms and achieved 90% true positive detection for clusters of microcalcifications with an average of four false positive clusters per image. In addition, an observer study was conducted in order to examine the effect of the computer-vision aid on radiologists' performance in the detection of subtle clustered microcalcifications in digital mammograms. It was found that there was statistically significant improvement in the radiologists' accuracy when the aid was used.

10.3.2   Chest Radiography

a.   Computerized Detection of Pulmonary Nodules

Radiographic detection of lung nodules is important in the diagnosis and management of patients. The best hope of cure in lung cancer depends upon early detection, while the tumor is still small and localized.[10.56, 10.57] Detection of malignant lung nodules in chest radiographs is one of the most difficult tasks performed by radiologists, because the visibility of nodules can be obscured by overlying ribs, bronchi, blood vessels, and other normal anatomical structures. The miss rate for radiographic detection of early lung nodules is currently about 30%.[10.38, 10.58] The observer error that causes such miss rates may be due to the camouflaging effect of the surrounding anatomical background on the nodule of interest, or to the subjective and varying decision criteria used by radiologists. Underreading of a radiograph may be due to a lack of clinical data,[10.59] lack of experience,[10.60] a premature discontinuation of the film reading because of a definite finding,[10.61] focusing of attention on another abnormality by virtue of a specific clinical question,[10.61] failure to review previous films,[10.62] distractions,[10.62] and "illusory visual experiences."[10.63] A computerized system that alerts radiologists to the locations of suspicious lung nodules should reduce the frequency of such errors by acting as a "second opinion" while leaving the final decision to the radiologist.

Computer-vision schemes are being developed[10.16, 10.64–10.65] for the detection of pulmonary nodules in digital chest radiographs. The computer-vision scheme for the detection of lung nodules is based on a difference image approach. A digital chest image is processed to yield nodule-enhanced and nodule-suppressed versions of the original image. This difference image approach involves an attempt to remove the structured anatomical background before application of feature-extraction techniques. Both linear and nonlinear filters have been employed in this image-processing step. The linear filters used are a "matched filter" for signal enhancement and a ring-shape averaging filter for signal suppression. The matched-filter kernel corresponds to the projected and detected profile of a spherical nodule. e.g., 9 mm in diameter. The ring-shape averaging filter replaces the gray level of each pixel by the average of the pixel values that lie along a ring of a given radius and width. The nonlinear filters include a circular morphological open operation for signal enhancement and a ring-shape median filter for suppression. Figure 10.5 illustrates an original chest image and its corresponding difference image. An arrow indicates the location of an actual pulmonary nodule.

After the difference of the signal-enhanced image and the signal-suppressed image has been obtained, global gray-level thresholding is performed in order to segment candidate nodules from normal anatomical background. Figure 10.6 shows an example of a gray-level histogram of (a) an original chest image and (b) one after preprocessing by the difference image step. In the original image, the pixel value of the nodule is similar to that of other anatomical structures, since the nodule can be superimposed on any structure in the chest. In the difference image, however, the pixel value of the nodule is at the high end of the histogram. Thus,

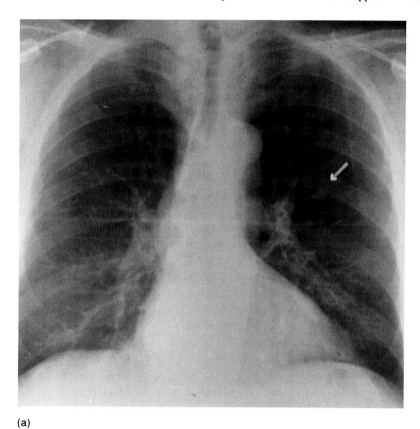

(a)

FIGURE 10.5. (a) Example of an original chest radiograph with a 1 cm left lung nodule (arrow) and (b) the difference image obtained using a 9 mm diameter, nodule-matched filter and an 18 mm, ring-shape averaging filter for enhancement and suppression, respectively. From Giger, et al.[10.64]

in this initial segmentation step, the difference image is thresholded by a cutoff value corresponding to a specific percentage of the area under the histogram of the difference image.

After multiple threshold images have been obtained, feature-analysis techniques (involving the size, contrast, and shape of candidate features) are performed by the computer to reduce the number of false positives arising from normal anatomical background. The sizes and shapes of candidate features that arise from actual nodules and non-nodules (i.e., normal anatomy) vary with threshold level in characteristically different ways. Thus, the size and shape of each feature are analyzed as functions of threshold level. Any remaining locations of suspected nodules are analyzed with respect to their contrast in the original chest image. This step often eliminates false positive detections in the perihilar region that arise from vascular

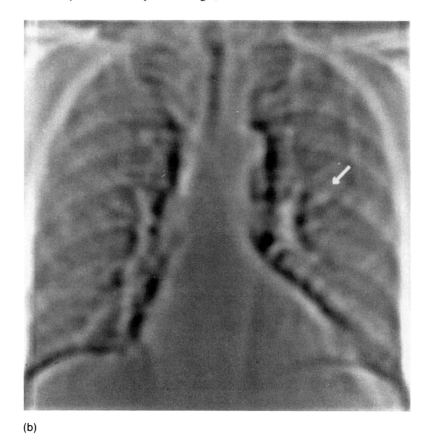

(b)

FIGURE 10.5. (*continued*)

structures. The computer results can be presented to a radiologist for the final decision in a manner demonstrated in Fig. 10.7.

Use of the detection scheme with only the linear filters achieves a detection sensitivity of 70% with an average of seven false positive detections per image. It was tested in a preliminary study in order to examine its utility as a diagnostic aid for radiologists. Results of receiver operating characteristics (ROC) analysis indicated that use of the aid slightly increased radiologists' detection performance $(p = 0.2)$.[10.54]

Nonlinear filters were employed both in the difference image step[10.66] and as part of a feature-extraction step.[10.65] Used as part of the feature-extraction step, a morphological opening operation (circular shape with a diameter of 9 mm) helped to eliminate false positives that arose from ribs. Use of the nonlinear filters in the difference image step showed that most nodules are detected by either the linear or nonlinear filtering methods. The false positive detections of these two techniques, however, generally occur at different locations. With a combination method that

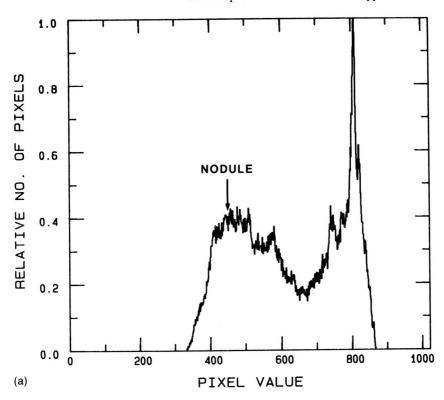

(a)

FIGURE 10.6. Gray level histograms of (a) the original image and (b) the difference image in Fig. 10.5. From Giger, et al.[10.64]

employs the results from both types of filters (i.e., the linear and the nonlinear), the false positive rate can be reduced by approximately 50%.

b.   Characterization of Interstitial Lung Disease

The evaluation of interstitial lung disease is a difficult task for radiologists. Katsuragawa, et al.[10.15, 10.27, 10.67–10.68] have developed a computerized scheme to isolate the fine details of lung texture, which are affected by interstitial disease, from the normal gross anatomy in digital chest images.

Initially, regions of interest (6.4 by 6.4 mm) are selected from intercostal (inter-rib) spaces on a digital chest image (0.1 mm pixel size), as demonstrated in Fig. 10.8. An initial preprocessing step involves estimating the background trend in each region of interest using a two-dimensional polynomial-surface-fitting technique. This is necessary since the lung field includes both that due to the gross anatomy of the lung and chest wall (background trend) and that due to the fine underlying texture which is related to interstitial diseases. The nonuniform background trend is subtracted from each ROI in order to yield the underlying fluctuations (i.e., the lung textures).

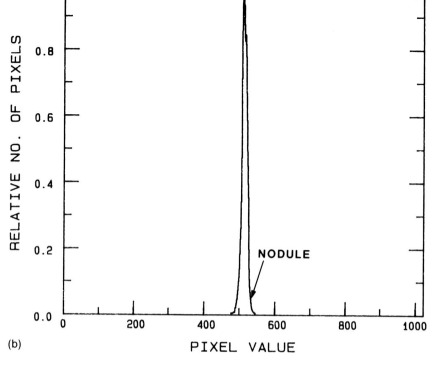

FIGURE 10.6. (*continued*)

Next, for each ROI, the square of the two-dimensional Fourier transform of the lung texture is calculated, referred to here as the "power spectrum." (It should be noted that strictly speaking , however, the power spectrum needs to be determined from an ensemble average of the square of the Fourier transform over an infinitely large area.) The power spectra of the lung textures may contain low frequency components due to some residual uncorrected background trend and very high frequency components due to radiographic mottle[10.69] in the original chest radiograph. Thus, the power spectra are filtered by the human visual system response function,[10.70] which is basically a bandpass filter.

Once each ROI has been preprocessed for background correction and its power spectrum filtered, it is subjected to a texture analysis process involving two measures: the root-mean-square (rms) variation (R) and the first moment of the filtered power spectrum (M) which corresponds to the magnitude and the coarseness of lung textures, respectively. Now

$$R = \sqrt{\int_{-\infty}^{\infty} \int_{-\infty}^{\infty} V^2(u, v) \mid F(u, v) \mid^2 du dv}$$

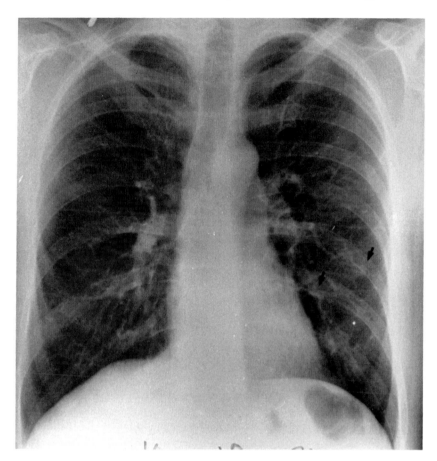

FIGURE 10.7. Chest radiograph with computer-reported locations of possible lung nodules indicated by arrows. The computer correctly detected a subtle nodule (right arrow) and one false positive (left arrow).

and

$$M = \frac{\int_{-\infty}^{\infty} \int_{-\infty}^{\infty} \sqrt{u^2 + v^2} V^2(u, v) \mid F(u, v) \mid^2 dudv}{\int_{-\infty}^{\infty} \int_{-\infty}^{\infty} V^2(u, v) \mid F(u, v) \mid^2 dudv},$$

where $V(u, v)$ and $F(u, v)$ correspond to the visual system response and the Fourier transform of the lung texture, respectively.

Automated classification for distinguishing between normal and abnormal lungs with interstitial disease is based on analysis of the two texture measures and on a database derived from clinical cases.[10.15] From normal lungs of 100 PA chest images, the average and the standard deviation of the rms variation and that of the first moment of the power spectrum were calculated. These values serve in the normalization process of measures for subsequent classifications. The two texture measures are combined into a single texture index by analysis of the cluster plot

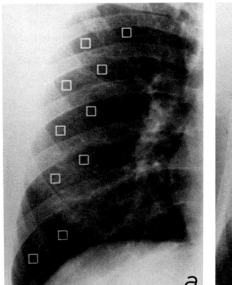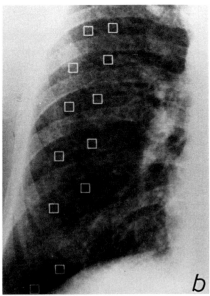

FIGURE 10.8. Digital chest radiographs with selected ROIs in intercostal spaces for determination of physical texture measures. (a) Normal lung and (b) abnormal lung with nodular pattern. From Katsuragawa, et al.[10.15]

of the two measures.[10.27] Figure 10.9 demonstrates the output of the computerized texture analysis scheme for a chest image exhibiting an abnormal lung pattern. Computer output is superimposed on the chest image using symbols to indicate the type and severity of the infiltrate and the probability of disease is noted. Comparison of ROC curves (Fig. 10.10) obtained by radiologists and by means of the computerized approach may provide performance similar to that of human observers in distinguishing lungs with mild interstitial disease from normal lungs.

c.   Detection of Cardiomegaly

Cardiac size is an important and useful diagnostic parameter in chest radiographs. Unsuspected abnormal enlargement of the heart (cardiomegaly) is often first diagnosed by radiologists during the interpretation of chest radiographs. The current method used by radiologists is manual measurement of the cardiothoracic ratio (CTR), which is the ratio of the transverse diameter of the cardiac shadow to the greatest transverse diameter of the thorax or the transverse diameter of the thorax at the highest level of the diaphragm. Figure 10.11 illustrates a schematic diagram of the chest with the various parameters indicated.

Nakamori, et al.[10.14] are developing an automated method for computing the various parameters related to cardiac size as an aid for radiologists in their diagnosis of heart disease. Initially, points along the rib cage edges are found using the second derivative of a horizontal profile obtained across the chest image [10.71]

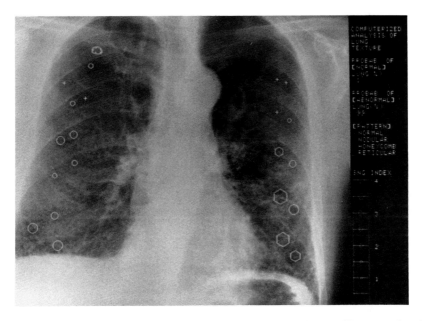

FIGURE 10.9. Chest image with computer output indicating a mixture of honeycomb and nodular patterns in the lung texture.

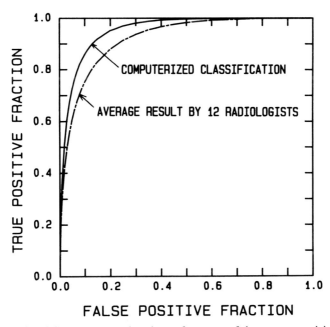

FIGURE 10.10. ROC curve comparing the performance of the computer vision scheme with that of radiologists. From Katsuragawa, et al.[10.15]

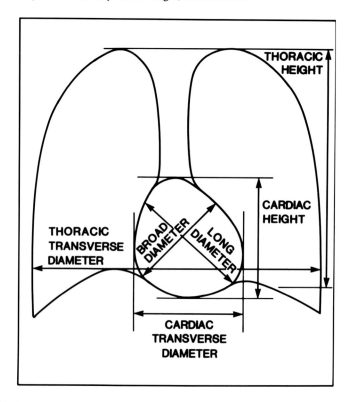

FIGURE 10.11. Schematic diagram of a chest image indicating various parameters related to the size and shape of the heart and lungs. After Nakamori, et al.[10.14]

and are fitted separately to polynomials. Points along the heart boundary and the diaphragm edge are detected based on analysis of edge gradients obtained in the horizontal and vertical directions on the chest image. From examination of heart contours from various chest radiographs (a clinical database). Nakamori, et al.[10.14] selected a model function to describe the contour of the heart as given by

$$r(\theta) = r_0 + r_1 \cos 2\{(\theta - \phi) - \alpha \cos(\theta - \phi)\},$$

where r_0, r_1, α, and ϕ are parameters to be determined by fitting the heart boundary points by means of a nonlinear least-squares method. Note that the detected heart boundary points are represented in polar coordinates as described in Fig. 10.12. Once the heart boundary points are fitted to the shift-variant cosine function, the entire contour of the heart can be described and used in determining the various parameters related to heart size. Figure 10.13 shows the detected edge points and the resulting fit that yields the heart size and other parameters.

The algorithm was applied to 60 PA chest radiographs and calculated CTRs correlated well (correlation coefficient of 0.91) with those from manual measurements. The computed outlines of the projected heart in the majority of the chest

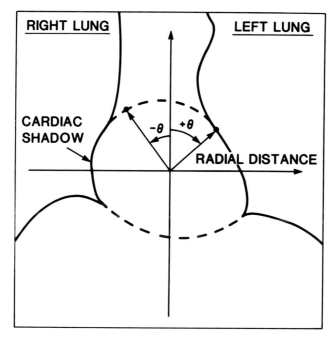

FIGURE 10.12. Polar coordinate system used for expressing the cardiac contour in terms of radial distance. After Nakamori, et al.[10.14]

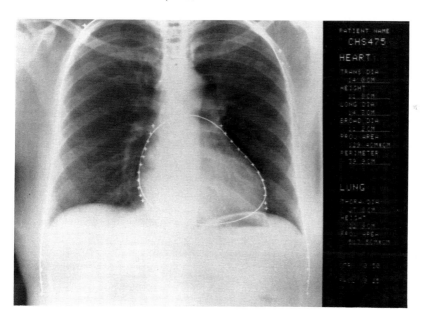

FIGURE 10.13. Chest image with computer-determined outline of the heart and rib cage edges. The cardiac-to-thoracic ratio, as well as other measurements, are listed.

radiographs were acceptable to most radiologists for estimation of the size and area of the cardiac shadow.

10.4   Decision Support

When an observer studies a medical image, a complex series of events may occur. First, normal anatomical structures are identified in order to orient the observer to the organ being imaged and the perspective from which the organ is viewed. This usually occurs very rapidly. Having identified expected normal structures and features, the observer then searches for unexpected findings. Errors in human interpretation of medical images may occur for a variety of reasons. The abnormal feature may not be seen (a cognitive error).[10.72] Both perceptual and cognitive components may be strengthened using conventional and computer-based techniques. If an observer fails to recognize the presence of abnormality, or recognizes an object but believes that it is not significant, reinspection of the image may result in correct recognition if the observer is given new knowledge to guide the search. For example, this new knowledge could come from the kind of computer-extracted features described earlier, from a computer-based decision support system, or from additional clinical knowledge.

In some medical imaging problems, the primary task is to determine the presence or absence of a particular abnormality (e.g., to rule out pneumonia on a chest X ray). Once this task is performed, the primary objective has been achieved. In other cases, recognition of abnormalities is only the beginning. Their implications may be wholly or partially unknown. They may be manifestations of an unfamiliar disease, or may be caused by a variety of different diseases. Sometimes, these distinctions can be made through further study of the image, through the consideration of other nonimage data (clinical findings, laboratory data), or from further characterization of the disease process through additional imaging perspectives of the same organ using other imaging modalities. Because knowledge in medicine is growing much faster than it can be absorbed by its practitioners, clinicians are increasingly unable to fully appreciate the significance of the data that they gather. The increasing presence of computers in medicine and the pronounced proliferation of computers to generate, display, and distribute images in diagnostic radiology make it very attractive to use computer-based decision support techniques to compensate for human limitations. Conversely, while computers have begun to achieve impressive decision-making capabilities, they cannot match the human's ability to make decisions when not constrained by memory limitations. If the relative strengths of humans and computers are combined, the result can represent a significant improvement over either acting alone.

A variety of different approaches to decision support in medicine have been taken, but only a few of these have been applied to diagnostic imaging. These include clinical algorithms, mathematical models of physical processes, statistical pattern matching, Bayesian statistical methods, decision theory, modeling, simu-

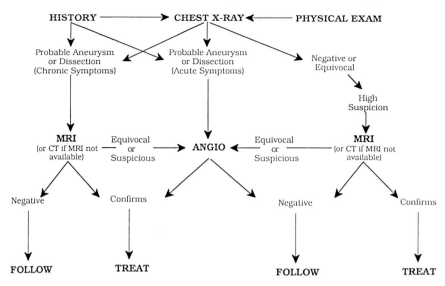

FIGURE 10.14. Clinical algorithm for the evaluation of patients suspected of having a thoracic aortic aneurysm. Depending on findings seen on chest X ray, a CT, or angiogram should be performed, or the patient should be followed clinically. Depending on the outcome of the CT scan or angiogram, surgical treatment, medical treatment, or no treatment is indicated. After Staub.[10.76]

lation, connectionism, and symbolic reasoning.[10.73] The end result of the use of these methods may be to *filter knowledge,* thus enabling the human decision-maker to function more effectively.

10.4.1   Clinical Algorithms

Clinical algorithms (or "decision trees") are flow charts to which a physician may refer in order to solve a specific diagnostic or therapeutic problem.[10.74] A typical clinical algorithm contains a series of clinical states and conditional branches which the clinician may follow to reach a diagnostic or therapeutic decision. Several of these have been applied to diagnostic imaging, particularly to guide the selection of appropriate imaging tests in the evaluation of clinical states and syndromes[10.75–10.77] (Fig. 10.14). These algorithms are often applied by simply consulting a reference book, although in some settings they may be effectively implemented by computer.[10.77–10.81] Hospitals are increasingly using large distributed computer systems to record clinicians' orders, convey test results, and so on. Integration of algorithm-driven decision support methods into the hospital's clinical care computer offers the potential for the use of computers to influence clinical care in real time.

A computer-based checklist has been used to help radiologists to interpret mammograms.[10.82] In this approach, radiologists were asked to explicitly comment on whether specific image features were present and how confident they were in their observations. This required the reading process to be systematized and the radiologist to consider all findings that could contribute to a malignant or benign diagnosis. This method was supplemented by the application of a computer-based classifier which suggested a likelihood of malignancy for observed localized lesions. A measurable improvement in performance was demonstrated.

10.4.2   Mathematical and Casual Models of Physical Processes

The physiological principles underlying some diseases can be represented by mathematical formulas and relationships. For example, there is a predictable relationship between various elements in blood. If one of these is depleted, corresponding decreases or increases in other blood elements occur. Therefore, knowledge of blood levels of individual components may lead to prediction of other values. These may further be used in casual models that can suggest resulting disease states. This approach has been used in areas such as acid-base/electrolyte disorders and respiratory care.[10.83]

10.4.3   Statistical Pattern-Matching Techniques

A mathematical relationship between measurable signs of disease and diagnosis or prognosis may be defined. Once this has been done, it may be possible to classify a patient into a specific diagnostic or prognostic category given the presence of a subset of features.[10.84] Bayes' theorem has provided a particularly popular approach. Its appeal is that it makes it possible to calculate the probability of a disease, given a set of observations, based on the frequency with which those observations have been associated with the disease in the past. If disease D_i is one of n mutually exclusive diagnoses under consideration and E is the evidence or observations supporting that diagnosis, then if $P(D_i)$ is the *a priori* probability of the ith disease[10.73]

$$P(D_i \setminus E) = \frac{P(D_i) \times P(E \setminus D_i)}{\Sigma_{j=1}^{n} P(D_j) \times P(E \setminus D_j)},$$

where $P(D_j \setminus E)$ is the probability of the ith disease (of a total of n diseases) given the evidence E and $P(E \setminus D_i)$ is the probability that the indicated piece of evidence occurs in the ith disease.

This theorem depends on the availability of accurate statistical data concerning the frequency of finding specific clinical states in a variety of diseases and on the independence of the observations. Bayesian analysis has been used in many areas of medicine. Early applications included the diagnosis of congenital heart disease[10.85] and acute abdominal pain.[10.86] An important early application of Bayes' law in diagnostic radiology was by Lodwick, et al.[10.87] in a system that dealt with computer diagnosis of primary bone tumors. This was a domain that was well suited to this method since sufficient statistical data were available and the variables are discrete and well-defined. Some tumors were always correctly evaluated in more than 80% of cases.

10.4.4   Decision Theory

Decision analysis goes beyond purely statistical approaches by attempting to add value judgements to clinical decision making. At each point in a decision tree, it is possible not only to represent the possible outcome of the decision taken, but the "cost" of the alternatives. This cost factor may take into account such issues as risk and discomfort of a diagnostic procedure, the value of life extension, or financial cost. When decisions are made that consider these broader, often more subjective issues, a more appropriate outcome may result. Decision theory provides a methodology for dealing with such circumstances.[10.88]

10.4.5   Connectionism

Artificial neural networks (ANN) have evoked excitement because in some ways they mimic the structure, function, and decision-making capabilities of the human brain.[10.89–10.90] ANN is a nonalgorithmic approach to information processing. Unlike many artificial intelligence techniques which require exhaustive knowledge engineering, ANNs learn directly from examples that are provided repeatedly. Once trained, a neural network can distinguish different input patterns according to its learning experience. The input data can correspond either to clinical data, or to human-extracted radiographic data, or to computer-extracted image features, or to any one of them. The input data are represented by numbers ranging from 0 to 1 and are supplied to the input units of the neural network. The output data from the neural network are then provided from output units through successive nonlinear calculations in the hidden and output layers. The calculation at each unit in a layer includes a weighted summation of all entry numbers, an addition of a certain offset number, and a conversion into a number ranging from 0 to 1 using a sigmoid-shape function such as a logistic function. Two different basic processes are involved in a neural network: these are a training process and a testing process. The neural network can be trained by such methods as a backpropagation algorithm[10.91] using pairs of training input data and desired output data. Once trained, the neural network accepts features of the abnormality in question and outputs a value from 0 to 1, which , e.g., can be related to a likelihood of malignancy or a possible course of clinical action.

ANNs have been most successfully used for pattern classification.[10.91] Asada, et al.[10.92–10.93] applied an ANN in the differential diagnosis of interstitial diseases. Gross, et al.[10.94] used an ANN in the interpretation of neonatal chest radiographs.

10.4.6   Symbolic Reasoning and Expert Systems

Most of the methods which have just been described employ numerical calculations to solve problems. The very subjective natures of many decisions in medicine lend themselves to less quantitative approaches using symbolical calculations rather than numerical ones. These programs typically use "knowledge" contained in a knowledge base. It is important to distinguish knowledge from data. An item of data in a medical database is a discrete value or fact (e.g., if the hematocrit is less than 37, then anemia is present). Such knowledge can be represented symbolically and conclusions may be reached by evaluating these knowledge structures using an inferencing mechanism (inference engine). The broad field which employs this kind of heuristic reasoning is known as *artificial intelligence*(AI).[10.73]

Expert systems are a special kind of artificial intelligence program which endeavor to solve problems in specific problem domains. In medicine, they usually contain knowledge about a category of disease along with problem solving approaches which are likely to be useful. As the name implies, expert system behavior is modeled after that of human experts and these programs typically produce as output a specific conclusion (e.g., a diagnosis or group of possible diagnoses), or advice that the human user may apply to facilitate problem solving. Expert systems can be quite effective at answering the questions that they have been programmed to answer, but usually do not "understand" their domain in depth and often cannot learn from their experiences.[10.95] Expert systems have been used for a variety of purposes outside of medicine including oil and mineral exploration, organic chemistry, automobile repair, space shuttle navigation, and as a source for political advice.[10.95]

Over the past 20 years, a variety of systems have been developed that apply artificial intelligence to medicine. Several surveys of this work exist.[10.96–10.99] Examples of systems that have focused primarily on diagnosis are EXPERT,[10.100] which has been applied to opthalmological and endocrine problems; PIP,[10.101] dealing with renal disease; ABEL,[10.102] which deals with acid-base and electrolyte disorders; MYCIN,[10.103] which deals with infectious disease; and INTERNIST,[10.104] which embraces the whole field of internal medicine. Some of these programs are very effective at making sophisticated diagnoses on a par with human experts, but in general they have not been accepted in real-world clinical practice. In part, this is because many expert medical systems require clinicians to enter into a time-consuming interaction with the computer, and also because they have not been designed to fit conveniently into current medical practice habits.

Many expert systems use *production rules* to guide decision making. A production rule is a conditional statement in the following form: IF premise assertions are true, THEN consequent assertions are true.[10.105] A weight (or confidence) factor is often associated with the rule to guide the use of the rule when it fires. An example of a production rule used in the MYCIN system is:

IF: 1. The stain of the organism is gram positive and
        2. The morphology of the organism is coccus and
        3. The growth confirmation of the organism is chains.

THEN: There is suggestive evidence that the identity of the organism is *Streptococcus.*[10.103]

Various systems use different global approaches to problem solving. Some systematically search for evidence (satisfied rules) supporting specific conditions or goals (backward chaining), while others are driven by the data available (forward chaining).

Heuristically based AI systems may use an ad hoc statistical scoring system to guide the selection of knowledge and ranking of candidate decisions. INTERNIST is an example of such a system.[10.104] This system assists with test selection and diagnosis of all diseases in internal medicine. It utilizes disease profiles consisting of diagnoses and associated findings or manifestations. For each disease, all manifestations of that disease are weighted according to the *frequency* that it is associated with the disease in question and the *evoking strength* or how strongly each disease manifestation suggests the indicated disease. These values range from 1 to 5 and are relative and subjective.

A similar semiquantitative approach has been used in several systems used in diagnostic radiology. Paraino, et al.[10.106] developed a system for the diagnosis of focal bone abnormalities that utilizes production rules and disease entities. This system evaluates findings provided by a physician concerning focal bone lesions. Production rules produce a preliminary reduction in the number of possible diagnoses. The scoring algorithm is then applied to this disease subset to produce a list of diagnoses in order of relative likelihood. Bramble[10.107] used similar techniques for a system that evaluated plain bone film findings observed by a clinician observer concerning arthritis.

Kahn, et al.[10.78] developed a rule-based expert system to guide radiological test selection. This system has been integrated with a commercially available radiology information management system and produces an algorithm or decision tree as its output. The decision tree is selected to show the radiological test options available to the user and the system can also provide information about the likely diagnostic possibilities along with workup goals. Another system focusing on radiological workup is DxCON[10.79] which structures advice about radiological test selection as a critique of the clinician's workup plan.

10.4.7 Critiquing Systems

Most medical expert systems have focused on making a diagnosis, while some have sought to give advice about treatment. Many of the early systems produced categorical statements as output (a diagnosis or list of possible diagnoses) without explanation of the rationale used by the expert system to reach that conclusion. A different approach was taken by Miller[10.108] who sought to cast the computer in the role of an intelligent assistant. Rather than delivering computer-based advice as categorical statements, critiquing systems discuss important aspects of a problem in the form of written prose critiques. These systems not only evaluate clinical data as input, but also ask the clinician to give a working diagnosis, or plan. This provides a frame of reference for the advisory system which then structures its advice around the clinician's own thinking. Importantly, critiquing systems also seek to explain the reasons for any recommendations that are made. This makes it possible for the clinician to decide whether the computer's advice is logical and makes sense in light of the broader clinical perspective. Critiquing systems have been built to give advice about many domains including the management of essential hypertension,[10.109] the evaluation of pheochromocytoma,[10.110] and the workup of obstructive jaundice.[10.79]

The critiquing approach has been applied to radiological differential diagnosis in the ICON system which gives advice about the evaluation of chest X rays in patients with a specific class of disease, the pulmonary lymphoproliferative disorders.[10.111] This system present English prose case discussions containing information about the significance of the clinical and radiographic findings of the case under consideration. ICON discusses the differential diagnosis of the major radiographic findings, evidence supporting the radiologist's working diagnosis, and evidence supporting other diagnoses which may not be under consideration. This style of advice proves useful for classical problems of differential diagnosis where the findings are clear cut, but the implications of the findings are in question. In image interpretation, however, decision making is often much more subjective and depends on the appreciation of the significance of visual or pictorial features which cannot be adequately expressed in words. As a result, the IMAGE/ICON system[10.112] was developed which supplements prose case discussions with display of images which are selected based on their likely relevance to the known or expected image content of a case. This approach was originally applied to the radiographic diagnosis of the pulmonary lymphoproliferative disorders and subsequently for breast cancer detection by mammography.[10.113-10.115]

10.4.8   Knowledge Filters

Because there is potentially a vast amount of medical knowledge that could be useful during any decision-making episode, one of the most effective uses of computers in support of clinical decision making in medicine is to *filter knowledge* for their human users. Provided with information about the problem that the clinician is dealing with (directly from the clinician, or from information systems containing the patient's electronic medical record), the computer may sift through the large amount of potentially useful information and deliver only specific information that is precisely relevant to the clinician's problem.[10.116] This process may continue iteratively as the computer receives information about the clinician's intermediate or provisional conclusions. Many of the methods just described may contribute to this kind of context-sensitive decision support.

10.5   A Decision Support Example: Mammography

Once a radiologist has identified abnormalities on a mammogram, it is necessary to determine their significance. The primary goal in interpreting mammograms is to determine whether there are findings present which are diagnostic or suspicious for breast cancer. There are many findings which suggest a malignant diagnosis including the presence of a mass, malignant forms of calcifications, abnormal pattern of blood vessels in the involved breast, thickening of the skin near the tumor, and retraction of the nipple. Once any of these findings have been identified, they must be characterized. For example, many masses in the breast are benign and it is often possible to determine whether a mass could be a tumor by studying morphological details such as the characteristics of its borders (shaggy masses tend to be malignant, while smoothly marginated masses are often benign). The significance of each individual finding is determined and then the collective significance of all findings is considered. Additional insight into the meaning of the observations may come from inspecting previous mammograms, noting how the appearances of the breasts have changed with time. In some cases, it is possible to make an absolute diagnosis of malignancy on the basis of such features. Often, however, the radiologist tries to determine whether the findings that are present are sufficiently suspicious to warrant a biopsy. Although the primary goal is the evaluation of breast cancer, there are other diseases of the breast which can be detected by mammography and it is important for the radiologist to correctly classify these conditions as well.

Some radiologists specialize in reading mammograms and have acquired enough experience so that they can accurately evaluate these issues in most patients. (Even the experts are perplexed by a few cases every day.) Many radiologists, however, cannot limit their practice to mammography and may not have had extensive training in reading these studies. For them, reading mammograms is difficult and prone to error. They may fail to recognize subtle abnormalities and they may not appreciate the significance of all the findings which they have observed. The MAMMO/ICON System[10.113–10.115] is an expert system that was developed to provide context-sensitive visual and cognitive feedback to help the radiologist function at a greater level of expertise than might otherwise be possible. MAMMO/ICON combines the critiquing approach just described with expert system-controlled image display. In order to use MAMMO/ICON, the radiologist provides the system with basic background information about the patient such as the family history of breast cancer, the menstrual history, and the presence of a palpable lesion within the breast. The radiologist records all observations extracted from the mammogram and proposes a working diagnosis that might explain these findings. The expert system uses these clinical and radiological findings as a basis for providing context-sensitive advice to the radiologist. To facilitate data entry, a voice recognition interface has been developed that allows the radiologist to dictate findings and automatically provide input data for MAMMO/ICON at the same time.[10.115]

The output produced by MAMMO/ICON is in two different forms designed to strengthen both preattentive and attentive phases of image interpretation. The preattentive phase is a purely visual form of pattern recognition and is the primary method used by experienced clinicians. For example, an experienced mammographer may immediately identify a mass and almost simultaneously "know" that it represents cancer. When the findings are less clear cut, the mammographer may switch to a more deliberate cognitive analysis (the attentive phase) in which the findings are carefully reviewed for distinguishing characteristics, and all diagnoses that could cause such findings are considered and ranked by probability. It is during this phase that additional knowledge may be helpful. The additional knowledge may represent a cognitive memory prompt (e.g., "Remember that this patient is at increased risk for developing breast cancer because of the following factors...”). Often a visual memory prompt is more helpful in diagnostic radiology where pattern recognition is so important. As a result, the MAMMO/ICON system supplements the cognitive advice embodies in a critique with the display of images selected to help to clarify the implications of the findings present. For example, if calcifications are observed, the expert system retrieves examples of similar calcifications from an image database. These images are grouped by benign and malignant etiologies and are presented in a way that facilitates classification.

Figures 10.15 and 10.16 show sample runs of the MAMMO/ICON system. Figure 10.15(a) shows a critique. Note that this English prose style critique has several sections. First, the system assesses *a priori* risk factors. Second, the system focuses on the radiologist's working diagnosis and looks for information supporting the conclusions. If there is information which suggests a different conclusion, this is presented next. Finally the system seeks to give advice about whether a biopsy may be indicated. Note that the system does not attempt to make a diagnosis. Rather it presents possible implications of clinical and radiographic findings. The final determination of a diagnosis or course of action is made by the radiologist.

MAMMO/ICON uses a variety of different criteria to select images for display. This is because there are a number of different reasons why a radiologist may wish to consult a reference image. It may be helpful to review proven cases of their patients that demonstrate the same findings that are present in the current case and are caused by a hypothesized diagnosis. Alternatively, the radiologist may wish to focus on an individual finding and be reminded of all conditions which could produce such a finding. If a particular diagnosis is being pursued, it may be useful to be reminded of the spectrum of all findings which might be seen in that condition, rather than just those thought to be present. It is also often useful to review variations of specific findings, or to be reminded of things which are often confused with the entity or finding under consideration. To accomplish this, the images are retrieved and arrayed along different "axes"[10.112–10.113] [see Fig. 10.15(b)], including:

1. Hypothesis axis (examples of the suspected diagnosis constrained by the findings present).

2. Spectrum axis (examples of the suspected diagnosis not constrained by the findings present).

3. Differential diagnosis axis (examples of all diagnoses which can produce the findings present).

4. Morphology axis (examples of variations of important morphologic features).

5. Clinical axis (examples of findings expected in the general clinical problem presented by the patient).

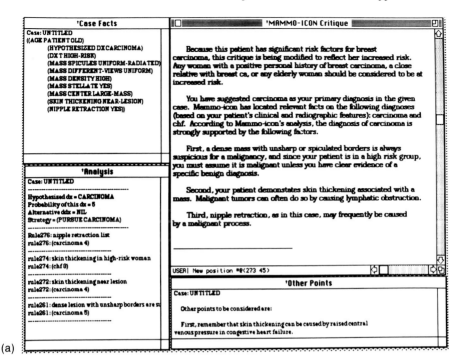

(a)

FIGURE 10.15. (a) Spiculated mass critique. The primary case discussion is presented in the window at the upper right. Below this, other points are discussed which may be less specifically relevant to the case under consideration, but may be helpful. Other windows summarize case facts, and the rules which fired to produce the critique. (b) Spiculated mass axis display. The large image at the upper left is selected from the image database because it closely matches the case under consideration. To the right are image axes where images have been retrieved in a variety of different ways, as described in the test. (c) Spiculated mass-recalculated axes. If the radiologist changes the working diagnosis, the criteria which govern selection of images are changed. Here, the radiologist has suggested a diagnosis which is unlikely to produce the findings which have been described. As a result, the hypothesis axis (second from top) is blank indicating that there were no examples in the image database of spiculated masses caused by the hypothesized diagnosis (fibroadenoma).

Each one of these axes contains a potentially large number of images arranged in order of their probable relevance to the current case. The physician can "scroll" through each axis to find those specific images which best answer whatever visual questions there may be.

In essence, the MAMMO/ICON system functions as a knowledge filter. Given basic information about the problem the radiologist is dealing with, the system sifts through a large amount of factual and pictorial data and endeavors to present a highly filtered subset of these data in a way that enables the radiologist to be a more informed decision maker.

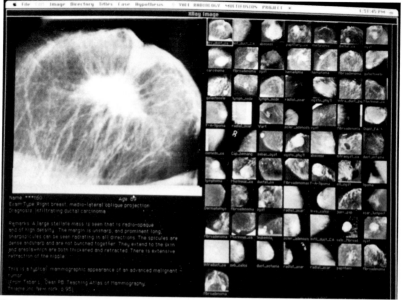

(b)

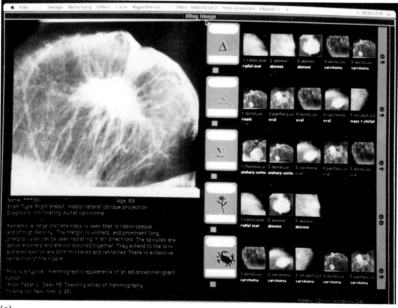

(c)

FIGURE 10.15. (*continued*)

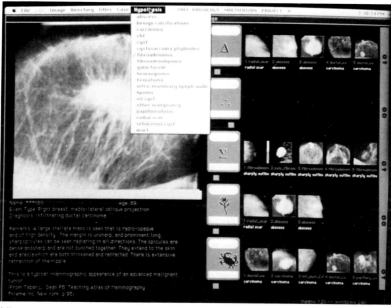

(d)

FIGURE 10.15. (*continued*)

10.6   Combining Decision Support and Computer Vision

Computer-vision and decision-support techniques may complement each other and work effectively together. Consider the examples that have been given concerning mammography. An "intelligent" mammography workstation might include computer-vision routines that would locate regions containing possible masses or microcalcifications for further inspection by a radiologist. As long as the computer's role is limited to alerting the radiologist to lesions that might be significant, it is not necessary for the computer to be completely accurate. If the computer system misses a lesion (false negative), the radiologist still has the opportunity to find the abnormality in the review of the case. Or, if the system incorrectly flags an abnormality as being significant (false positive), the radiologist can overrule the computer. For a radiologist to use such a computer system, however, the number of false positives must be low. The acceptable rate of false positives varies with the type of lesion being considered and depends on: (a) the costs and benefits of the subsequent follow-up procedure if a lesion is suspected by the radiologist and (b) the subjective tolerance of the radiologist to the average number of false positives reported by the computerized system.

Once the radiologist (or computer, or both) has located an abnormality present on the image, various decision-support tools may be used to yield a better understanding of the diagnostic significance of the abnormality. For example, a

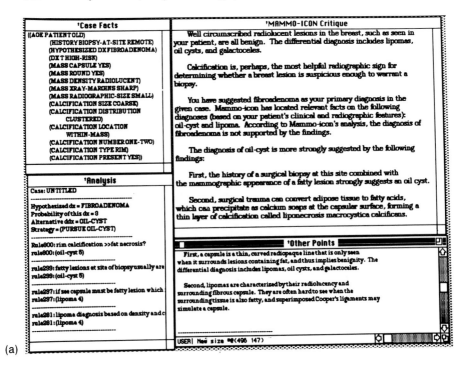

(a)

FIGURE 10.16. (a) Oil cyst critique. In this example, the radiologist has suggested a diagnosis which is unlikely to cause the findings present in the case under consideration. The critique indicates that it cannot find evidence in the knowledge base to support that diagnosis, but suggests other diagnoses which are more likely. (b) Oil cyst axis display. The images which have been retrieved also reflect the fact that the proposed diagnosis has not been responsible for the described findings of any cases in the image database. The differential diagnosis axis contains more likely possibilities and the large image which is displayed is an example of one of them.

computer-vision routine might identify a mass in a mammogram and also provide some information on the characteristics of the lesion such as its size, degree of speculation, and so on. An image-based expert system like MAMMO/ICON could then retrieve examples of similar masses from an image database. This would allow the radiologist to automatically have ready access to similar reference images to facilitate the decision to accept or reject the flagged abnormality. In addition, rule-based expert systems or artificial neural networks could then merge human- or computer-determined descriptions (or both) of the lesion with other available data (family history and *a priori* risk factors, for example) to help the radiologist to evaluate those abnormalities with respect to the diagnosis.

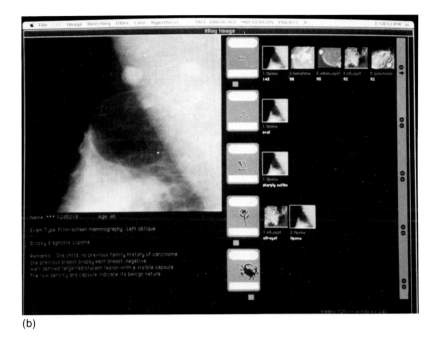

(b)

FIGURE 10.16. (*continued*)

10.7   References

[10.1]   Pratt W.K. *Digital Image Processing.* New York: Wiley; 1978.

[10.2]   Serra J. *Image and Analysis and Mathematical Morphology.* New York: Academic; 1982.

[10.3]   Ballard D., Brown C. *Computer Vision.* Englewood Cliffs: Prentice-Hall; 1982.

[10.4]   Gonzales R.C., Wintz P. *Digital Image Processing.* New York: Addison; 1982.

[10.5]   Hall E.L. *Computer Image Processing and Recognition.* New York: Academic; 1979: 173–181.

[10.6]   Jain A.K. *Fundamentals of Digital Image Processing.* Englewood Cliffs: Prentice-Hall; 1989.

[10.7]   Watanabe S. *Pattern Recognition: Human and Mechanical.* New York: Wiley; 1985.

[10.8]   Bracewell R.N. *The Fourier Transform and its Application.* New York: McGraw-Hill; 1978.

[10.9]   Bartoo G.T., Kim Y., Haralick R.M., Nochlin D., Sumi S.M. Mathematical morphology techniques for image processing applications in biomedical imaging. *Proc SPIE* 1988; 914: 466–475.

[10.10] Barba J., Yuan L., Gel J. Edge detection in cytology using morphological filters. *Proc SPIE* 1989; 1075: 311–318.

[10.11] Herman S. Feature-size dependent selective edge enhancement of X-ray images. *Proc. SPIE* 1988; 914: 654–659.

[10.12] Hall D.H., Lodwick G.S., Kruger R.P., Dwyer S.J., Townes J.R. Directed computer diagnosis of rheumatic heart disease. *Radiology* 1971; 101: 497–509.

[10.13] Kruger R.P., Townes J.R., Hall D.L., Dwyer S.J., Lodwick G.S. Automated radiographic diagnosis via feature extraction and classification of cardiac size and shape descriptors. *IEEE Trans. Biomed. Eng.* 1972; BME-19: 174–186.

[10.14] Nakamori N., Doi K., Sabeti V., MacMahon H. Image feature analysis and computer-aided diagnosis in digital radiography: Automated analysis of sizes of heart and lung in digital chest images. *Med. Phys.* 1990; 17: 342–350.

[10.15] Katsuragawa S., Doi K., MacMahon H. Image feature analysis and computer-aided diagnosis in digital radiography. Detection and characterization of interstitial disease in digital chest radiographs. *Med. Phys.* 1988; 15: 311–319.

[10.16] Toriwaki J., Suenaga Y., Negoro T., Fukumura T. Pattern recognition of chest X-ray images. *Comput. Graph. Image Proc.* 1973; 2: 252–271.

[10.17] Hashimoto M., Sankar P.V., Sklansky J. Detecting the edges of lung tumors by classification techniques. *Proc. IEEE Int. Conf. Patt. Recogn.* 1982; CH-1801: 276–279.

[10.18] Giger M.L., Doi K., MacMahon H. Image feature analysis and computer-aided diagnosis in digital radiography. 3. Automated detection of nodules in preipheral lung fields. *Med. Phys.* 1988; 9: 635–637.

[10.19] Ballard D.H. Generalizing the Hough transform to detect arbitrary shapes. *Patt. Recogn.* 1981; 13(2): 111–122.

[10.20] Ballard D.H., Sklansky J. A ladder-structured decision tree for recognizing tumors in chest radiographs. *IEEE Trans. Comput.* 1976; C-25: 503–513.

[10.21] Lampeter W.A., Wandtke J.C. Computerized search of chest radiographs for nodules. *Invest. Radiol.* 1986; 21: 384–390.

[10.22] Sanada S., Doi K., MacMahon H. Image feature analysis and computer-aided decision in digital radiography: Automated delineation of posterior ribs in chest images. *Med. Phys.* 1991 (in press).

[10.23] Pratt W.K., Faugeras O.D., Gagalowicz. Applications of stochastic texture field models to image processing. *Proc. IEEE* 1981; 69(5).

[10.24] Fu K.S. *Sequential Methods in Pattern Recognition and Machine Learning.* New York: Academic; 1968.

[10.25] Tou J.T., Gonzalez R.C. *Pattern Recognition Principles.* Reading: Addison-Wesly; 1974.

[10.26] Fukunaga K. *Introduction to Statistical Pattern Recognition.* New York: Academic; 1972.

[10.27] Katsuragawa S., Doi K., MacMahon H. Image feature analysis and computer-aided diagnosis in digital radiography: Classifications of normal and abnormal lungs with interstitial disease in chest images. *Med. Phys.* 1989; 16: 38–44.

[10.28] Sutton R.N., Hall E.L. Texture measures for automatic classification of pulmonary disease. *Ieee Trans. Comput.* 1972; C-21: 667–676.

[10.29] Kruger R.P., Thompson W.B., Turner A.F. Computer diagnosis of pneumoconiosis. *IEEE Trans. Syst. Man and Cybern.* 1974; SMC-4: 40–49.

[10.30] Turner A.F., Kruger R.P., Thompson W.B. Automated computer screening of radiographs for pneumoconiosis. *Invest. Radiol.* 1976; 11: 258–266.

[10.31] Tully R.J., Conners R.W., Harlow C.A., Lodwick G.S. Towards computer analysis of pulmonary infiltration. *Invest Radiol.* 1978; 13: 298–305.

[10.32] Revesz G., Kundel H.L. Feasibility of classifying disseminated pulmonary diseases based on their Fourier spectra. *Invest. Radiol.* 1973; 8: 345–349.

[10.33] Kimme C., O'Loughlin B.J., Sklansky K. Automatic detection of suspicious abnormalities in breast radiography. In: Klinger A., Fu K.S., Kunii T.L., eds. *Data Structures, Computer Graphics and Pattern Recognition.* New York: Academic; 1975: 427–447.

[10.34] Hand W., Semmlow J.L., Ackerman L.V., Alcorn F.S. Computer screening of zeromammograms: A technique for defining suspicious areas of the breast. *Comput. Biomed. Res.* 1979; 12: 445–460.

[10.35] Semmlow J.L., Shadaopappan A., Ackerman L.V., Hand W., Alcorn F.S. A fully automated system for screening xeromammograms. *Comput. Biomed. Res.* 1980; 13: 350–362.

[10.36] Magnin L.E., Cluzeau F., Odet C.L., Bremond A. Mammographic texture analysis: An evaluation of risk for developing cancer. *Opt. Eng.* 1986; 25: 780–784.

[10.37] Caldwell C.B., Stapleton S.J., Holdsworth D.W., Jong R.A., Weiser W.J., Cooke G., Yaffe M.J. Characterization of mammographic parenchymal pattern by fractal dimensions. *Phys. Med. Biol,* 1990; 35: 235–247.

[10.38] Naidich D.P., Zerhouni E.A., Siegelman S.S. *Computed Tomography of the Thorax.* New York: Raven; 1984.

[10.39] Shapiro S., Venet W., Strax P.H., Venet L., Roeser R. Ten to fourteen year effect of screening on breast cancer mortality. *J. Nat. Cancer Inst.* 1982; 69: 349–355.

[10.40] Verbeek A.L.M., Hendricks J.H., Holland R., Mravunac M., Sturmans F., Day N.E. Reduction of breast cancer mortality through mass screening with modern mammography. *Lancet* 1984; 1: 1222–1224.

[10.41] Collette H.J.A., Day N.E., Rombach J.J., deWaard F. Evaluation of screening by means of a case-control study. *Lancet* 1984; 1: 1224–1226.

[10.42] Tabar L., Gad A., Holmberg L.H., Lungquist U., Fagerberg C.J.G., Baldetorp L., Gontoft O., Lundstrom B., Manson J.C., Eklund G., Day N.E., Petterson F. Reduction in mortality from breast cancer after mass screening with mammography. Randomized trial from the breast screen working groups of the Swedish National Board of Health and Welfare. *Lancet* 1985; 1: 829–832.

[10.43] Martin J.E., Moskowitz M., Milbrath J.R. Breast Cancer missed by mammography. *Am .J. Roentg.* 1979; 132: 737–739.

[10.44] Hillman B.J., Fajardo L.L., Hunter T.B., Mockbee B., Cook C.E., Hagaman R.M., Bjelland J.C., Frey C.S., Harris C.J. Mammogram interpretation by physician assistants. *Am. J. Roentg.* 1987; 149: 907–911.

[10.45] Kalisher L. Factors influencing false negatives rates in xeromammography. *Radiology* 1979; 133: 297–301.

[10.46] Bassett L.W., Bunnell D.H., Jahanshahi R., Gold R.H., Arndt R.D., Linsman J. Breast cancer detection: One versus two views. *Radiology* 1987; 165: 95–97.

[10.47] Moskowitz M. Benefit and risk. In: Bassett L.W., Gold R.H., eds. *Breast Cancer Detection* (2nd ed.). New York: Grune and Stratton; 1987: 131–142.

[10.48] Baines C.J., Miller A.B., Wall C., McFarbane D.V., Simor I.S., Jong R., Shapiro B.J., Audet L., Petitclerc M., Onimet-Olivia D., Ladonceur J., Herbert G., Mihuk T., Hardy G., Standing H.K. Sensitivity and specificity of first screen mammography in the Canadian national breast screening study: A preliminary report from five centers. *Radiology* 1986; 160: 295–298.

[10.49] Haug P.J., Tocino I.M., Clayton P.D., Bair T.L. Automated management of screening and diagnostic mammography. *Radiology* 1987; 164: 295–298.

[10.50] Giger M.L., Yin F.F., Doi K., Metz C.E., Schmidt R.A., Vyborny C.J. Investigation of methods for the computerized edtection and analysis of mammographic masses. *Proc. SPIE* 1990; 1233: 183–184.

[10.51] Giger M.L., Doi K., Yion F.F., Yoshimura H., MacMahon H., Vyborny C.J., Schmidt R.A., Metz C.E., Montner S.M. Computer-vision schemes for lung and breast concer detection. In: Bregan D., Carr K., Gale A.G., eds. *Proceedings of the Second International Conference on Visual Search*. London: Taylor and Francis; 1991.

[10.52] Chan H.P., Doi K., Galhotra S., Vyborny C.J. MacMahon H., Jokich P.M. Image feature analysis and computer-aided diagnosis in digital radiography. 1. Automated detection of microcalcifications in mammography. *Med Phys.* 1987; 14: 538–548.

[10.53] Chan H.P., Doi K., Vyborny C.J., Lam K.L., Schmidt R.A. Computer-aided detection of microcalcifications in mammograms: Methodology and preliminary clinical study. *Invest. Radiol.* 1988; 23: 664–671.

[10.54] Chan H.P., Doi K., Vyborny C.J., Schmidt R.A., Metz C.E., Lam K.L., Ogura T., Wu Y., MacMahon H. Improvement of radiologists' detection of clustered microcalcifications on mammograms: The potential of computer-aided diagnosis. *Inves. Radiol.* 1990; 25: 1102–1110.

[10.55] Nishikawa R., Giger M.L., Doi K., Schmidt R.A., Vyborny C.J. Automated detection of microcalcifications on mammograms: New feature extraction techniques with morphological filters. *Radiology* 1990; 288: 177.

[10.56] Metz C.E. ROC methodology in radiologic imaging. *Invest. Radiol.* 1986; 21: 720–733.

[10.57] Silverberg E., Boring C.C, Squires T.S. Cancer statistics 1990. Ca-A Cancer J. Clin. 1990; 40: 9–27.

[10.58] Forrest J.V., Friedman P.J. Radiologic errors in patients with lung cancer. *West J. Med.* 1981; 134: 484–490.

[10.59]   Johnson M.L. Observer error, its bearing on perception. *Lancet* 1955; 2: 422–424.

[10.60]   Vernon D.M. *The Psychology of Perception.* Middlesex: Penguin; 1962.

[10.61]   Tuddenham W.J. Visual Search, image organization, and reader error in Roentgen diagnosis. *Radiology* 1962: 78; 694–703.

[10.62]   Smith M.J. *Error and Variation in Diagnostic Radiology.* Springfield: Thomas; 1967.

[10.63]   Abercrombie M.L. *The Anatomy of Judgement.* London: Hutchinson; 1960.

[10.64]   Giger M.L., Doi K., MacMahon H., Metz C.E., Yin F.F. Pulmonary nodules: Computer-aided detection in digital chest images. *Radiographics* 1990; 10: 41–51.

[10.65]   Giger M.L., Ahn N., Doi K., MacMahon H., Metz C.E. Computerized detection of pulmonary nodules in digital chest images. Use of morphological filters in reducing false positive detections. *Med. Phys.* 1990; 37: 1300–1307.

[10.66]   Yoshimura H., Giger M.L., Doi K., Ahn N., MacMahon H. Use of morphologic filters in the computerized detection of lung nodules in digital chest images. *Radiology* 1989; 347: 73.

[10.67]   Katsuragawa S., Doi K., Nakamori N., MacMahon H. Image feature analysis and computer-aided diagnosis in digital radiography: Effect of digital parameters on the accuracy of computerized analysis of interstitial disease in digital chest radiographs. *Med. Phys.* 1990; 17: 72–78.

[10.68]   Katsuragawa S., Doi K., MacMahon H., Nakamori N., Sasaki Y., Fennessy J.J. Quantitative analysis of lung texture in the ILO pneumoconioses standard radiographs. *Radiographics* 1990; 10: 257–269.

[10.69]   Doi K., Holje G., Loo L.N., Chan H.P., Sandik J.M., Jennings R.J., Wagner R.F. MTFs and Weiner spectra of radiographic screen-film systems. In: *HSS Publication 82-8187.* Rockville: FDA; 1982.

[10.70]   Chan H.P., Metz C.E., Doi K. Digital image processing: Optimal spatial filter for maximization of the perceived SNR based on a statistical decision theory model for the human observer. *Proc SPIE* 1985; 535: 2–11.

[10.71]   Powell G.F., Doi K., Katsuragawa S. Localization of inter-rib spaces for lung texture analysis and computer-aided diagnosis in digital chest images. *Med. Phys.* 1988; 15: 581–587.

[10.72] Kundel H.L., Nodine C.F., Krupinki M.A. Computer-displayed eye position as a visual aid to pulmonary nodule interpretation. *Invest. Radiol.* 1990; 25: 890–889.

[10.73] Shortliffe E.H., Buchanan B.G., Feigenbaum E.A. Knowledge engineering for medical decision making: A review of computer-based clinical decision aids. In: Clancey W.J., Shortliffe E.H., eds. *Readings in Artificial Intelligence—The First Decade.* Reading: Addison-Wesley; 1984: 34–71.

[10.74] Sherman H., Reiffen B., Komaroff A.L. Ambulatory care systems. In: Driggs M.F., ed. *Problem Directed and Medical Information Systems.* New York: International Medical; 1973: 143–171.

[10.75] Kuhns L.R., Thornbury J.R., Fryback D. *Decision Making in Imaging.* Chicago: Year Book Medical; 1989.

[10.76] Staub W.H. *Manual of Diagnostic Imaging.* Boston: Little Brown; 1984.

[10.77] McNeil B.J., Abrams H.L., eds. *Brigham and Women's Hospital Handbook of Diagnostic Imaging.* Boston: Little Brown; 1986.

[10.78] Kahn C.E., MesserSmith R.N., Jokich M.D. PHOENIX: An expert system for selecting diagnostic imaging procedures. *Invest. Radiol.* 1987; 22: 978–980.

[10.79] Swett H.A., Rothschild M.A., Weltin G.G., Fisher P.R., Miller P.L., Optimizing radiologic workup: An artificial intelligence approach. *J. Dig. Imag.* 1989; 2: 15–20.

[10.80] Greenes R.A. Computer-aided diagnostic strategy selection. *Radiol. Clin. N. Am.* 1986; 24: 105–120.

[10.81] Menn S.J., Barnett G.O., Schmechel D., Owens W.D., Pontoppidan H. A computer program to assist in the care of acute respiratory failure. *J. Am. Med. Assoc.* 1973; 223: 308–312.

[10.82] Getty D.J., Prickett R.M., D'Orsi C.J., Swets J.A. Enhanced interpretation of diagnostic images. *Invest Radiol.* 1988; 23: 240–252.

[10.83] Patil R.S. Review of casual reasoning in medical diagnosis. In: *Proceedings of the Tenth Annual Symposium on Computer Application in Medical Care.* Washington: IEEE; 1986: 11–16.

[10.84] Armitage P., Gehan E.A. Statistical methods for the identification and use of prognostic factors. *Int. J. Cancer* 1974; 13: 16–36.

[10.85] Warner H.R., Toronto A.F., Veasy L.G. Experience with Bayes' theorem for computer diagnosis of congenital heart disease. *Ann. N. Y. Acad. Sci.* 1964; 115: 2.

[10.86]   duDombal F.T., Leaper D.J., Staniland J.R., Horrocks J.C. Computer-aided diagnosis of acute abdominal pain. *Br. Med. J.* 1964; 2: 9–13.

[10.87]   Lodwick G.S., Cosmo L.H., Smith W.E., Keller R.F., Robertson E.D. Computer diagnosis of primary bone tumors: A preliminary report. *Radiology* 1963; 80: 273–275.

[10.88]   Pauker S., Kassirer J. Decision analysis. *N. Engl. J. Med.* 1987; 316: 250–258.

[10.89]   Makhoul J. Artificial neural networks. *Invest. Radiol.* 1990; 25: 748–750.

[10.90]   Boone J.M., Groso G.W., Creco-Hunt V. Neural network in radiologic diagnosis. I. Introduction and illustration. *Invest. Radiol.* 1990; 25: 1012–1019.

[10.91]   Rumelhart D.E., Hinton G.E., Williams R.J. Learning internal representation by error propagation. In *Parallel Distributed Processing* (Vol. 1). Cambridge: MIT Press; 1968: 318–362.

[10.92]   Asada N., Doi K., MacMahon H., Montner S., Giger M.L., Abe C., Wu Y. Neural network approach for different diagnosis of interstitial lung diseases. *Proc SPIE* 1990; 1233: 45–50.

[10.93]   Asada N., Doi K., MacMahon H., Montner S.M., Giger M.L., Abe C., Wu Y. Potential usefulness of artificial neural network for differential diagnosis of interstitial lung disease: A pilot study. *Radiology* 1990; 177: 857–860.

[10.94]   Gross G.W., Boone J.M., Greco-Hunt V., Greenberg B. Neural networks in radiologic diagnosis. II. Interpretation of neonatal chest radiographic. *Invest. Radiol.* 1990; 25: 1017–1023.

[10.95]   Schank R.C. The cognitive computer. In: *On Language, Learning, and Artificial Intelligence.* Reading: Addison-Wesley; 1984: 33.

[10.96]   Shortliffe E.H., Buchanan B.G., Feigenbaum E.A. Knowledge engineering for medical decision making: A review of computer-based clinical decision aids. *Proc. IEEE* 1979; 67: 1207–1224.

[10.97]   Kulikowski C.A. Artificial intelligence methods and systems for medical consultation. *IEEE Trans. Patt. Anal. Mach. Itell.* 1980; PAMI-2: 464–476.

[10.98]   Szolovits P., Ed. *Artificial Intelligence in Medicine.* Boulder: Westview; 1982.

[10.99]   Clancey W.J., Shortliffe E.H., eds. *Readings in Medical Artificial Intelligence—The First Decade.* Reading: Addison-Wesley; 1984.

[10.100] Weiss S., Kulikowski C. EXPERT: A system for developing consultation models. In: *Proceedings of the Sixth International Joint Conference on Artificial Intelligence.* Stanford: Stanford University, Department of Computer Science; 1979: 942–947.

[10.101] Pauker S.G., Gorry G.A., Kassirer J.P., Schwartz W.B. Towards the simulation of clinical cognition: Taking a present illness by computer. *Am. J. Med.* 1976; 60: 981–996.

[10.102] Patil R.S., Szolovits P., Schwartz W.B. Casual understanding of patient illness in medical diagnosis. In: *Proceedings of the Seventh International Joint Conference on Artificial Intelligence.* Vancouver; 1981: 893–899.

[10.103] Shortliffe E.M.: *MYCIN: Computer-Based Medical Consultations.* New York: Elsevier; 1976.

[10.104] Miller R.A., Pople H.E., Myers J.D. INTERNIST-1, an experimental computer-based diagnostic consultant for general internal medicine. *N. Engl. J. Med.* 1982; 307: 368–476.

[10.105] Kulikowski C. Artificial intelligence methods and systems for medical consultation. In: Clancey W.J., Shortliffe E.H., eds. *Readings in Medical Artificial Intelligence—The First Decade.* Reading: Addison-Wesley; 1984: 86–87.

[10.106] Piraino D.W., Richmond B.J., Uetani M., Luetkehaus T., Rockey D., Belhobek G., Armistead J., Jones F. Problems in applying expert system technology to radiographic image interpretation. *J. Dig. Imag.* 1989; 2: 21–26.

[10.107] Bramble J.M. Computer-aided diagnosis of arthritis. In: *Proceedings of the Ninth Conference on Computer Applications in Radiology.* Hilton Head; 1988: 229–235.

[10.108] Miller P.L. *Expert Critiquing Systems.* New York: Springer; 1986.

[10.109] Miller P.L., Black H.R. Medical plan analysis by computer: Critiquing the parmacologic management for essential hypertension. *Comp. Biomed. Res.* 1984; 17: 38–54.

[10.110] Miller P.L., Blumenfrucht S.J., Black H.R. An expert system which critiques patient workup: modeling conflict expertise. *Comp. Biomed. Res.* 1984; 17: 554–569.

[10.111] Swett H.A., Miller P.A. ICON: A computer-based approach to differential diagnosis in radiology. *Radiology* 1987; 1963: 555–558.

[10.112] Swett H.A., Fisher P.R., Cohn A.I., Miller P.I., Mutalik P.G. Expert system controlled image display. *Radiology* 1989; 172: 487–493.

[10.113] Mulatik P.G., Fisher P.R., Swett H.A. Radiologic knowledge representation and axis display in the IMAGE/ICON system. In: Arenson R.L., Friedenberg R.M., eds. *Computer Applications to Assist Radiology.* Carlsbad: Symposia Foundation; 1990: 185–190.

[10.114] Swett H.A. Decision support in diagnostic radiology. In: Arenson R.L., Freidenberg R.M., eds. *Computer Applications to Assist Radiology.* Carlsbad: Symposia Foundation; 1990: 654–657.

[10.115] Swett H.A., Fisher P., Mutalik P., Miller P.L., Wright L. The IMAGE/ICON system; voice activated intelligent image display for radiologic diagnosis. In: *Proceedings of the Thirteenth Annual Symposium on Computer Applications in Medical Care.* Washington: IEEE Computer Society; 1989: 977–978.

[10.116] Wright K. The road to the global village. *Sci. Am.* 1990; 84–94.

11

Architecture and Ergonomics of Imaging Workstations

Paul S. Cho and H.K. Huang

11.1 Architecture of Imaging Workstation

Display workstations transform the binary representation of an image to visual signals and thus serve as the link between invisible pixels and the eye. The bit pattern stored in computer memory is converted to video signals, which in turn stimulate the light-emitting phosphors on the display screen. Therefore, what the eyes see is ultimately dictated by the display system's ability to faithfully reproduce the original image. Unless it becomes possible to transfer visual information directly from computer memory to the visual cortex, this mode of signal transfer which relies on light photons will remain vital to image presentation.

An imaging workstation consists of three major components: image processing hardware, display monitor, and storage device. The image processing hardware generates and transforms pixel data for their visualization on the display monitor. Different types of storage devices are used to meet the high-capacity, high-performance requirements of imaging applications. These include image memory, magnetic disk drives, and optical disks.

11.1.1 Image Processing Hardware

a. Special-Purpose Image Processing Hardware

For high resolution, high speed image processing and display applications, special-purpose image processing hardware is required. The image processors may be board-level units that plug directly into the computer's general-purpose bus, or they may be chassis-level products that communicate with the computer via a bus adapter. A typical image processing system consists of an image memory (or frame buffer), a pixel processor, and a video output processor (Fig. 11.1). These components usually share a common image transfer bus to realize high speed transfer of data.

To store images, either a part of the CPU memory or a separate image memory can be used. Ideally, the image memory should be addressable in linear, two-dimensional, or three-dimensional modes to eliminate address-calculation overhead otherwise necessary for pixel and voxel data access. The pixel processor performs arithmetic operations on the data copied from the image memory. These operations include point functions such as image addition, subtraction, and

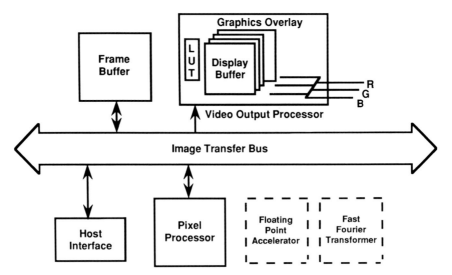

FIGURE 11.1. A simplified hardware configuration of a typical image processing system controlled by a host computer.

merge; geometric functions such as magnification; statistical functions such as histograms; and transformation functions such as lookup table operations. Often, optional hardware components are available to speed up the image processing computation. These components may include a floating-point accelerator, fast Fourier transform coprocessor, and spatial convolver. The video output processor normally contains three channels of image output to provide either one 24-bit full-color image or three 8-bit gray-scale images. In addition to providing the image channels, a well-designed system also supplies an alpha channel for graphics overlay.

b.   General-Purpose Hardware Platform

The idea of adding special-purpose imaging hardware to an existing computer is one approach to an imaging workstation, but it has some serious limitations. For instance, the lack of a universal device driver requires the user to learn a new library of subroutines for each new image processing hardware configured. Also, the programmer must become familiar with the way the host-computer's operating system interacts with these special-purpose devices. This use of multiple software can significantly delay the development of application software. In addition, the application software is likely to have a short life span because of its device-dependent nature.

With recent advances in CPU speed, memory size, and the ability of built-in graphics devices to handle large amounts of pixel data, it is possible to design an upper-performance imaging workstation around a standard hardware platform. Application programs are written using the standard software tools such as the X-window library.[11.1–11.2] The X-window has gained popularity because of its portability from one computer hardware to another. It is designed as a network communications protocol and therefore permits network-transparent image transfer and display. This is a powerful concept which has helped usher in the present Internet revolution.

11.1.2 Display Monitor

The video output processor generates picture signals which are used to drive a display device. Today, the display monitor using the cathode ray tube (CRT) is still the predominant display device because its overall image quality is superior to other types of display devices such as liquid crystal screens, light valves, and flat-panel displays. Since the image quality of the visualized data can affect the interpretation of the data, CRT display monitors play a crucial role in the formation of visual information. Characteristics of the video display monitor that affect visual quality include scanning technique, resolution, phosphor characteristics, luminance-contrast, color display, and geometric distortion.

a. Scanning Technique

Early vector graphic CRT systems used a random scanning technique in which the electron beam is moved randomly to draw lines. In this method, only the endpoint coordinates of the line segments had to be stored. In the days when computer memory was expensive and slow, this method proved expedient although it was unable to display complex objects.

The emergence of image processing technology in which the data are represented as a rectangular array of pixels has required a raster-scan CRT display. In the raster scanning method the electron beam scans the CRT screen in a fixed pattern of closely spaced parallel lines. These lines are usually horizontal and scanned in succession from the top to the bottom of the screen. The beam current is varied either as "on-off" or continuously to produce a black-white or gray-scale picture, respectively. There are two modes of scanning: interlaced and progressive. In interlace scanning, a complete frame is displayed in two steps. For example, in the American standard the even lines are painted (1/60), then the odd lines fill the gaps (1/60) to comprise a complete picture frame in 1/30. The most common interlaced scanning standards are the American standard of 525 lines at 30 interlaced frames per second and the European standard of 625 lines at 25 interlaced frames per second. It is generally assumed that a 25 to 30 Hz interlaced refresh rate is sufficient to produce a picture without causing the sensation of brightness variation or flicker. Although this assumption may be valid for commercial TV standards, higher refresh rates are often required for imaging workstation applications. In the progressive scan, the entire image is painted continuously from left to right and top to bottom. For this type of scan the trend is to use a monitor that is operating between 60 and 72 frames per second (Hz). It is commonly accepted that in order to eliminate flicker, a 72 Hz progressive scan should be used in medical imaging applications.

b.   Phosphor

A visible light image from the monitor screen is produced by the phenomenon of *cathodo-luminescence* in which high energy electrons excite phosphors deposited on the screen plate. As the excited electrons in the phosphor return to the ground state, light is emitted whose intensity is proportional to the electron beam current. Over 50 different types of phosphor are available as listed in the P number register regulated by the U.S. Electronic Industries Association. Crystalline materials most commonly used as phosphors are based on zinc sulphide, although rare earth activated oxides and oxisulphides have been increasingly used. The important factors to consider in selecting phosphors are energy conversion efficiency, decay time, color, and longevity. A large number of phosphors can convert 10% to 20% of electron beam energy into light. These efficiencies are remarkably high, an order of a magnitude above those used in other types of display devices. This is one reason for the continued predominance of CRT technology.

Phosphor decay times may vary from less than a microsecond to a few seconds. Phosphors with longer persistence can be used to reduce flicker by allowing significant integration of temporal variations in luminance. This works well for stationary images but causes smearing in dynamic images. Phosphors of almost any emission color may be selected from the visible light spectrum. The conversion efficiency degrades with the amount of excitation. Most phosphors have a practical lifetime of several thousand hours. To prolong phosphor longevity, the screen-saver feature is often enabled in workstations, in which the electron beam is automatically shut off after a certain period of inactivity.

The selection criteria for the phosphor depend on the imaging applications and involve the following considerations (see Table 11.1).

1. Phosphor steady-state color (color of phosphor during excitation should appear with a slight blue tint to resemble the hue of the radiograph). The C.I.E.[Commission Internationale de l'Eclairage (International Commission on Illumination)] chromaticity diagram can be used to find the relative percentage of the primary colors presented in a given color.

2. Phosphors with fluorescence decay times that are shorter than the field refresh rate of an interlacing scan (16.7 ms), or the frame refresh rate of a progressive scan (33.3 ms), will contribute an annoying flicker effect to the static image. To minimize this flicker fluorescence decay times chosen for the phosphor should be longer than the scanning refresh rate.

3. Light output efficiency provides a scale of efficiency for a given input beam current. A phosphor with a low efficiency (relative to P4) needs a larger beam current to produce the same light intensity. As a result, it reduces the life expectancy of the monitor.

TABLE 11.1. Specifications of three popular phosphors used in display monitors.

Phosphor	Primary (Red	color Green	percent Blue)	Decay time (milliseconds)	Efficiency (relative to P4)
P4	27	30	43	.100	1.00
P40	26	33	41	>100.000	0.76
P164	27	29	44	>100.000	0.60

c.   Resolution

The display screen contains a layer of phosphor grains $10\mu m$ or less in size. The pixel size is limited by the spot size of the electron beam. The spot has a Gaussian-like current distribution over a cross-sectional area. It is usually defined as the width between points where the beam current has dropped to a certain fraction such as $1/e$ (about 37%) of its maximum value. The beam spot diameter increases as the square root of the beam current. The minimum diameter to which the spot size can be reduced is dictated by the beam current required to produce an acceptable phosphor brightness. Therefore, there is a tradeoff between resolution and image brightness.[11.3]

The resolution of a display monitor is most commonly specified in terms of the number of scan lines. For example, the term "1k monitor" means that the monitor has 1024 raster lines. In the strict sense of the definition, however, it is not sufficient to specify spatial resolution simply in terms of raster lines because the actual resolving power of the monitor may be less. Consider a digitally generated line pair pattern (black and white lines in pairs). The maximum displayable number of these line pairs on a 1k monitor is 512. However, the monitor may not be able to resolve 1024 alternating black and white lines in the vertical direction if the electron beam spot is out of focus or the contrast/brightness is set too high, causing the adjacent raster lines to overlap. Horizontal resolution has no relation to the number of raster lines; rather, it is limited by the beam spot size and dependent on how quickly the beam current changes according to the driving video signals.

There are several techniques for measuring spatial resolution. The simplest and most commonly used method used a test pattern which consists of varying widths of line pair objects in both vertical, horizontal, and sometimes radial directions. It should be noted that this visual approach measures the resolution of the total display-perception system, including the visual acuity of the observer, and therefore is prone to subjective variations.

Other techniques include the shrinking raster test, the scanning spot photo-meter, the slit analyzer, and the measurement of the modulation transfer function (MTF).[11.4] Another noteworthy fact is that resolution is a function of the location of the electron beam on the screen. In general, the defocusing effect increases as the beam moves away from the center of the screen. Therefore, resolution specifications must be accompanied by data concerning the location on the screen where the measurement was taken as well as the luminance of the screen.

d. Luminance and Contrast

Luminance measures the brightness in candelas per square meter (cd/m^2) or in foot-lamberts (ft L; 1ft L $= 3.426cd/m^2$). The concept of luminance applies when the luminous source cannot be treated as a point source, i.e., it has a finite area. Luminance of the display monitor is a function of the electron beam current and the conversion efficiency of the phosphor.

There are various definitions for contrast and contrast ratio. Contrast is most often defined as a ratio: the difference between two luminances divided by one of the two luminances, usually the larger. Thus

$$C = (L_s - L_0)/L_s$$

or

$$C = (L_0 - L_s)/L_0,$$

where L_0 is the luminance of the object and L_s is the luminance of the background. The contrast ratio C_r is frequently defined as the ratio of the luminance of an object to that of the background. This is expressed by

$$C_r = L_{max}/L_{min},$$

where L_{max} is the luminance emitted by the area of greatest intensity and L_{min} is the luminance emitted by the area of least intensity.

Because of the particulate nature of the phosphor and its optical transparency, the screen acts as a high efficiency reflector such that 25% to 75% of the incident light is scattered back. The light emitted from the phosphor is viewed against this reflected light. The contrast ratio, therefore, depends on both the luminance of the CRT image and the intensity of the ambient light. For instance, in bright sunlight the display surface can have an apparent luminance of $3 \times 10^4 cd/m^2$. To achieve a contrast of 10, the luminance of the cathodo-luminescence must be $3 \times 10^5 cd/m^2$, which is extremely high even for a high efficiency phosphor.

Since perceived contrast depends on the human observer, the psychophysical parameter called just-noticeable-difference (JND) is sometimes used to characterize a CRT. JND is the smallest luminance difference that can be perceived by the human observer in the presence of noise caused by temporal fluctuations in luminance output and nonuniformity in the CRT phospor layer. Other sources of noise are found in the observer's visual system, such as the random firings of neurons.[11.5]

e. Color CRT

The oldest and still most widely used design of a color CRT is the shadow mask. It consists of three electron guns, a shadow mask with circular holes, and a phosphor

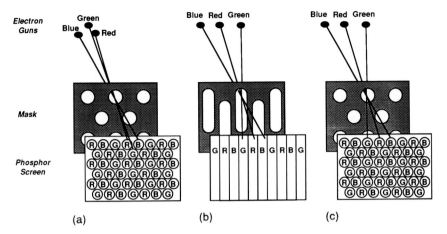

FIGURE 11.2. (a) Delta gun with circular hole mask. (b) In-line gun with slotted mask. (c) In-line gun with circular hole mask.

screen. The phosphor screen has arrays of "triads," where each triad consists of the phosphors of the three primary emission colors, namely, red, green, and blue. The guns are positioned in a triangular or delta form and therefore are referred to as the delta gun. The geometrical relationship of the gun, shadow mask, and screen is such that the beam from each gun strikes only one type of the phosphor in the triad. The other two types of phosphor are masked from the gun by the shadow mask. The beams are made larger than the holes so that the beam completely fills the hole to transfer maximum energy to the phosphor. Three guns can independently modulate the beam current to produce the desired color mixture for each triad [Fig. 11.2(a)].

In recent years the design of the shadow mask and guns has improved to meet the increasing demands of the high brightness, high resolution market in such applications as large-scale PC board design, detailed CAD/CAM, and computer graphics. The current trend is to place guns in line to achieve better tracking of the beam, thus improving the hit rate of the proper color phosphor. The shadow mask may be either slotted [Fig. 11.2(b)] or circular [Fig. 11.2(c)], with color triads laid down as vertical stripes or as circular dots, respectively.

A classic problem in shadow mask technology is associated with the significant dissipation of electron beam energy by the shadow mask. Apart from the loss of luminance, the absorbed energy can cause local thermal expansion which may result in color distortion by dealigning the geometry between the shadow mask and the phosphor screen. This has led to the development of flat pretensioned shadow masks in conjunction with flat screens as described by Dietch, et al.[11.6]

Advancements in spot shaping and focusing techniques as well as video amplifier technology have enabled the recent development of a large screen (20-inch square), high resolution (2048 × 2048) color CRT which operates at 60 Hz noninterlaced refresh rate. The spot size is 0.5 mm at the center and 0.6 mm at the corner.[11.7]

f.   Geometric Distortion

Geometric distortions of various types may be present in a CRT. These include pincushion, barrel, nonlinearity, hook and flagging, line-pairing, and ringing.[11.8] Pincushion distortion is common in a large-size screen and is exhibited by non-linear inward edges. Barrel distortion consists of nonlinear outward edges, which can result from overcorrection of pincushion distortion. Nonlinearity, if present, is observable when a large circle is displayed as a stretched or flattened oval. Hook or flagging usually manifests itself as a bending of the upper-left corner of the screen on the edge of the raster scan. Line-pairing is a bunching of horizontal scan lines that show up as bright and dark regions. This type of distortion occurs most often in interlaced monitors. Ringing is seen as a series of dark and light shaded bands at the left side of the screen that disappear a small distance away from the edge. Geometric distortion on a monitor may occur from time to time. Calibration during preventive maintenance with a standard phantom such as an SMPTE pattern[11.9] should be performed to detect distortion.

11.1.3   Image Storage Devices

Stringent requirements are placed on the storage device in imaging applications. The sheer volume of data calls for a large capacity in excess of the gigabyte range, and increasing demands for image processing and graphics at interactive speeds, require a very high throughput capacity. There are two categories of image storage. One type is random access memory (RAM), which is directly connected to the image processor or the computer system bus. This type of image storage has a very fast input/output rate but is volatile and expensive. The second type of storage is slower, permanent, and less expensive. Today, high capacity magnetic disk drives dominate the primary storage arena while optical disks are widely used for secondary storage.

a.   Random Access Memory

Random access memory is used as a buffer in an imaging workstation. Typically, workstations are configured with 32 to 256 megabytes of RAM. One 2k × 2k × 8-bit image takes up four megabytes of memory. The RAM is used as a buffer in the sense that a set of images is first loaded from disk storage into the RAM. From there, one image at a time is transferred to the video random access memory (VRAM). The VRAM has higher input/output rates than the RAM and is connected to the display monitor through a fast digital-to-analog (D/A) converter. This architecture allows an image to be displayed rapidly on the video monitor.

b.   Magnetic Disk Drives

The storage capacity of magnetic hard disk drives has increased dramatically in recent years. For example, the capacity of the 5.25-inch drive, which was first introduced in 1980 (Seagate ST 506), increased from its initial five megabytes to 800 megabytes in 1990 and to 9,000 megabytes (or 9 gigabytes) in 1995.

Currently, the speed of data transfer is dictated by the disk-to-computer interface. The disk drive reads out bit information faster than the interface can accept it. In applications where display speed becomes critical, a parallel transfer disk (PTD) or disk array is used to overcome this limitation. The PTD allows multiple read/write heads to transfer data simultaneously. The disk array, on the other hand, configures multiple conventional disks in parallel. This scheme is often referred to as RAID (redundant array of independent disks). Two common approaches for the disk array are software striping and hardware paralleling. In software striping, the disks are connected to the system bus through traditional controllers. Blocks of data are segmented and moved to and from the disk drives in parallel. Data segmentation and recombination are handled by the software. In the hardware approach, a parallel drive array controller simultaneously manages multiple disk drives.

The useful life cycle of parallel architecture may be nearing an end as the new generation of fast disk-to-computer interfaces emerges. These are Ultra SCSI (small computer system interface), Fiber Channel, and SSA (serial storage architecture) interfaces, all of which exceed the disk drive performance.[11.10]

c.   Optical Disk Drives

Optical disks provide a convenient means for storing and archiving a large volume of data. Unlike magnetic tape, which requires sequential data access, the optical disk allows random access of recorded data. There are two types of optical disks: WORM (write once, read many) optical disks and erasable MO (magneto-optical) disks. These disks are available in different sizes: 3.5-inch (WORM and MO), 5.25-inch (WORM and MO), 12-inch (WORM), and 14-inch (WORM), ranging in capacity from 180 megabytes to 10 gigabytes. In addition, the capacity of on-line storage can be increased to the terabyte range by configuring optical disk jukeboxes.

11.2   Examples of Imaging Workstation

Today, imaging workstations are used in many different fields, including medicine, engineering, scientific visualization, defense, and entertainment. A wide range of specifications are seen. The spatial resolution requirements range from 1k to 3k and the pixel representation can be monochrome, gray scale, or color. The number of configurable monitors is generally limited by the available hardware address space and physical bus slots for video output boards. The examples of imaging workstations presented here are from radiological applications.

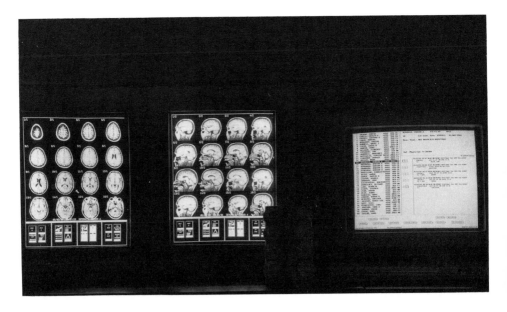

FIGURE 11.3. A dual-monitor 2k imaging workstation showing two sets of MR neuro images. The separate text monitor shows patient directory and study list. Image processing functions are controlled by the icons located at the bottom of the 2k screens and the external dials.

11.2.1   Diagnostic Workstation

The first workstation example is the diagnostic workstation used by radiologists for making primary diagnoses.[11.11–11.12] Components in this type of workstation should be of highest quality. If the workstation is used for displaying projection radiographs, then a number of 2k monitors are needed; if used for CT and MR images, multiple 1k monitors are sufficient. The diagnostic workstation requires a rapid (about 1 to 2 seconds) image display time and a digital dictating phone to report findings. The workstation provides software to append the reports with the images. Figure 11.3 shows a dual-monitor 2k imaging workstation at the University of California, San Francisco (UCSF). This 2k station is based on the SUN SPARCserver 470 computer and two 21-inch diagonal 2k portrait-mode monitors (UHR-4820P MegaScan display system, E-Systems, Littleton, MA). Each 2k station has a parallel transfer disk with 5.2 gigabytes of formatted storage which can display a 2k × 2k × 12-bit image in 1.5 seconds (Storage Concepts, Irvine, CA).

11.2.2   Review Workstation

Review workstations are used by radiologists and referring physicians to review cases in the hospital wards.[11.13–11.14] The transcribed report should be available with the corresponding images. The review workstation does not require 2k mon-

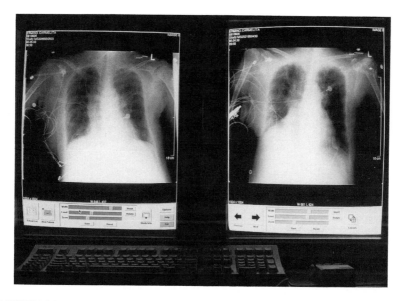

FIGURE 11.4. A dual-monitor 1k (1,600-line) imaging workstation for intensive care units showing two images. The left-hand monitor shows the current image, and the right monitor is used to show previous images by clicking the two lower-left icons (Previous and Next). Image processing functions are controlled by the icons located at the bottom of each screen.

itors because referring physicians are not looking for minute detail. Figure 11.4 shows a dual-monitor 1k (1,600-line) review workstation used at the ICUs (intensive care units) at UCSF. This workstation consists of a SUN SPARC 20 workstation with 2 gigabytes of disk storage, two GXTurbo video display boards and two 1,600 × 1,024 display monitors.

11.2.3   Analysis Workstation

Analysis workstations differ from diagnostic and review workstations in the sense that they are used to extract useful parameters from images. Some parameters are easy to extract from a simple region of interest (ROI) operation. Others such as blood flow measurements from DSA and three-dimensional reconstructions from sequential CT images, are computationally intensive. These operations require an analysis workstation with a more powerful image processor and high performance software. Figure 11.5 shows an analysis workstation, Onyx, developed at Silicon Graphics (Mountain View, CA).

11.2.4   Digitizing and Printing Workstation

Digitizing and printing workstations are designed for radiology department technologists or film librarians to digitize historical films and films from outside the department. The workstation is also used to convert softcopy images to hardcopy.

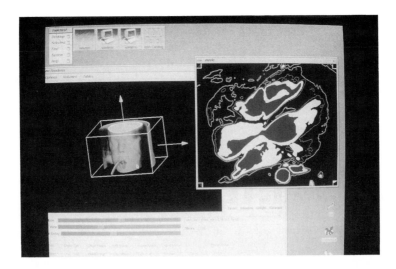

FIGURE 11.5. An analysis workstation (Onyx) at UCSF showing a three-dimensional rendering of a CT examination of a child, on the left, and a simulated blood flow (dark gray) in the four chambers of the heart, on the right. (Courtesy of Silicon Graphics Computer Systems, Mountain View, CA., S. Wong and E. Grant.)

In addition to the standard workstation components, this workstation also requires a laser film scanner, a laser film imager, and a paper printer. The paper printer is used for pictorial report generation from the diagnostic, review, and editorial and research workstations. A 1k display monitor for quality control purposes would be sufficient for this type of workstation.

11.2.5   Interactive Teaching Workstation

The teaching workstation is used for interactive teaching. It emulates the role of a teaching library but has more interactive features. Figure 11.6 shows a digital mammography teaching workstation from VICOM (Fremont, CA).

11.2.6   Editorial and Research Workstation

The editorial and research workstation is used for simple data analysis as well as for generation of slides, teaching materials, and illustrated reports. Macintosh, PC, or UNIX-based workstations with imaging software are commonly used. Figure 11.7 is an example of a Macintosh-based editorial-research workstation. Ideally, the software should include both ready-to-use imaging functions and research development utilities which enable researchers to assemble new research tools in a timely fashion. The latter can be accomplished through the use of object-oriented software modules.[11.15]

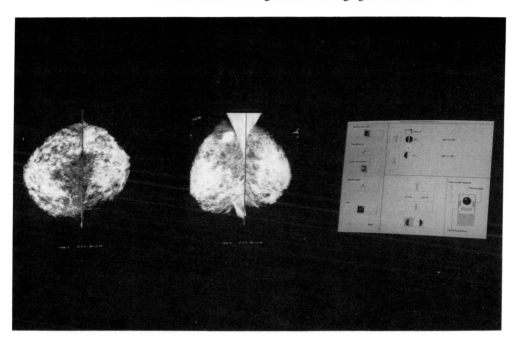

FIGURE 11.6. A 2k digital mammography teaching workstation showing four mammograms. Left: left and right craniocaudal views; Middle: left and right mediolateral oblique views; Right: text monitor with icons for image display and manipulation.

11.3   Ergonomics of Imaging Workstation

The ergonomics analysis of an imaging workstation addresses a variety of human engineering factors, including interactive devices, acoustic noise, lighting conditions, and workstation layout.[11.16] This section focuses on ergonomic issues that affect perceived image quality, including glare, ambient illuminance, user interface, and display monitor configuration.[11.17]

11.3.1   Glare

According to the study conducted by Stammerjohn, et al., glare is the most frequent complaint among workstation users.[11.18] Glare is defined as the sensation produced by luminance within the visual field that is sufficiently greater than the luminance to which the eyes are adapted that can cause annoyance, discomfort, or loss in visual performance and visibility.[11.19]

Glare can be caused by reflections of electric light sources, windows, and light-colored objects such as furniture and clothing. The magnitude of sensation of glare is a function of the size, position, and luminance of a source, the number of sources, and the luminance to which the eyes are adapted at the moment. Glare may be categorized according to its origin as either direct or reflected. Direct

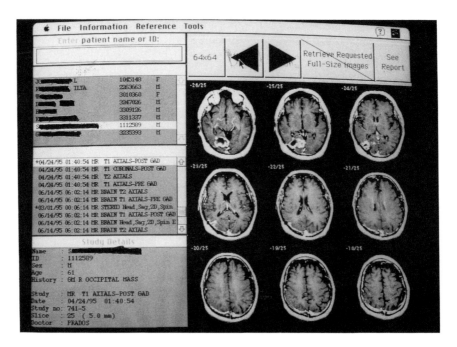

FIGURE 11.7. A Macintosh-based editorial-research workstation. The image window shows a set of MR T_1 images subsampled to 128 × 128 pixels. The text windows contain patient list (top), study list (middle), and demographic information of the currently selected patient (bottom).

glare may be caused by bright sources of light in the visual field of view (e.g., sunlight and light bulbs). Reflected glare is caused by light reflected from the display screen. If the reflections are diffuse, they are referred to as veiling glare.

Image reflections are both distracting and annoying because the eye is induced to focus alternately between the displayed and reflected images. Reflected glare can be reduced by increasing display contrast, wearing dark clothing, correctly positioning the screen with respect to lights, windows, and other reflective objects, and adjusting the screen angle.

Further reduction of glare can be achieved by etching the screen glass, treating the glass with antireflective coatings, or superimposing antireflection filters on the screen, such as micromesh, circular polarizers, and neutral density filters with anti-reflective coatings.[11.20] Etching the glass reduces reflection by scattering reflected light, thus giving a more diffuse and less conspicuous appearance of reflected objects. Since the light emitted from the display phosphors is also dispersed, image sharpness may be compromised. Coating the screen with a thin film of antireflective agent such as Lambda-4 usually does not diminish image sharpness.[11.21] The coating layer, however, can be easily and permanently smeared with fingerprints. Antireflection filters reduce perceived glare by attenuating reflected light more than the light emitted from the CRT, since the reflected light must pass through the filter twice. Morse[11.22] has evaluated some of these antireflective techniques. Objective measurements were made of luminance, modulation transfer function, contrast, and glare, as well as subjective evaluations of the brightness, sharpness, contrast, color, and glare. On average, the order of preference was: gray antireflection filter, micromesh filter, circular polarizer, blue antireflection filter, etched plastic filter, and no filter.[11.22]

11.3.2   Ambient Illuminance

An important issue related to the problem of glare is the proper illumination of the workstation area. Excessive lighting can increase the readability of paper documents but may also increase reflected glare, while dim lighting can reduce glare but make reading at the work surface difficult. Illuminance refers to the light flux incident on a unit area of surface. A common unit of illuminance is the meter-candle or lux.

Ergonomic guidelines for a traditional office environment recommend a high level of lighting of 700 lux or more.[11.23] According to a study conducted by van der Heiden, et al.,[11.24] among 38 CAD operators allowed to adjust the ambient lighting, median illumination levels were around 125 lux, with 90% of the readings falling between 15 and 505 lux. These levels are optimized for CRT viewing but certainly not for reading documents. An illuminance of 200 lux is normally considered inadequate for an office environment. Cushman recommends a reasonable compromise.[11.23] He suggests a lower range of 150 to 400 lux for tasks that do not involve information transfer from paper documents. At these levels, lighting is sufficiently subdued that good display contrast is usually obtainable. A higher range of 400 to 550 lux is suggested for tasks that require reading of paper documents. Increasing ambient lighting above 550 lux reduces display contrast appreciably. If paper documents contain small and low contrast print, 550 lux may not provide adequate lighting. In such cases, special task lighting directed only at the document surface may be recommended.

In radiology, where optimal viewing conditions are most critical, the level of ambient illuminance should be equal to the average luminance of displayed images. In addition, lighting for the viewing room containing display monitors should be lower than that in a conventional film viewing room. The luminance of the X-ray film placed on the lightbox ranges from 3 to 200 cd/m^2, while that of a display monitor varies from 0.1 to 50 cd/m^2.[11.25]

11.3.3   User Interface

The main goal of the user interface is to provide efficient tools with which the user can call up and manipulate images. This is accomplished through software switches that appear on the screen as graphic icons and menus. The mouse is widely used in today's workstations as the point-and-select input device. A well-designed user interface should provide ergonomic efficiency to a degree that the interaction becomes second nature to the user.

One design aspect of the user interface that can seriously affect image viewing is the window format. In applications where only a single image needs to be displayed, this may be straightforward, but the window design becomes complicated for applications where multiple images must be displayed. This situation is prevalent in medical imaging applications such as examination of CT/MRI slices, side-by-side comparison of projection radiographs, and movie loops of dynamic studies. The image windows can be formatted to display images in a tiled format in which all the windows are fully visible simultaneously or as stacked images which require the user to select one window to be placed on top of the stack for full, unobstructed display.[11.26]

The windows may or may not have a frame or scroll bars. If they do, they should not interfere with the task of image viewing. The width of the frame bars should be narrow and dark so that potential sources of glare and distraction are minimized.

11.3.4 Multiple Display

Depending on the size of the image set to be displayed, it may be necessary to configure multiple monitors. Comfortable viewing with a single imaging workstation is provided by tilting the monitors to form a mosaic of concave surfaces focused at the user while maintaining a viewing distance of 50 to 90 cm. For group viewing, a flat arrangement is preferred since this provides fewer restrictions for people located peripherally.[11.25]

To facilitate the visual integration of images displayed on multiple monitors, the idle space between screens should be minimized. It may be necessary to remove each CRT from its original chassis which tends to add extra width around the screen. The bare CRTs can then be spliced more closely together, and a custom face plate installed to form a multiple-display workstation.

Because the lifetime of the phosphor is a function of active use, a set of CRTs commissioned at the same time may degrade at different rates over time. Therefore, it is important to maintain equality among the monitors in terms of contrast, brightness, and dynamic range. It is a good policy to bring up a test pattern on a daily basis to ascertain that all the monitors are optically in tune. Adjustments can be made using the analog brightness-contrast controls. The brightness control adjusts the beam current and provides a means to offset brightness loss. The contrast control changes the gain of the video amplifier. A comprehensive quality control program of CRT displays is provided by Roehrig, et al.[11.26]

11.4 References

[11.1]   Jones O. *Introduction to the X Window System.* Englewood Cliffs, NJ: Prentice Hall; 1989.

[11.2]   Weinberg W.S., Hayrapetian A.S., Cho P.S., Valentino D.J., Taira R.K., Huang H.K. X-window based 2k display workstation. *Proc. SPIE* 1991; 1446: 35–39.

[11.3] Glasford G.N. *Fundamentals of Television Engineering.* New York: McGraw-Hill; 1974: 64–74.

[11.4] Sherr S. *Fundamentals of Display System Design.* New York: Wiley; 1970: 399–411.

[11.5] Ji T., Roehrig H., Blume H., Seeley G., Browne M. Physical and psychophysical evaluation of CRT noise performance. *Proc. SPIE* 1991; 1444: 136–150.

[11.6] Dietch L., Palac K., Chiodi W. Performance of high-resolution flat tension mask color CRTs. *SID Dig.* 1986; 322–323.

[11.7] Awata Y., Sumiya H., Shibata Y., Umemura S. A new large-screen high-resolution Trinitron color display monitor for computer graphics application. *SID Dig.* 1986; 459–462.

[11.8] Keller P.A. Cathode-ray tube displays for medical imaging, *J. Dig. Imag.* 1990; 3: 15–25.

[11.9] Lisk K.G. SMPTE test pattern for certification of medical diagnostic display devices. *Proc. SPIE* 1984; 486: 79–82.

[11.10] Mayer J.H. Revolutionizing mass storage. *SunExpert* 1995; 6: 45–50.

[11.11] Dwyer S.J., Cox G.G., Cook L.T., McMillan J.H., Templeton A.W. Experience with high resolution digital gray scale display systems. *Proc. SPIE* 1990; 1234: 132–138.

[11.12] Taira R.K., Simons M., Razavi M., Kangarloo H., Boechat M.I., Hall T., Chuang K.S., Huang H.K., Eldredge S. High resolution workstations for primary and secondary radiology readings. *Proc. SPIE* 1990: 1234: 18–25.

[11.13] Arenson R.L., Seshadri S.B., Kundel H.L., DeSimone D., Van der Voorde F., Gefter W.B., Epstein D.M., Miller W.T., Aronchick J.M., Simon M.B., Lanken P.N., Khalsa S., Brikman I., Davey M., Brisbon N. Clinical evaluation of a medical image management system for chest images. *Am. J. Roentg.* 1988; 150: 55–59.

[11.14] Cho P.S., Huang H.K., Tillisch J., Kangarloo H. Clinical evaluation of a radiologic picture archiving and communication system for a coronary care uinit. *Am. J. Roentg.* 1988; 151: 823–827.

[11.15] Ehricke H-H., Grunert T., Buck T., Kolb R., Skalj M. Medical workstations for applied imaging and graphics research. *Comp. Med. Imag.* 1994; 18: 403–411.

[11.16] Horii S.C., Horii H.N. An eclectic look at viewing station design. *Proc. SPIE Med. Imag. II* 1988; 914: 920–928.

[11.17] Horii S.C. Electronic imaging workstations: ergonomic issues and the user interface. In: *Syllabus: Special course in computers for clinical practice and education in radiology.* Radiological Society of North America; 1992: 125–134.

[11.18] Stammerjohn L.W., Smith M.J., Cohen B.G.F. Evaluation of work station design factors in VDT operations. *Hum Fact.* 1981; 23(4): 401–412.

[11.19] Kaufman J.E., ed. *IES Lighting Handbook.* (5th ed.). New York: Illuminating Engineering Society; 1972: 1–9.

[11.20] Rancourt J., Grenawalt W. Approaches to enhancing VDT viewability and methods of assessing the improvements. *Proc. SPIE* 1986; 624: 8–13.

[11.21] Grandjean E. Design of VDT workstation. In: Salvendy G., ed. *Handbook of Human Factors.* New York: Wiley; 1987: 1359–1397.

[11.22] Morse RS. Glare filter preference: influence of subjective and objective indices of glare, sharpness, brightness, contrast, and color. *Proc. Hum. Fact. Soc.* (29th Ann. Mtg.); 1985: 782–786.

[11.23] Cushman W.H. Illumination. In: Salvendy G., ed. *Handbook of Human Factors.* New York: Wiley; 1987: 670–695.

[11.24] van der Heiden G.H., Brauninger U., Grandjean E. Ergonomic studies on computer-aided design. In: Grandjean E., ed. *Ergonomics and Health in Modern Offices.* London: Taylor & Francis; 1984: 119–128.

[11.25] Arenson R.L., Charkraborty D.P., Seshadri S.B., Kundel H.L. The digital imaging workstation. *Radiology* 1990; 176: 303–315.

[11.26] Ho B.K.T., Ratib O., Horii S.C. PACS workstation design. *Comp. Med. Imag.* 1991; 15: 147–155.

[11.27] Roehrig H., Blume H., Ji T-L., Browne M. Performance tests and quality control of cathode ray tube displays. *J. Dig. Imag.* 1990; 3: 134–145.

12

Virtual Reality and Augmented Reality in Medicine

David Hawkes

12.1   Introduction

Virtual reality (VR) and associated technologies will have far-reaching effects on all aspects of our lives in the next 10 to 20 years. VR technology is well established in the aerospace industry, architecture and building design, the entertainment industry, and increasingly in medicine. While many claims might seem far-fetched, rapid technological advances in computer graphics and animation, primarily for the entertainment industry, have resulted in very realistic virtual environments. The computer power to generate these graphics is becoming widely available and, as a result, the number of potential applications for virtual reality is increasing very rapidly.

This chapter provides an overview of what is being done currently and what is achievable in the near future with particular emphasis on medical applications. After a brief description of the technologies and applications that have emerged we show how virtual and real environments are being combined in planning and guiding surgery and other interventions and look forward to what might be practically and usefully achieved in the next few years. First, however, it is useful to clarify terminology in a way that reflects different types of applications of this technology.

Virtual reality is usually defined as the process by which a computer and associated peripherals replace the sensory stimuli of sight, sound, or touch (or even smell) that would be provided by the natural environment to give an illusion of existing within a synthetic or virtual world. The closer the experience is to the real world, the more convincing is the virtual reality. As in the real world, visual effects are the most important stimulus provided by VR, but an additional key component which separates VR from computer graphics is the provision of the facility to interact with and change the virtual world.

A closely related area, termed *augmented reality*, combines synthetic information with stimuli obtained directly from the real environment to provide additional information pertaining to the real world that would not otherwise be available.

In *tele-presence*, sensory information from the real environment is relayed to a remote location and the operator at that location is able to manipulate the environment at the original site while experiencing the illusion of immersion in that environment.

12.1.1   Virtual Reality

The components of a VR system include a means to display visual information, a means of three-dimensional location of the view-point in the virtual environment, a means of manipulating the virtual environment with hand held tools or manipulators and, if necessary, the provision of force feedback and other tactile or sensory feedback of the virtual environment. A useful survey of the current status of VR technology is provided in Burdea and Coiffet's textbook.[12.1]

Real-time or near real-time displays can make significant demands on computational resources and effective VR systems require powerful graphics processors. Displays can be provided on a computer workstation with 3D cues provided by surface shading and motion parallax. Stereo vision can be provided with polarized glasses and polarizable shutters switched in synchrony with display of right and left stereo pairs. Most VR work, however, uses head-coupled displays. There are two types. The first and most common is a head-mounted display (HMD) which uses two video screens, one for each eye, and associated lenses to allow the user to fuse the two images into a single stereo image. In the second type, the head is tracked and the image displayed on the screen is updated to give the appropriate view. In the autostereoscopic display, multiple views are generated and as the head moves an illusion of both stereo and motion parallax is provided by displaying the appropriate views to each eye.[12.2] These systems generally allow the user much less freedom of movement, but perception of stereo can be very effective. Head-mounted displays usually incorporate liquid crystal displays because of their light weight and limited size. The limited spatial resolution of head-mounted displays, in particular with color displays, can have a significant impact on the perception of the virtual scene.[12.3]

Devices such as the Data Glove (VPL Research, Redwood City, California) record hand movements which in turn can be used to manipulate the virtual environment. Mechanical devices have been designed to provide force feedback and devices have even been proposed to detect and transmit olfactory information (smell) for applications in medicine.[12.4]

12.1.2   Augmented Reality

In augmented reality additional input information is required from the real scene. The virtual information can be displayed aligned with the real-world scene in semi-transparent HMDs or can be projected onto a glass screen through which the real scene is visible. This technology has applications in heads up displays (HUDs) in the aerospace industry[12.5] and in medicine.[12.6–12.7] In our work, described below, we project virtual 3D images of a patient's anatomy aligned with a stereo view through the binocular operating microscope.[12.8] All three methods require correspondence between the virtual world and the real world. Methods of establishing and maintaining this registration are still under development. Display strategies which provide the operator with appropriate visual cues of the combined 3D environment require significant further research.

12.1.3   Tele-Presence

Tele-presence requires coupling of the input information provided by the remote operator with a device to manipulate the real environment, as well as immersion of the operator at the remote site in as "real" a virtual environment as is necessary for the application. Remote manipulation is usually provided by devices with sufficient degrees of freedom to reproduce the operator's movements together with appropriate end-effectors such as grabbers and cutters. In addition to medical applications, tele-presence has applications in hostile environments such as space research; deep sea exploration; security operations, such as defusing suspect devices; and inspection and maintenance work in nuclear reactors.[12.9]

12.1.4   Real Time

In virtual reality, augmented reality, and tele-presence, it is useful to be a bit more specific about what we mean by the term "real time." Formally, "real time" describes processes for which the timing of the output is critical to the application. This could mean an immediate or instantaneous response, but no transfer of information can ever achieve this. In more practical terms, "real time" means "fast enough for the application." When using head-coupled displays, our sensory system expects temporal coupling of movement with a visual stimulus. Even a very small time delay of 40 ms can lead to nausea, headaches, and disorientation, especially when the viewer is immersed in VR for more than a few minutes.[12.10] The experience is very similar to motion or travel sickness where a similar decoupling of visual and motion cues occurs. Updating displays in this very short time interval requires significant computational resources which can make many applications commercially unviable. Predictive algorithms and Kalman filtering techniques are being investigated as a means of overcoming this problem.

If augmented reality is to be available continuously in head-coupled displays or is required to give the 3D cue of motion parallax, the constraints on the speed of image update will be the same as for VR. On the other hand, some environments change relatively slowly. If the projection of the image of the virtual scene follows the optical path of the image of the real scene and head-coupled displays are not used, then the update rate is that required to keep pace with the changing environment. For example, the outline of a runway or target in a pilot's HUD, for even a fast-moving aircraft, will usually not change that rapidly. In our application of augmented reality in the surgical binocular microscope, the microscope is often kept in position for several hours and even when moved an update rate of a few seconds is perfectly adequate. As a result modest PC-based technology is sufficient for our application with the binocular microscope, reducing computational costs by up to two orders of magnitude.

If tele-presence is used for critical tasks such as might be envisaged in surgery, the response needs to be sufficiently fast for the users to perform the procedure safely and deal with emergencies.[12.9] The operator might adopt a "move and look" strategy in which the operator moves an instrument a small distance and then waits for the display to update. For continuous motion, delays should not be longer than half a second in total.[12.9] On the other hand in mentoring or supervisory control, where the expert provides advice but does not directly interact with the scene, much longer time delays could be tolerated, depending on the application.

Taking these comments to their logical extreme, annual monitoring of, say, benign tumor growth could still be considered as "real time" in that visualization is "fast enough for the application."

12.2   Medical Applications of VR Technology

Virtual reality technology is being developed for use in the following medical application areas:

- As a teaching aid;

- As a training aid;

- To plan and simulate interventions;

- To guide interventions;

- To guide remote manipulation and telesurgery;

- To provide remote consultation;

- To aid in rehabilitation and provide an enhanced environment for the disabled; and

- To aid in design of clinical facilities.

This list is by no means exhaustive and more application areas are likely to arise in all areas of medicine and health care.

12.2.1   Teaching

Three-dimensional computer models of anatomical structures and software that simulates the physical properties of tissue are under development to teach anatomy. These tools may partially replace the use of human cadavers for dissection. Although it will be a while before the dissecting room is completely obsolete, such "virtual cadavers" have the advantage of providing a quick aide memoir as well as being indestructible, i.e., they can be used again and again while the cadaver can be used only once. Several systems are now available on CD ROM or CDi, but these usually only provide 2D displays. An example of a fully 3D system is the Voxel Man system from Professor Hoehne's Group in Hamburg,[12.11] which provides 3D graphics and interaction to allow simulation of dissection or removal of 3D anatomical objects revealing structures underneath. Such systems require painstaking labeling and editing of every voxel in the 3D datasets by skilled anatomists. Considerable effort is still required to complete validation of this labeling process.

The Visible Human Project, whose images are distributed by the National Library of Medicine (Washington DC, USA), provides a rich source of data for this approach.[12.12] This data set consists of photographs of cut sections and corresponding MR and CT data of a male cadaver. Data from a female cadaver is also now available. The full Visible Human dataset is extremely large (approximately 15 gigabytes) and complete labeling is a daunting task, but the data set is now widely available and there is considerable activity preparing this data set for teaching purposes.

12.2.2 Surgical Simulation and Training

The opportunities for VR technology to provide effective surgical simulation and training environments is now widely recognized.[12.13] Sophisticated surgical simulators providing a significant degree of realism are becoming available (e.g., Cine Med, VPL Research). These usually incorporate mechanical manipulators to mimic the feel and dexterity of real surgical tools. Research is underway to incorporate effective force feedback. Visual feedback is provided by head-mounted displays in simulation of open surgical procedures, or by simulating endoscopic or microscopic views in minimal access surgery or microsurgery. In the former, the user's head movements are tracked with a 3D localizer, while in the latter the movement of the endoscope or microscope is simulated with a mechanical manipulator. Currently, the realism of the intervention is limited by the lack of realistic deformation of the virtual structures. Significant effort is currently underway to develop more complete computer models incorporating the physical properties of different tissue structures.[12.14–12.17] Virtual endoscopy is emerging as a new way of viewing CT or MR data. The volumetric data are viewed as if seen through an endoscope. Provision of additional information may enhance conventional endoscopic examinations and in certain circumstances may obviate them entirely.

12.2.3 Planning and Simulation of a Specific Procedure

With regard to a procedure on a particular individual, the image data and the simulation are specific to that individual, allowing detailed planning and rehearsal of difficult or complex procedures. This process is probably most advanced in planning the correction of growth defects in maxillo-facial surgery, repairing trauma to the skull, neurosurgery, and cosmetic surgery. Outlines defined in 3D images can be used to generate physical models by the process known as stereo-lithography.

The surgical procedure can be rehearsed on these models. Alternatively the surgical process can be simulated on the computer display as described above,[12.11, 12.18–12.19] except that patient-specific models are generally less detailed than those used to teach anatomy.

12.2.4   Image-Guided Surgery

In the emerging application of image-guided surgery, the 3D medical image acquired on a particular patient before an operation is brought into correspondence with the patient's anatomy during the procedure. A 3D localizer can then be used rather like a 3D mouse to interrogate the preoperative images in correspondence with the patient's anatomy during surgery. Recently these techniques have been developed to provide a combination or fusion of synthetic preoperative data and visual information obtained during the intervention, either by superimposition on a computer workstation,[12.20–12.21] see-through head-mounted displays,[12.22] or by combination with endoscope or operating microscope images.[12.8] More detail is provided below.

The concepts behind these techniques are far from new. The first case of image-guided surgery was reported only eight days after Roentgen published his discovery of X-rays in 1896. A radiograph of a patient with a needle embedded under the skin of the hand was used to direct the incision to remove the offending object.[12.23] In the 1930s an ingenious device was described for back projecting a radiograph of a bullet or shrapnel fragment onto the patient's skin to guide surgical removal.[12.24] For 20 years the stereotactic frame has provided a means to establish a coordinate system fixed to the skull for guiding biopsy or needle placement within the brain. Markers, usually "N," "X," or "W" shapes, are attached at known locations with respect to the base ring of the frame, which is fixed in place prior to scanning. The location of these markers in the images provides a reference from image coordinates to surgical instrument (for recent reviews, see Maciunas[12.25]).

12.2.5   Remote Manipulation, Tele-Surgery, and Mentoring

Tele-presence technology in principle allows the surgeon to be removed an arbitrary distance from the patient. This is often termed *tele-surgery*. The surgeon might be in the room next door, e.g., to avoid radiation exposure when placing radioactive seeds within a tumor in interstitial radiotherapy, or maybe many hundreds or even thousands of miles away. The patient is real, the interaction is in "real time," but the operation is controlled from a remote site. Visual, audio, and touch feedback could be provided to the surgeon at the remote site, while the surgeon's actions are duplicated at the operation site. In tele-surgery, input devices are designed to mimic the size and feel of real surgical instruments. Television cameras, microphones, and mechanical sensors provide visual, audio, and tactile feedback of the procedure.

Interesting demonstrations have been presented at the recent meeting on Medical Robotics and Computer Assisted Surgery (MRCAS 95) and a case has been made for provision of skilled surgical expertise in hostile or remote environments. The surgical profession in general remains sceptical of the ability of remotely controlled devices to perform complex surgical procedures. Nevertheless there is significant research effort worldwide and in the more convincing demonstrations the surgeon at the remote site acts as mentor or teacher to an individual who already has sufficient skills to perform the basic tasks of cutting, dissection, and sewing. These tasks require more dexterity than can, at present, be provided easily by remote manipulators. Scarce skills in training might be provided by mentoring or provision of remote assistance in areas such as minimal access surgery, where it is widely recognized that current training is often inadequate. Effective tele-surgery is currently still limited by problems of time delay and limited bandwidth for communicating the very large volumes of data necessary to relay a realistic environment in apparent real time. Real-time remote manipulation is therefore kept to an absolute minimum. The problem of time delay is much reduced when transmission distances are kept small.

Tele-manipulation with dexterity enhancement allows gearing so that large movements of the surgeon's hand are translated into the fine movements required in microsurgery of, for example, the eye or ear. Tele-manipulation also allows the delivery of radiation in radiosurgery, without exposing the surgeon, or therapy such as heat, vacuum, high-speed motion, laser energy, or light, which cannot easily be delivered by hand.[12.26]

12.2.6   Remote Consultation and Rehabilitation

In remote consultation, as in tele-surgery, remote expertise might be more effectively utilized by providing more complete information to the consultant than would otherwise be possible. There is some evidence in psychiatric consultations that patients are less inhibited when interacting indirectly via television cameras than directly face to face. Virtual reality can also provide an enhanced environment for the disabled and can aid in rehabilitation. The recent conferences entitled "Medicine Meets Virtual Reality" provide an interesting update of these possibilities.[12.27]

12.3   Augmented Reality in Image-Guided Surgery

In the remainder of this chapter, specific issues related to image-guided surgery and augmented reality will be discussed in more detail, but first it is useful to review the limitations of current 3D medical imaging and to describe how 3D information is displayed on computer workstations.

12.3.1   Three-Dimensional Medical Imaging

Medical imaging has provided exquisite tools for peering into every part of the body without recourse to cutting with a knife or the insertion of endoscopes. The process started just over 100 years ago with the discovery of X-radiation, and its penetrative power through human tissue, by Wilhelm Roentgen. With the advent of 3D imaging by nuclear medicine in the 1960s, X-ray-computed tomography in 1972, and magnetic resonance imaging later in the 1970s, true three-dimensional representations of the patient's anatomy became available. (A good text on the different image acquisition technologies is provided by Steve Webb's book, *The Physics of Medical Imaging.*[12.23])

With current state-of-the-art spiral CT technology, it is possible to collect data to reconstruct 30 or more slices in one acquisition at the rate of one slice per second. This significantly reduces the effect that patient movement may have on data integrity, in particular in the abdomen. For a high resolution acquisition, with each slice having a thickness of 1.5 mm, this corresponds to 45 mm or more of data. For scans of this type a typical in-plane resolution is .5 mm. Larger volumes can be constructed by concatenation of adjacent acquisitions or by increasing the slice thickness. The radiation dose to the patient has to be kept as low as reasonably achievable and in practice this limits contrast resolution and spatial resolution for the particular volume scanned.

In MR, 3D acquisitions are now commonplace and volumes corresponding to 200 mm axial length with voxel dimensions of $|x|x|$ mm are usual. These acquisitions still take several minutes and patient movement can be a problem in the abdomen and thoracic region. Very fast echo planar acquisition sequences are available on some machines, allowing acquisition times of 50 to 100 ms per slice although tissue contrast may be significantly compromised and image artifact from susceptibility effects remains a problem compared with conventional MR scanning. Very fast volume imaging is also under development.

In both MR and CT, great care has to be taken during data acquisition to ensure sufficient image integrity for image-guided surgery. Routinely acquired diagnostic scans may not be of sufficient quality for image guidance. Slice thickness may be 5 mm or more and there may be significant patient movement during data acquisition which will cause artifact or distortion of the resulting 3D representations.

With MR and CT images, and nuclear medicine images if they are used, the design of the scanner ensures that data are delivered as a set of parallel slices of known spacing. It is therefore computationally straightforward to build a true 3D volume from these scans. Errors in slice spacing and orientation, geometric distortion inherent in the imaging modality, and patient movement during the acquisition all conspire to reduce the integrity of this representation. A careful quality control of post-processing techniques can reduce these errors to a minimum. Patient movement can be a significant problem, although careful radiographic technique and faster scans reduce these errors. Generally, it is now reasonably straightforward to produce images without movement artifact in the head and extremities. The quality of 3D volumes has improved significantly with faster scanning techniques in the chest and pelvis while the abdomen can still present problems.

In many ways ultrasound is an ideal tool for image guidance, as it is cheap, noninvasive, the output is effectively instantaneous, and the ultrasound transducer is easy to manipulate in the surgical environment. Unfortunately, the data produced are difficult to interpret. Considerable further research is necessary before computerized segmentation and image interpretation techniques become available for routine use. In order to build a 3D representation, the orientation of the ultrasound probe is found using some form of 3D localizer. Outlines of corresponding structures are defined; currently this is done interactively and requires considerable skill, and an interpolation scheme is then used to build the 3D representation. These techniques have been used very effectively to build 3D models of the fetal organs[12.28] and have been incorporated in an augmented reality system.[12.29]

While these images are predominantly still viewed as 2D slices or cuts through the 3D volume of the patient, methods have been available for many years to build a full 3D representation of the patient.[12.30] Unlike conventional optical images, these imaging systems record the 3D distribution of some physical parameter such as change in acoustic impedance, concentration of injected radionuclide, X-ray attenuation coefficient, proton density, and so on. We sense the 3D world that we inhabit by visual perception of reflected light from surfaces which are generally opaque. We have a problem sensing and experiencing the 3D virtual world created by medical imagery. Some attempts have been made to create a visual representation of these true 3D distributions of physical parameters, through the use, for example, of holography or the vari-focal mirror. Unless the images are very sparse (for example, images of vascular networks), the eye is confused and we do not perceive the full richness of the 3D data. As a result, in 3D visualization we usually simplify the data significantly to produce displays on a computer screen which simulate the 3D cues of surface shading and motion parallax. Stereopsis can be exploited with synchronized polarizable shutters and display of images for the left and right eyes. Significant preprocessing is necessary to delineate the boundary regions of relevant structures and to compute the rendered images for display.

a.   Image Segmentation

Accurate and automated delineation of anatomical structures remains a significant challenge to the computer vision community but a detailed review is inappropriate here. Image-guided surgery makes significant demands on computer-aided image interpretation. The task for image guidance is to identify the relevant structures in the image and to delineate their boundaries as accurately as possible from one or more of the imaging modalities used. Relevant structures include the target lesion, critical structures to avoid, such as nerve fibers, blood vessels or organs with critical function, and structures to be used for navigation purposes, such as bony landmarks. Certain structures, such as bone from CT or the skin surface from either MR or CT, are relatively straightforward to delineate using simple thresholding. The contrast between soft tissue and bone in CT and soft tissue and air in either MR or CT is sufficient to identify a single intensity which will correspond reasonably accurately to the boundary between these structures. Separate bony structures or other structures, such as large blood vessels in MR angiograms or contrast enhanced tumors, may be identified by region growing, intensity thresholding, or by pixel classification schemes.[12.31] Interactive editing and delineation when these processes fail is very time-consuming. In certain circumstances, boundaries can be found by examining image gradients or high-order differentials of image intensity (see Chap. 10). Multi-scale approaches (see Chap. 4) are proving effective at understanding image structure, but are yet to yield practical, fully automated image segmentation tools. Hierarchical multi-scale image representations do, however, provide a means of simplifying image structure and, together with appropriate "point and click" user interfaces, can provide an effective interactive tool for structure delineation.[12.32]

b. Registration

When information for image guidance is provided by two or more different 3D images, these images are registered into a common co-ordinate system and the structures within the combined data set are accurately delineated. Registration must be sufficiently accurate so as not to degrade the integrity of data from the individual modalities. Point landmarks attached to the patient are used for registration but identification of corresponding anatomical landmarks can provide reasonably accurate registration.[12.33] In higher resolution MR and CT images of the skull base, twelve or more anatomical landmarks can lead to a registration accuray of between 1 and 2 mm. Semi-automated methods based on matching corresponding surfaces or other visible feature have been used effectively,[12.34–12.35] but these need prior identification which may entail significant manual interaction and editing. Recently, statistical measures based on maximizing mutual information between the two images have proved very effective and have the potential for very accurate and robust image registration.[12.36-12.39] Combining registration and segmentation into a single process based on these statistical methods is showing great promise.

The processes of registration and object delineation are now sufficiently accurate to focus attention back to the image acquisition process, where residual MR distortion effects and small scaling errors become dominant sources of error in the construction of accurate patient representations for image-guided surgery. If we use the complete affine transformation, which includes scaling and skew or nonlinear deformation, it may be possible to correct for at least some of these errors.

We can now produce patient representations with spatial integrity of the order of 1 mm in the head and extremities and 3 mm in the neck, thorax, and pelvis.

c. Visualization

An alternative to the display of slice information is the generation of displays which give the illusion of three dimensions. These displays are particularly important in guiding interventions where knowledge of the precise 3D spatial relationships between structures are important yet views of the patient in interventional images are frequently far from those of established radiography. Disorientation and difficulties in interpretation can slow down procedures and in extreme cases could lead to a risk of errors. These 3D displays are rarely used for diagnosis but can be very helpful during planning and guidance. The task in effective 3D display is to provide the illusion of 3D structure on a 2D computer screen or head-mounted display.

We can utilize the stereo cue with the aid of polarized screens and glasses. Either each eyepiece of the glasses contains active polarized shutters or the screen can switch polarization rapidly between left and right circular polarization, while the glasses have static polarizing filters. Images are displayed in synchrony with the shutter such that the left view is displayed to the left eye and the right view to the right eye. It must be possible to refresh the screen and switch polarization sufficiently rapidly to avoid flicker (i.e., at least 50 frames per second). The perception of depth is very good, although there may be some slight distortion as the observer moves, with the virtual image appearing to rotate slightly with movement of the observer's head. Also, although this affords good perception of the relative depths of different objects, there is an effect known as "cardboarding" in which the objects themselves appear flat and lacking in depth, rather like cardboard cut-outs in children's toys. Wearing polarizing glasses can be inconvenient and precluded during certain surgical and interventional procedures. The polarized glasses do not eliminate all the light from the contralateral view, resulting in some cross talk which can give confusing cues.

Depth from the parallax cue is achieved by dynamically changing the viewpoint of the rendering. This is usually done by precomputing and displaying a movie sequence, typically at about 1° steps of rotation giving side-to-side rocking of about 10° over a period of a second or two. Head-coupled displays, described previously, also provide appropriate parallax cues.

Depth from shading provides a strong visual cue of three dimensions. It attempts to provide, in a virtual scene, the same 3D cues provided in a photograph or still image of a natural scene. To understand how this works, we need to look at the concepts of surface shading and lighting models. A good review is provided by Hoehne, et al.[12.40] Methods are broadly classified into surface rendering and volume rendering methods. In 3D computer graphics where an explicit model of a structure exists, points on the surface are triangulated and each triangle or "facet" is rendered according to an appropriate shading model. Smoothing of the resulting tiled appearance can be achieved with various interpolation routines. This is usually termed polygon rendering. With 3D voxel data, in which a threshold of image intensity can be used to define a surface between two tissues, effective rendering can be achieved by depth gradient shading, by gray-level gradient shading, or marching cubes.[12.41] In depth gradient shading, the orientation of the surface is determined by looking at the relative depth of adjacent voxels in the surface. Gray-level gradient shading improves the appearance of these images by defining surface orientation directly from the image gray-level gradients.[12.42] In the marching cubes algorithm, all possible intersections of the surface with each voxel are defined and a look-up table is used to compute the resulting shaded surface intensity.[12.41]

All these methods require explicit delineation of the surface. Often this is a difficult task with noisy voxel data. An alternative approach is volume rendering in which the concept of a "partial" surface is introduced and defined as a surface of fractional strength.[12.43–12.44] Objects to be visualized are constructed as "occupancy maps," with zero corresponding to no evidence of the object at that voxel and ranging to the maximum value when the voxel is completely occupied by that object. Hence, one voxel can contain contributions from many structures. In certain circumstances, the occupancy map is a direct, linear remapping of the original voxel intensity data, for example bone and soft tissue from CT. Usually, however, some "fuzzy" segmentation of the voxel volume is necessary to assign appropriate occupancy values.

d. Registration and Augmented Reality in Image-Guided Interventions

In augmented reality, in which a virtual model specific to the patient is superimposed on the real-world scene, we need to establish the spatial correspondence between the real world and the virtual environment. This process of registration is a difficult yet vital component of image-guided surgery. There have been significant advances in this area in the last few years, but it is not yet clear which of the many different schemes will be adopted as standard, as the number of alternative technologies bears witness.

The basic task can be split into two components. First, the correspondence between 3D preoperative images and the patient must be established. Once this is done then locating a point in a surgical scene allows visualization of that same co-ordinate within the preoperative images. Likewise, a feature in the preoperative images can be mapped onto the surgical scene. The second task is to maintain the integrity of the virtual patient model as the operation proceeds. Any movement or distortion of the patient's anatomy is monitored and used to update the model.

In the first task, registration is established by locating corresponding features. These are usually external fiducials attached to a sterotactic frame, pins drilled into bone,[12.45] visible anatomical landmarks, or markers fixed to the skin. The latter is less accurate because of skin movement between scan and surgery. These fiducials must provide sufficient contrast to be located accurately in the preoperative scans and in the operating room. A wide range of localizer technologies, which are capable of delivering 3D co-ordinates of a point in space to better than a millimeter, have been demonstrated. These include mechanical articulated arms,[12.46–12.47] sonic localizers,[12.48–12.49] and devices based on localization of light-emitting diodes with cameras.[12.50]

The process of registration establishes correspondence of patient and image co-ordinate systems with that of the localizing system. In theory, three non-co-linear points are sufficient to establish this correspondence; in practice, many more (six to twelve) are used to reduce errors in point localization. If the fiducials are larger than voxel dimensions then their centers can be defined with sub-pixel accuracy by determining the center of gravity of the distribution of image intensities. It is important that the same co-ordinate in space is located in both images and operating room. The Vanderbilt system,[12.45] in particular, incorporates an ingenious design of removable landmarks and an interlocking ball-and-socket design of the pointer, which has the potential for very high accuracy.

Alternative systems based on visible surfaces have been developed in which pattern light[12.51] or a scanned laser beam[12.20] is used to reconstruct a visible surface, usually the skin surface in the operating room. Surface reconstruction accurate to about 1 mm is now achievable in surgical conditions. This surface is then registered to the corresponding surface delineated in the preoperative images, using chamfer matching[12.52] or related algorithms. The accuracy and robustness of this process is very dependent on the total surface area available and surface symmetry and curvature properties, but clinically usable registration is achievable.[12.53] Outstanding problems include the consistency of skin surface co-ordinates, which are known to change with anesthesia, temperature, and the position of the patient.

Tracking motion and deformation during interventions often follows the same procedure but, having established correspondence by one method, it can be appropriate to insert other more stable, if more invasive, fiducials at the operating site and to use these for tracking movement and deformation. Investigations are under way to use ultrasound[12.54] or X-ray images[12.28] to establish registration and monitor any movement. As stated above, ultrasound probes are easier to manipulate, but the image is difficult to interpret. On the other hand, X-rays deliver an additional radiation dose to the patient and theater personnel unless care is taken to avoid unnecessary exposure, yet X-rays potentially provide very accurate spatial information.

Interventional MRI (iMRI) is emerging as a subdiscipline of MR imaging and more open and accessible scanners are being designed which allow access to the patient during scanning. Access on all currently available systems is still very restricted and significant development is necessary to ensure MR compatibility of all surgical instruments and monitoring equipment. Nevertheless it is likely that iMRI will emerge as a major technology in image-guided interventions in the next decade. Whereas the last two decades have been dominated by a revolution in the quality and versatility of medical imaging technology, the next decade will see a similar revolution in using medical imaging in surgery and other interventions in conjunction with graphics, virtual reality, mechanized tools, and robotics.

12.3.2 Applications in Skull Base Surgery and Neurosurgery

This section describes some of the surgical applications that have made use of this technology as part of collaborations in London between the Image Processing Groups in Radiological Sciences and Neurology at UMDS, Guy's and St. Thomas's Hospitals, the Maudsley Hospital, Kings College Hospital, and The National Hospital for Neurology and Neurosurgery.

Three-dimensional combined representations have been used extensively over the last three years in the joint Skull Base Clinic of Guy's and the Maudsley hospitals in London. Images were generated prior to surgery and evaluated by a surgeon and neuroradiologist. The accuracy of these representations was then assessed during the surgical procedure. Full details are published elsewhere.[12.19] We were able to show that combined images displayed as slices significantly improved the perception of spatial relationships between the lesion and bone and between the lesion and other soft tissue structures. The 3D displays provided improved clarity and accuracy of interpretation of spatial relationships between vascular structures and the lesion. This is of particular importance in planning the resection of lesions of the skull base.

Figure 12.1 shows a 3D view of registered MR, CT, and MR angiographic (MRA) images generated using a multimodal 3D volume rendering technique.[12.44] The vascular structures in the MRA images were obtained by region growing and intensity thresholding. The 3D rendered images provide greater clarity of the relationships between vascular structures and the tumor than was possible from the separate tomographic images. Figure 12.2 shows two views from a movie sequence of another patient in which the bone is cut away to reveal the precise relationship between bony structures, a meningioma, and related blood vessels. Figure 12.3 shows the precise relationship between the optic nerve and a meningioma of a third patient. The perception of three-dimensions in all these images is best when they are viewed as a movie on a high quality graphics workstation. This effect cannot be reproduced on the printed page (interested readers might like to peruse our Web pages http://www-ipg.umds.ac.uk/ for a movie display, although for full effect the original movies should be viewed on a high quality workstation). The multimodal 3D volume renderer provides a reasonable visual perception of the inherent uncertainty in object boundaries due to image noise, limited image contrast, and the partial volume effect. These image were generated on a SUN SPARC 10 workstation using software developed at UMDS.

The VISLAN project (*Vis*ible *Lan*dmarks in image-guided neurosurgery)[12.18, 12.21, 12.51, 12.53] has integrated a neurosurgical guidance system based on video matching and tracking technology with a multimodal preoperative planning system. The intra-operative system includes a pair of video cameras and a small projector. Patterned light is projected onto the exposed skin surface and a 3D representation of this surface is reconstructed from the two video images. This surface is then matched to the corresponding surface delineated from preoperative MR and CT scans. This establishes patient-to-image correspondence in the operating room. The 3D location of a hand-held localizer, with a distinctive visible pattern, is reconstructed and tracked with images from the same two video cameras. The localizer can then be used to interrogate the preoperative images in the coordinates of the operating room. Hence, tasks such as craniotomy and resection can be guided using the preoperative images and surgical plan. A facility is provided to insert visible landmarks attached to pins which can be drilled into the skull to allow correspondence to be maintained if the patient (or camera system) moves during the procedure. Figure 12.4 shows the camera system mounted on a stand or ceiling gantry above the patient. Figure 12.5 shows an "augmented reality" display. A live video frame shows the region of interest around the operating site. The gray-scale image is from the live video, while the outline is a computer overlay of the planned craniotomy window. The surgeon placed dots on the (real) scalp so that they were in correspondence with the (virtual) plan, by observing his own hand and the pen at the same time as the virtual data. Figure 12.6 shows one

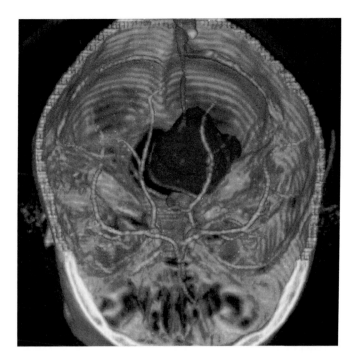

FIGURE 12.1. Three-dimensional multimodal volume rendered view[12.42] of accurately registered MR, CT, and MRA images of a patient with a large acoustic neuroma. In a color display, bone from CT is shown in gray, tumor from gadolinium enhanced MR is shown in green, and blood vessels from MRA in red.

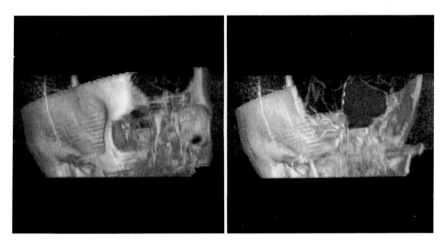

FIGURE 12.2. Two multimodal volume rendered scenes from a movie sequence of a second patient, showing bone cut away to reveal the precise relationship between bony structures, a meningioma, and related blood vessels.

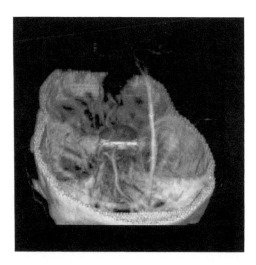

FIGURE 12.3. A multimodal volume rendered image of a third patient with a meningioma, showing the relationship between the optic nerve from MR (yellow in a color display) tumor and blood vessels.

of the localizers and Fig. 12.7 shows its use in surgery. Figure 12.7(a) shows the VISLAN system in use during an operation. The surgeon has positioned the tip adjacent to tissue that he is dissecting around a deep-seated tumor in the medial temporal lobe. By inspecting the "virtual reality" graphics display, shown in Fig. 12.7(b), he can see the tip of the localizer in relation to a rendered view, which can be computed to show any arbitrary viewpoint, and in relation to the original MR scan data (orthogonal slices which intersect at the current tip position are shown). These displays are updated in real time as he moves the localizer. Studies have shown that the localization accuracy is 0.15 mm to 0.5 mm in 3D camera coordinates, that, using surface registration, localization of features initially defined in pre-operative coordinates is accurate to 0.4 to 0.9 mm. A further facility is provided for visible marker pins to be screwed into the skull at the beginning of surgery, to allow updating of registration if the patient or camera system are moved relative to each other during the operation. The system has been tested so far in 15 surgical cases with good results.

In another project we are cooperating with Leica, Switzerland, to project images of critical structures, pertinent to the surgical procedure, into the stereo microscope.[12.8] The left and right optical projections of the microscope are each accurately calibrated. The patient coordinates, microscope coordinates, and 3D preoperative representations are brought into correspondence by identifying corresponding landmarks and tracked with the aid of an Optotrak 3020 localizer (Northern Digital, Toronto, Canada). The relevant anatomical and pathological structures are projected into each optical path of the microscope. Figure 12.8 shows a general view of the modified operating microscope in position during an operation to remove an epidermoid cyst. The projectors are clearly shown on each

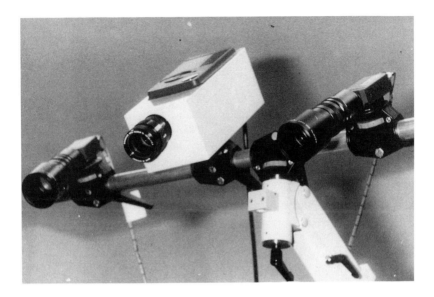

FIGURE 12.4. Cameras and projector of the VISLAN system mounted on a stand or ceiling gantry above the patient.

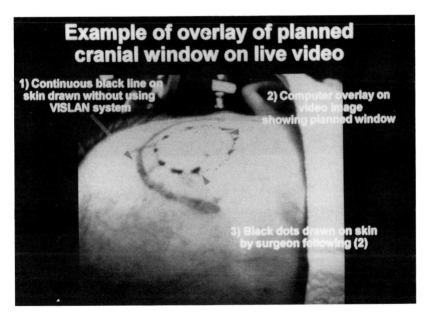

FIGURE 12.5. "Augmented Reality" display showing a live video image with superimposition of graphics of the planned craniotomy window.

FIGURE 12.6. The VISLAN localizer, showing the distinctive pattern which is tracked automatically in "real-time" by the video system.

side of the microscope assembly and the Optotrak localizer is seen on the far side of the room to the left. Figure 12.9 shows a skull phantom used to test the system. Figure 12.10 shows a view of the phantom, taken through one binocular eyepiece of the microscope with a hand-held camera, with overlays from projection of the 3D virtual model generated from the CT images. Figure 12.11 is a photograph taken through one eyepiece during the operation to remove the epidermoid cyst. Figure 12.12 is a photograph of a similar view with overlays projected from the 3D model created from registered MR, CT, and MRA images. These overlays are available in real time. Figure 12.13 shows a similar view with overlays but with captions describing various structures. Subsequent surgical exposure revealed accurate positioning of the overlay of the carotid artery, lesion, and facial nerve canal. The text captions were added later and currently do not appear in the overlay. The overlay is updated about once per second in the current implementation.

These overlays require very different types of display from the 3D rendered scene depicted above. Perception of the operative scene through the overlay is necessary and therefore the overlay should be a simplified, symbolic representation of the structure of interest. In this example, the tumor is displayed as a wire mesh and the vascular structures and optic chiasma as colored ribbons. Although much work has still to be done to optimize the contrast and quality of the display overlays, these results demonstrate the principle of the system.

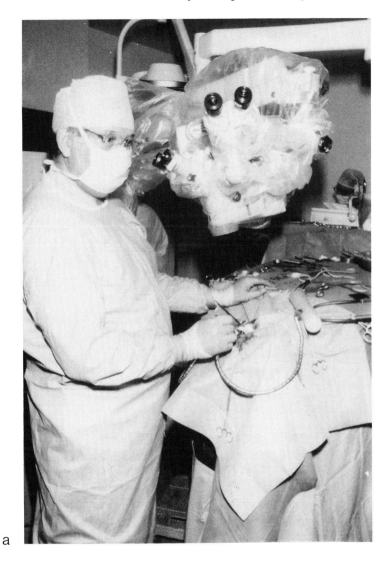

a

FIGURE 12.7. (a) The VISLAN system in use during an operation; (b) The corresponding graphical display on the workstation in theater.

12.3.3 Visual Perception in Augmented Reality

A number of research issues remain unresolved regarding the fusion of 3D visual information derived from the 3D virtual world with direct visualization of the real world. A related area is the use of head-up displays in the aerospace and automotive industries and recent findings are summarized in Weintraub and Ensing.[12.5] The main recommendations in this review were that displays should be kept simple and uncluttered, color should not be used, and regions should not be filled in. The main

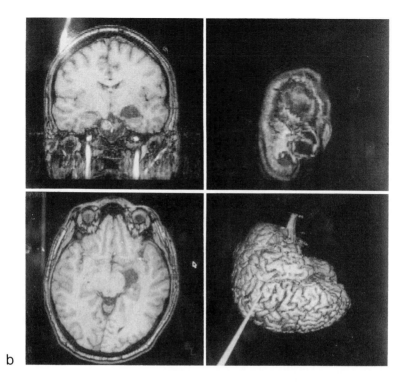

b

FIGURE 12.7. (*continued*)

emphasis was on the display of alphanumerics and "symbology" associated with aircraft cockpit instruments (artificial horizon, airspeed, roll, flight path, runway, and so on).

In our applications, instead of symbology we view a detailed 3D model of relevant anatomical and pathological structures registered with the real scene. In our first functional model, described previously, we used ribbons to outline tubular structures and meshes to outline blobby objects such as bone or tumor. Little research has been done into the use of color and perception of depth from stereo in overlaid displays. Visual cue efficiency is a function of size, luminance, color, shape, position, and possibly image texture. Although these parameters may be fixed in the production of any given cue, their visual perception will depend on the background. The apparent brightness[12.55] and visual acuity-luminance function[12.56] will vary according to the background luminance and color respectively. The perceived color of the overlaid cue also varies depending on the background color.[12.57] Conflicting cues between the real scene and the synthetic scene will create problems of perception. Color in HUDs is not recommended as perceived contrast varies with background color. In the surgical applications the predominant color is red and therefore this should be less of a problem. Filled regions are not recommended as they may obscure detail in the real scene.

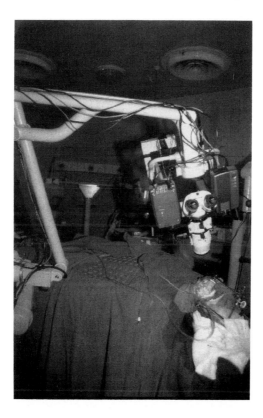

FIGURE 12.8. General view of the modified operating microscope with projectors and localizer in position during an operation to remove an epidermoid cyst. Reprinted with permission from Wiley-Liss, Inc., a subsidiary of John Wiley & Sons, Inc.[12.8]

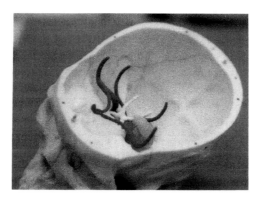

FIGURE 12.9. Photograph of a skull phantom used to test the microscope overlay system. Reprinted with permission from Wiley-Liss, Inc., a subsidiary of John Wiley & Sons, Inc.[12.8]

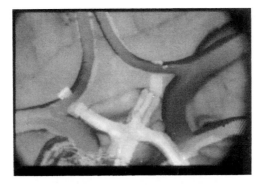

FIGURE 12.10. Photograph taken through one of the binocular microscope eyepieces of the phanton with overlays derived from a CT scan of the phantom. Reprinted with permission from Wiley-Liss, Inc., a subsidiary of John Wiley & Sons, Inc.[12.8]

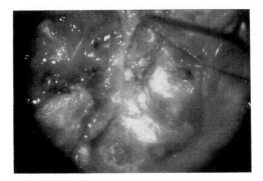

FIGURE 12.11. Photograph taken through one of the binocular microscope eyepieces during the operation to remove the epidermoid cyst. Reprinted with permission from Wiley-Liss, Inc., a subsidiary of John Wiley & Sons, Inc.[12.8]

One problem of perception is that the synthetic scene appears to be "pasted" on top the real scene.[12.29] The attempt has been made to overcome this problem by drawing a synthetic rectangular hole with appropriately rendered sides to give the illusion of "pulling" the synthetic scene down into the patient. Although we wish to render the real tissue translucent to perceive the synthetic structure beneath, we get conflicting cues from surgical instruments which we know are opaque. Ideally our system should know where such instruments are and render them accordingly. A structure in the virtual world may be:

- Currently invisible in the real world, as it is hidden by overlying structure or indistinguishable from adjacent structure;

- Visible, as it lies on the surface; or

- Invisible, because it has been surgically removed.

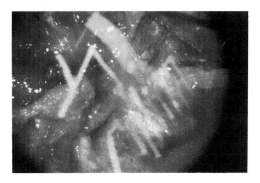

FIGURE 12.12. A view similar to that in Fig. 12.11, through one of the binocular microscope eyepieces, showing overlayed structures derived from MR, CT, and MRA. These were available in real time during the operation. Reprinted with permission from Wiley-Liss, Inc., a subsidiary of John Wiley & Sons, Inc.[12.8]

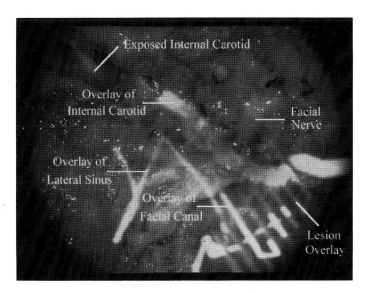

FIGURE 12.13. A view similar to that in Fig. 12.12, through one of the binocular microscope eyepieces showing the overlays with text annotation which was added later. Reprinted with permission from Wiley-Liss, Inc., a subsidiary of John Wiley & Sons, Inc.[12.8]

Different graphical cues are appropriate for these three cases. This implies that we need to know the location of the visible surface in the real world in 3D space corresponding to the currently exposed surgical scene. Surface reconstruction using range finders or patterned light, as described above, might be appropriate for this task.

Problems of time lag are particularly acute in head-coupled displays when the head is free to move, but are less of an issue with the binocular microscope in which the viewpoint can be fixed for significant lengths of time. In the latter case, the cue of motion parallax is not available.

This technology then raises the possibility of being able to update the match between the synthetic 3D world and the real scene by identifying corresponding surfaces such as bone or exposed blood vessels. Research is underway to explore this possibility using surface matching techniques or the statistical measure of mutual information.

Considerable further work is required to test the "human factors" associated with these displays:

- The perception of stereo in these displays, and in particular the perception of the correct depth of different real and synthetic structures, i.e., perception of what is in front and what is behind;

- The appropriate use of color; and

- Whether displayed virtual scenes should be simple line drawings or more realistic volume renderings.

Finally, the ergonomics of these systems must not become so complicated that their control distracts from the surgical task they are meant to be assisting. The overall system must be integrated effectively into the operating room environment, along with other technologies being introduced. The floor around the surgeon's feet is becoming a confusing tangle of different foot-operated controls (diathermy, suckers, light switches, brakes for mechanical devices, etc.). An integrated environment incorporating reliable and safe voice recognition might provide appropriate control.

12.4   Conclusions and Future Work

In all the applications described above, the rigid body is assumed to be an adequate representation of the patient. In practice, anatomical and pathological structures will move and distort both as part of normal tissue function and activity and during surgical or therapeutic interventions.

Many sensors and imaging devices are potentially available during interventions including ultrasound, X-ray fluoroscopy, CT, endoscopy, the operating microscope, and conventional video. Interventional MR is now also being introduced as a near real-time device, with at least one manufacturer providing a system in which surgical procedures can be performed within the imaging volume.

Work is under way to use intraoperative sensor and image information to update detailed preoperative representation. This process requires accurate calibration and specification of each of the in-theater imaging devices together with tracking of imager coordinates. Rigid structures, such as the skull or individual vertebrae, can be tracked by X-ray or ultrasound. Realistic computer models of the physical attributes of tissue can be used with finite element analysis to simulate tissue distortion and predict the location of relevant structures currently hidden from view. Alternatively, simple interpolants such as radial basis functions may be used to predict tissue location between measured points.

The construction of accurate 3D representations from multiple modalities, assuming the validity of the rigid body assumption, is finding clinical application in our laboratory in neurosurgical guidance, particularly in the skull base. The work on deformation and tracking is at a very early stage of development at present and computational requirements are daunting, but the computational methodology is evolving very rapidly.

Careful specification of the clinical problem and subsequent clinical evaluation must be integral components of all research to ensure that these rapid developments are clearly directed toward solving real clinical problems. Success in this area relies on close cooperation between scientist, engineer, and surgeon, with a tight loop of development by the scientist and engineer and evaluation and feedback from the medical or surgical user. A typical project would pursue the following path:

- An initial understanding of the medical or surgical problem by the scientists and development engineers;

- Demonstration of concepts;

- Detailed user specification;

- Technical specification;

- Construction of functional models;

- Clinical evaluation; and

- A period of rapid and intense development with clinical feedback resulting in a prototype for clinical testing.

Only at this stage can clinical trials be contemplated to establish clinical utility. Finally, the full process known as health technology assessment (HTA) is used to determine whether the system is clinically effective and cost effective.

Other applications of augmented reality and virtual reality in teaching, training, tele-surgery, remote consultation, rehabilitation, and aids for the disabled are developing rapidly. Computer hardware of sufficient power for many of these applications is becoming available at a reasonable cost. We should be planning for advances that will exploit computer power predicted for three, five, or ten years from now.

12.5   References

[12.1]   Burdea G., Coiffet P. *Virtual Reality Technology.* New York: Wiley; 1994.

[12.2]   Dodgson N.A., Wiseman N.E., Lang S.R., Dunn D.C., Travis A.R.L. Autostereoscopic 3D display in laparoscopic surgery. In: Lemke H.U., Inamura K., Jaffe C.C., Vannier M.W., eds. *Computer Assisted Radiology CAR'95.* Berlin: Springer-Verlag; 1995: 1139–1144.

[12.3]   Rosen J.M., Lasko-Harvill A., Satava R. Virtual reality and surgery. In: Taylor R.H., Lavallee S., Burdea G.C., Moesges R., eds. *Computer-Integrated Surgery.* Cambridge: MIT Press; 1995: 231–243.

[12.4]   Keller P.E., Kouzes R.T., Kangas L.J., Hashem S. Transmission of olfactory information in telemedicine. *Medicine Meets Virtual Reality, MMVR'95;* 1995.

[12.5]   Weintraub D.J., Ensing M. *Human Factors Issues in Head-up Display Design: the Book of HUD.* University of Michigan: CSERIAC (Crew Systems Information Analysis Center); 1992.

[12.6]   Peuchot B., Tanguy A., Eude M. Virtual reality as an operative tool during scoliosis surgery. In: Ayache N., ed. *Computer Vision, Virtual Reality and Robotics CVRMed'95.* Berlin: Springer-Verlag; 1995: 549–554.

[12.7]   Uenohara M., Kanade T. Vision-based object registration for real-time image overlay. In: Ayache, N., ed. *Computer Vision, Virtual Reality and Robotics, CVRMed'95.* Berlin: Springer-Verlag; 1995: 13–22.

[12.8]   Edwards P.J., Hawkes D.J., Hill D.L.G., Jewell D., Spink R., Strong A.J., Gleeson M.J. Augmentation of reality using an operating microscope for otolaryngology and neurosurgical guidance. *J. Imag. Guid. Surg.* Vol 1, No. 3. New York: John Wiley and Sons, 1996.

[12.9]   Sheridan T.B. Human factors in telesurgery. In: Taylor R.H., Lavallee S., Burdea G.C., Moesges R., eds. *Computer Integrated Surgery.* Cambridge: MIT Press; 1995: 223–229.

[12.10]   Regan E.C., Price K.R. The frequency of occurrence and severity of side-effects of immersion virtual reality. *Aviat. Space Envir. Med.* 1994; 65: 527–530.

[12.11]   Tiede U., Bomans M., Hoehne K.H., Pommert A., Reimer M., Schiemann T., Schubert R., Lierse W. A computerised 3D atlas of the human skull and brain. *Am. J. Neurorad.* 1993; 14: 551–559.

[12.12]   V.M. Spitzer, D.G. Whitlock, R.F. Kilcoyne, A.L. Scherzinger, D. Rubinstein, P. Russ. The visible human male for teaching and reference in radiology. In: Lemke H.U., Inamura K., Jaffe C.C., Vannier M.W., eds. *Computer Assisted Radiology, CAR'95.* Berlin: Springer-Verlag; 1995: 677–683.

[12.13]   Wickham J.E.A. Future developments: minimally invasive surgery. *Br. Med. J.* 1994; 308: 193–196.

[12.14]   Cutting C., Grayson B., Bookstein F.L., Fellingham L.L., McCarthy J.G. Computer aided planning and evaluation of facial and othognathic surgery. *Comput. Plas. Surg.* 1986; 13: 449–462.

[12.15] Terzopoulos D., Metaxas D. Dynamic 3D models with local and global deformations: deformable superquadrics. *IEEE-PAMI* 1991; 13: 703–714.

[12.16] Water K. A physical model of facial tissue and muscle articulation derived from computer tomographic data. In: Robb, R.A., ed. *Visualisation in Biomedical Computing*. Washington D.C.: SPIE; 1992: 574–583.

[12.17] Edwards P.J., Hill D.L.G., Little J.A., Sahni V.A.S., Hawkes D.J. Medical image registration incorporating deformations. In: Pycock D., ed. *Proceedings of the British Machine Vision Conference (BMVC'95)*. BMVA 1995; 691–699.

[12.18] Zhao J., Colchester A.C.F., Henri C.J., Hawkes D.J. Visualisation of multimodal images for neurosurgical planning and guidance. In: Ayache N., ed. *Computer Vision, Virtual Reality and Robotics in Medicine*. Berlin: Springer-Verlag; 1995: 40–46.

[12.19] Gandhe A.J., Hill D.L.G., Studholme C., Hawkes D.J., Ruff C.F., Strong A.J., Cox T.C.S., Gleeson M.J. Combined and 3D rendered multimodal data for planning skull base surgery: a prospective evaluation. *Neurosurgery* 1994; 35: 463–471.

[12.20] Grimson W.E.L., Ettinger G.J., White S.J., Gleason P.L., Lozano-Perez T., Wells W.M., Kikinis R. Evaluating and validating an automated registration system for enhanced reality and visualisation in surgery. In: Ayache N., ed. *Computer Vision, Virtual Reality and Robotics in Medicine, CVRMed'95*. Berlin: Springer-Verlag; 1995: 3–12.

[12.21] Colchester A.C.F., Zhao J., Holton-Tainter K.S., Henri C.J., Maitland N., Roberts P.T.E., Harris C.G., Evans R.J. Development and preliminary evaluation of VISLAN, a surgical planning and guidance system using intra- operative video imaging. *Med. Imag. Anal.* 1996; 1.

[12.22] Wagner A., Ploder O., Enislidis G., Schumann B., Ewers R. Semi-immersive artificial environments in maxillofacial surgery. *Comput. Aid. Surg.* 1995; 1: 19–22.

[12.23] Webb S. *The Physics of Medical Imaging*. Bristol: Institute of Physics; 1988.

[12.24] Steinhaus H. Sur la localisation au moyen des rayons X. *C. R. Acad. Scis* 1938; 206: 1473–1475.

[12.25] Maciunas R.J. *Interactive Image Guided Neurosurgery*. AANS; 1993.

[12.26] Charles S. Dexterity enhancement for surgery. In: Taylor R.H., Lavallee S., Burdea G.C., Moesges R., eds. *Comput. Integ. Surg.* Cambridge: MIT Press; 1995: 467–471.

[12.27] Weghorst S.J., Seiburg H.B., Morgan K.S. *Medicine Meets Virtual Reality: Healthcare in the Information Age—Future Tools for Transforming Medicine.* Proceedings of MMVR4. San Diego: IOS Press; 1996.

[12.28] D'Arcy T., Ruff C.F., Hawkes D.J., Hughes S., Bhalerao A., Chiu W.C., Maxwell D., Saunders J. Volume estimation and shape characterisation of fetal organs. *5th Int. Conf. Fetal Neonatal Phys. Meas.* 1995; Abstract.

[12.29] Bajura M., Fuchs H., Ohbuchi R. Merging virtual objects with the real world: seeing ultrasound imagery within the patient. In: Taylor R.H., Lavallee S., Burdea G.C., Moesges R., eds. *Computer Integrated Surgery.* Cambridge: MIT Press; 1995: 245–254.

[12.30] Herman G.T., Liu H.K. Three-dimensional display of human organs from computed tomograms. *Comput. Graph. Imag. Proc.* 1979; 9: 1–21.

[12.31] Bezdek J.C., Hall L.O., Clarke L.P. Review of MR segmentation techniques using pattern recognition. *Med. Phys.* 1993; 20: 1033–1048.

[12.32] Griffin L.D., Robinson G.P., Colchester A.C.F. Multi-scale hierarchial segmentation. In: Illingworth J., ed. *British Machine Vision Conference, BMVC'93.* Sheffield: BMVA Press; 1993: 289–298.

[12.33] Hill D.L.G., Hawkes D.J., Gleeson M.J., Cox T.C.S., Strong A.J., Wong W.L., Ruff C.F., Kitchen N.D., Thomas D.G.T., Sofat S., Crossman J.E., Studholme C., Gandhe A.J., Green S.E.M., Robinson G.P. Accurate frameless registration of MR and CT images of the head: applications in planning surgery and radiation therapy. *Radiology* 1994; 191: 447–454.

[12.34] Pelizzari C.A., Chen G.T.Y., Spelbring D.R., Weichselbaum R.R., Chen C. Accurate three dimensional registration of CT, PET and/or MR images of the brain. *J. Comput. Asst. Tomog.* 1989; 13: 20–26.

[12.35] Jiang H., Robb R.A., Holton K.S. New approach to 3-D registration of multimodality medical images by surface matching. *SPIE* 1992; 1808: 196–213.

[12.36] Collignon A., Maes F., Delaere D., Vandermeulen D., Suetens P., Marchal G. Automated multimodality image registration using information theory. In: Bizais Y., Barillot C., Di Paola R., eds. *Information processing in medical imaging.* Dordrecht: Kluwer; 1995: 263–274.

[12.37] Wells W.M., Viola P., Kikinis R. Multimodal volume registration by maximisation of mutual information. In: Taylor R.H., Lavallee S., eds. *Medical Robotics and Computer Assisted Surgery MRCAS'95* 1995; 55–62.

[12.38] Studholme C., Hill D.L.G., Hawkes D.J. Multiresolution voxel similarity measures for MR-PET registration. In: Bizais Y., Barillot C., Di Paola R., eds. *Information Processing in Medical Imaging (IPMI'95)*. Dordrecht: Kluwer; 1995: 287–298.

[12.39] Studholme C., Hill D.L.G., Hawkes D.J. Automated 3D registration of truncated MR and CT images of the head. In: Pycock D., ed. *Proceedings of the British Machine Vision Conference (BMVC'95)*. BMVA 1995; 27–36

[12.40] Hoehne K.H., Bomans M., Pommert A., Riemer M., Tiede U., Wiebecke G. Rendering tomographic volume data: adequacy of methods for different modalities and organs. In: Hoehne K.H., Fuchs H., Pizer S.M., eds. *3D Imaging in Medicine*. Berlin: Springer-Verlag; 1990: 197–216.

[12.41] Lorensen W.E., Cline H.E. Marching cubes: a high resolution 3D surface construction algorithm. *Comput. Graph.* 1987; 21: 163–169.

[12.42] Hoehne K.H., Bernstein R. Shading 3D images from CT using grey level gradients. *IEEE-MI* 1986; 1: 45–47.

[12.43] Levoy M. Methods for improving the efficiency and versatility of volume rendering. In: Ortendahl D.A., Llacer J.,eds. *Information Processing in Medical Imaging*. New York: Wiley-Liss; 1990: 473–488.

[12.44] Ruff C.F., Hill D.L.G., Robinson G.P., Hawkes DJ. Volume rendering of multimodal images for the planning of skull base surgery. In: Lemke J.U., Inamura K., Jaffe C.C., Felix R., eds. *Computer Assisted Radiology, CAR'93*. Berlin: Springer-Verlag; 1993: 574–582.

[12.45] Maciunas R.J., Fitzpatrick J.M., Galloway R.L., Allen G.S. Beyond stereotaxy: extreme levels of application accuracy are provided by implantable fiducial markers for interactive image-guided neurosurgery. In: Maciunas R.J., ed. *Interactive Image Guided Neurosurgery*. AANS 1993: 259–270.

[12.46] Galloway R.L., Maciunas R.J., Edwards C.A. Interactive image guided neurosurgery. *IEEE-BME* 1992; 39: 1226–1231.

[12.47] Watanabe E., Mayanagi Y., Kosugi Y., Manaka S., Takakura K. Open surgery assisted by the neuronavigator, a stereotactic, articulated, sensitive arm. *Neurosurgery* 1991; 28: 792–800.

[12.48] Reinhardt H.F., Zweifel H.-J. Interactive sonar-operated device for stereotactic and open surgery. *Stereotact. Funct. Neurosurg.* 1990; 54+55: 393–397.

[12.49] Friets E.M., Strohbehn J.W., Hatch J.F., Roberts D.W. A frameless stereo-taxic operating microscope for neurosurgery. *IEEE-BME* 1989; 36: 608–617.

[12.50] Bucholtz R.D., Smith K.R. A comparison of sonic digitisers versus light emitting diode based localisation. In: Maciunas R.J., ed. *Interactive Image Guided Neurosurgery.* AANS; 1993: 179–200.

[12.51] Colchester A.C.F., Zhao J., Henri C., Evans R.J., Roberts P., Maitland N., Hawkes D.J., Hill D.L.G., Strong A.J., Thomas D.G.T., Gleeson M.J., Cox, T.C.S. Craniotomy simulation and guidance using a stereo video based tracking system (VISLAN). In: Robb R.A., ed. *Visualization in Biomedical Computing* SPIE 1994; 2359: 541–551.

[12.52] Borgefors G. Hierarchial chamfer matching: a parametric edge matching algorithm. *IEEE-MI* 1988; 10: 849.

[12.53] Henri C.J., Colchester A.C.F., Zhao J., Hawkes D.J., Hill D.L.G., Evans R.L. Registration of 3-D surface data for intra-operative guidance and visualisation in frameless stereotactic neurosurgery. In: Ayache N., ed. *Computer Vision, Virtual Reality and Robotics in Medicine (CVRMed '95).* Berlin: Springer-Verlag; 1995: 47–69.

[12.54] Hamadeh A., Lavallee S., Szeliski R., Cinquin P., Peria O. Anatomy based registration for computer integrated surgery. In: Ayache N., ed. *Computer Vision, Virtual Reality and Robotics in Medicine, CVRMed'95.* Berlin: Springer; 1995: 212-218.

[12.55] Padgham C.A., Saunders J.E. *The Perception of Light and Colour.* London: Bell; 1975.

[12.56] Domenech B., Segui M., Capilla P., Illueca C. Variation of the visual acuity-luminance function with background colour. *Ophthal. Physiol. Opt.* 1994; 14: 302–305.

[12.57] Klinker G.J. *A Physical Approach to Colour Image Understanding.* Massachusetts: Peters; 1993.

13

Problems and Prospects in the Perception of Visual Information

Peter N. T. Wells

The study of the perception of visual information does not have a logical hierarchical structure; rather, it consists of a number of parallel threads, more or less related to each other. In this book, these threads are dealt with separately, although the adroit reader will have noticed a degree of tangling! In this chapter, the same separations are adopted and there are some rather general conclusions about the problems and prospects in the perception of visual information.

13.1 Aspects of Visual Perception

13.1.1 Physiological Optics

The eye is the window through which the mind perceives the world around it. It is also a window through which to discern the workings of the brain.[13.1] The retina converts and produces the signals that the brain interprets as the visual image. In doing this, the retina outperforms the most powerful supercomputers. Yet individual neurons in the retina are about a million times slower than electronic devices and consume about one ten-millionth as much power. Clearly, biological computation must be very different from its digital counterpart.

The photochemical events in the eye are initiated by the conversion by light of the 11-cis form of retinal to the all-trans form within the rhodopsin molecule (Sec. 1.4). In the dark-adapted state, sodium ions can freely enter the rod outer segment; this influx is reduced by light, causing graded hyperpolarization of the cell. It now seems possible that a protein called "recoverin"[13.2] may be involved in the regulation of the cyclic guanosine monophosphate (c-GMP) which keeps the sodium channel open.[13.3] This is particularly relevant to an understanding of some inherited visual defects. Abnormalities of the rhodopsin gene account for about 20% of autosomal dominant disease. The many forms of retinitis pigmentosa unrelated to rhodopsin[13.4] may be due to abnormalities of other proteins, of which recoverin is the first to be identified. What is needed now is research into the gene sequences of the transduction proteins as the first step in renewed studies of the retinal dystrophies.

At present, the most likely model for color vision seems to be one containing elements of both the component and opponent theories (Sec. 1.6). Thus, it is believed that the red, green, and blue receptor signals are processed separately up to the level of the lateral geniculate nucleus where they combine through the lateral inhibitory networks. After this, three pairs of detectors act in opposition. One pair deals with red and green, another with yellow and blue, and the third pair is a black-and-white opponent system dealing with brightness and saturation. Exactly how the color signals are processed and added, however, remains to be determined.

Display brightness needs to be optimized for the dynamic range of the eye (Sec. 1.7). It is important to remember that vision at low light levels primarily uses the rods, with a consequent loss in acuity.

Ultimately, it is the achievable information transfer rate that limits the performance of the eye in providing data for visual image processing. Thus, the eye could provide more data than can be accommodated within the bandwidth of the optic nerve (Sec. 1.8). This means that there is scope for improving data transfer by appropriate compression of the displayed image (Sec. 13.2.8).

13.1.2   Detection of Visual Information

One of the most appealing aspects of vision science is that much can be learned from optical illusions. Some illusions appear to be innate from the evolutionary process. For example, vertical lines are less well discriminated than horizontal ones, but this is specific for straight lines and not wavy ones (Sec. 2.3). The moon illusion, in which the moon appears larger when on the horizon than when overhead, is another example; so far, no really satisfactory explanation has been found. Certainly, useful models exist: bottom-up and top-down approaches (Sec. 2.8) give helpful insights. The speed of visual perception is indeed remarkable. Although we are beginning to understand how some kinds of targets may be identified (through features such as textons,[13.5] for example), much more research is needed.

13.1.3   Interpretation of Visual Images

An individual's Snellen visual acuity is at best a rough-and-ready index of performance (Sec. 3.2). It does not necessarily correlate very well with performance in routine visual tasks such as night driving. The vision contrast test system (Sec. 3.7) provides a quick, simple, inexpensive and accurate way to measure contrast sensitivity and to evaluate the transmission of information through an observer's visual system. Further research with this system should be useful for testing and improving imaging systems as well as for evaluating the abilities of human observers.

13.1.4   A Multiscale Geometric Model of Human Vision

A crucial aspect of human perception is that we can move around in the three-dimensional world in which we live. No matter how we move around, the objects in our world are perceived as stable in a three-dimensional environment. When we are interested in scene structure, all we can do is to make an image and investigate the structure of the image as completely as possible. Making many images on different scales increases our knowledge about the scene, but this knowledge can never be complete.

There is evidence that the visual system determines spatial derivatives of images, up to at least the fourth order. In Chap. 4, this model is elucidated in detail. The visual system is seen as a "geometry engine," that determines image structure by measuring partial derivatives simultaneously at each point and at many resolutions. Some of the mathematics is far from trivial (in particular, see Sec. 4.6) and, indeed, some of the mathematical tools that have to be used are vague and incomplete. It is clear, however, that invariance theory provides a powerful language for the description of local image properties and that it is intimately related to differential geometry.

It is interesting to speculate that there might be a mathematical reason for the fact that receptive fields with a structure higher than the fourth order are rare or possibly even absent (Sec. 4.8). We have only just begun to understand the relationships between structure and scale space. Temporal behavior of invariants has not been discussed. It is clear, however, that visual perception relies heavily on dynamic changes of the visual scene induced by object or observer motion, or by eye movements.

13.1.5 Human Response to Visual Stimuli

There are currently three different techniques by which detection thresholds are commonly measured (Sec. 5.2). These are the method of limits, the method of adjustment, and the method of constant stimuli. Unfortunately, thresholds can vary with psychological factors such as expectation and motivation. The situation becomes more complicated when complex stimuli are involved (Sec. 5.3). Nowadays, it is accepted that there is an essentially dual visual system. Parallel processing occurs across the stimulus, and there is serial processing of selected stimulus information. Thus, it is possible to identify preattentive and attentive stages in feature extraction and object identification.

It is obviously sometimes of great importance optimally to match the characteristics of the image to those of the observer (Sec. 5.3). This minimizes the possibility of errors occurring in image analysis. At the very least, the size of the image must be optimized. Focal attention and observer factors are beginning to be understood; it is important to realize that interpretation errors may be more important than incomplete search.[13.6] A perplexing research problem, however, is that accurate measurement of where an observer is looking does not mean that the observer is actually attending there.[13.7] There are some equally paradoxical aspects of fixation relating both to position and time. Perhaps observers think when they are not necessarily looking at what they are thinking about. This is a serious problem, full of experimental difficulty. The role of training is obviously of great importance; whether or not scan paths help in cognition is still the subject of debate.

The relevance of psychology to visual stimulus response cannot be overemphasized. For example, in an experimental search task using chest radiographs,[13.8] the receiver operator characteristic sensitivity dropped after lunch. Everyone is familiar with postprandial lapses in wakefulness!

13.1.6   Cognitive Interpretation of Visual Signals

The mechanisms of perception and cognition have puzzled philosophers for thousands of years. Indeed, the subject was completely philosophical until psychologists began to apply experimental methods to the study of vision. It may be that the shapes of objects are perceived by a three-stage process (Sec. 6.2). First, a "primal sketch" is formed, allowing major features and intensity variations to be identified. Next, more subtle characteristics such as surface discontinuities and depth clues are absorbed. Finally, a three-dimensional model of the object is mentally constructed. Target identification, and particularly texture analysis, involves a two-stage process of preattentive and attentive vision. The higher the statistical order of the texture difference, however, the greater is the attentive effort necessary for cognition and the task may be beyond the ability of the observer. It is in texture analysis that computer-assisted advisory systems may have an important place in radiological and other types of image interpretation (Sec. 3.2.10).

Recent research[13.9] has revealed that imagined "images" of different sizes, created in the "mind's eye" by a closed-eye "observer", generate responses in different areas of the brain. This implies that all images lie close to the brain, before the brain has time to attach language-based labels to them. Moreover, it is evidence that knowledge can fundamentally bias what is seen, or that an observer is likely to see whatever it is expected will be seen.

The perception of color is still far from being understood (Sec. 6.5). Of course, there is no such thing as absolute color. Color perception is a property of the observer, often complicated by individual color vision defects. Underlying this is the fact that the spatial resolution limitation of the eye is better for luminance than for color.[13.10] This means that closely spaced color fringes, for example, can appear to the observer to be in black and white.

13.1.7   Visual Data Formatting

The information processing tasks performed by the retina and the visual cortex of the brain are still poorly defined and the biological computations of vision are hardly understood at all. The human brain can provide many interpretations of a single image. As pointed out in Sec. 7.1, the problem of vision is highly underconstrained and ill-conditioned.

Aliasing and noise are unavoidable in image formation. It is true that the location and shape of an image histogram give clues about the quality and character of an image (Sec. 7.2). Each image has a unique histogram; obviously, the inverse is not true. Adaptive histogram equalization has an important role in image processing.[13.11] Texture, which can be thought of as an "organized area phenomenon,"[13.12] is a statistical property of surface features. The application of fractals to the understanding of texture is a promising area of image analysis.

How are details perceived in images? Although the experimental facts are quite well known (Sec. 7.3), the conditions under which the higher cognitive centers can "fill in" missing information have not been properly worked out. Moreover, filling in of missing information can presumably work well only when the observer is preconditioned at least to the image class. Even when this is the case, there is a danger that what is "filled in" is wrong.

Edge detection is an important part of many image processing tasks. So far, automatic systems have significant limitations and advances using traditional filtering methods must have residual problems. Much more research is needed, since the inadequacy of existing methods means that there is often no substitute for the human observer even when the task is repetitive or massive. This is particularly so in medical imaging, in which contrast and detail are mostly dictated by physical processes, technology, and algorithms and, to a lesser extent, by image processing techniques and display media (Sec. 7.4). The visual detection of "minimal" morphological changes is often impossible. Despite the generally disappointing results that have so far been obtained in this area, it seems a particularly worthwhile subject for research.

Neural network models of the human brain (Sec. 7.5) have led to better understanding of visual detection processes. The concept of distributed memories and the massive parallelism and lateral interconnections of biological neural networks are beginning to reveal the ways in which the human visual system extracts features from images and recognizes their relevance. This is indeed an exciting area and one which may soon be much better understood.

13.1.8   Image Manipulation

Adjustments in gray-scale mapping and spatial filtering have well-understood effects on image perception (Secs. 8.3 and 8.4). There is scope, however, for developments in interpolation algorithms (Sec. 8.2) although, as already noted, there are risks in producing what seem to be esthetically pleasing pictures.

Geometric image processing is becoming more practicable as small computers become more powerful and less expensive. Image warping is particularly important in temporal studies; what can be achieved is in practice limited only by the availability of suitable fiducial points in the images.

It is image segmentation that usually poses the most intractable problems in image manipulation. Gray-level segmentation has only limited clinical applications. A promising technique of segmentation involves artificial neural networks and training sets (Sec. 8.7). Although most of the three-dimensional displays derived from two-dimensional scans that are already in clinical use depend on image segmentation, the ability to store three-dimensional blocks of image data and rapidly to extract and display any plane or surface is likely to become increasingly important.[13.13] Moreover, these techniques are opening up new opportunities in treatment planning and minimally invasive surgery and other therapeutic procedures (Chap. 12).

The process by which data are compressed determines the acceptability of the method. A really satisfactory implementation of picture archiving and communications systems using existing technologies for storage and transmission depends on reducing the data to numbers that can be economically handled. Using a form of adaptive block cosine transform coding,[13.14] it has been shown that compression ratios as high as 25 to 1 may be acceptable for primary diagnosis in chest radiology.

Color seems to be most useful when it is used to code a feature that is complementary to gray-scale image representation of anatomical spatial relationships. In addition to its use in nuclear medicine (Sec. 8.8), color is widely used to display blood flow in ultrasonic imaging systems.[13.15]

13.1.9   Physical and Psychophysical Measurement of Images

Much of the research into image evaluation tacitly assumes that human perception and pattern recognition are only affected in ways predictable from the substitution of one set of image parameters for another (Sec. 9.1). Another fundamental assumption is that information carried from the object through an image to the observer can be degraded, but never improved.

Diagnostic systems require human observers for pattern recognition because humans are extraordinarily efficient classifiers of visual information and because procedures that could possibly be used for computerizing the process remain largely unknown. In the future, imaging will rely more and more on currently underutilized capabilities of observers, such as motion detection and three-dimensional appreciation. The reality is that real-time ultrasonic imaging and computer-synthesized image projections are important not so much because they allow observers to see things that they have never seen before, but because they make use of human information channels that have previously been unused.

Although all image noise results in loss of information, correlated noise interferes more because its signal falls into the same spatial frequency band as the image features.[13.16] In any event, the signal-to-noise ratio (SNR) of an imaging system is directly related to image perception and designers should strive to minimize the SNR. In fact, spatial and contrast resolutions are linked through the SNR.[13.17]

The concept of the ideal observer is traditionally central to image evaluation, primarily because of the extreme difficulty of modelling the human observer. The use of models of perception developed by experimental psychologists and computer scientists, however, may be more valuable than the ideal observer in trying to understand how images and observers can be more effectively linked (Sec. 9.3). The method of receiver operating characteristic (ROC) turns out to be a versatile and powerful tool (Secs. 9.4 and 9.5). Ultimately, however, ROC analysis depends on independent standards of truth: radiological and clinical truth may differ. Because ROC studies almost always involve "selected" cases, it is difficult to make them representative. Often, it is prudent to carry out preliminary evaluation of an imaging system by means of a non-ROC pilot experiment to prepare for ROC studies that are appropriate but consume more time and resources.

Essentially, there are two ways to improve diagnostic accuracy in radiology (Sec. 9.5). Either the image can be improved so that abnormalities are more evident, or the interpretive process can be improved. The latter involves an understanding of psychological factors. The main difficulty is in generalizing to populations of cases and observers from the results of experiments of limited sizes. It is always important to be aware that observers can simulate the behavior of less well trained individuals. It is not satisfactory to think of observers simply as detectors; rather, they are recognizers and the measurement of performance in image perception needs more and better methodology than is currently available.

13.1.10 Computer Vision and Decision Support

Observers sometimes fail to recognize and classify objects present in an image, but it may be possible to strengthen the attentive phase of perception (Sec. 10.1). One way to do so is by computer vision, in which a computer extracts and determines meaningful descriptions of image features. In a clinical setting, the computer vision system can be arranged to provide the radiologist with a "second opinion" in detecting and characterizing abnormalities; success depends on an appropriate database. Filters of various kinds, both linear and nonlinear, have already been developed for template matching, and for signal enhancement and suppression (Sec. 10.2). Image segmentation remains rather primitive in terms of what can be achieved, despite the complexity of the algorithms. Nevertheless, feature extraction performance has reached a level that makes it practically useful in some well-defined clinical image perception and analysis tasks. Good examples are in mammography and chest radiography (Sec. 10.3). These tasks are rather exceptional, however, since it is difficult to imagine many other image perception problems that are so well constrained or widely practised. It is in decision support (Sec. 10.4) that computer vision is likely to be first introduced into widespread clinical practice. Even here, however, there are unsuspected obstacles; most notably, radiologists will have to change their attitudes to the use of workstations (Sec. 13.2.11) and they will have to be persuaded that this kind of help really is useful.

13.1.11 Architecture and Ergonomics of Image Workstations

Image workstations are still at a somewhat idiosyncratic stage of development. Their introduction on a wide scale, will require greater standardization, if only because of the economics. Nevertheless, the fundamental architecture is now quite well defined (Sec. 11.1) in terms of hardware, display, data storage, and the communication network. It is a truism, however, that the situation would be better if the instrumentation were better! This is a fast-moving area of technological development; large-screen high-resolution color monitors and fiber optic cables are still to have their full impact.

The importance of ergonomics in image workstation design cannot be overemphasized. The change in radiological culture that will be necessary to exploit the potential benefits of advances in image processing, manipulation, storage, and display, together with computerized advisory systems, is indeed profound. The workstation is the interface with the radiologist and the key to rapid acceptance and success. Overall, it is the image quality that is important.[13.18]

13.1.12   *Virtual Reality and Augmented Reality in Medicine*

Virtual reality is the process by which a computer and associated peripherals replace the sensory stimuli provided by the natural environment to give an illusion of existing within a synthetic or virtual world. Augmented reality combines synthetic information with stimuli obtained directly from the real environment to provide insight that otherwise would not be available.

The production of real-time, or near real-time, displays is computationally very demanding. Moreover, the virtual or augmented realities of today are far from ideal as far as visual display is concerned, and even less adequate for stimuli such as a force and motion. It is already clear, however, that the technology will have a profound affect on many aspects of medicine (Sec. 12.2), ranging from teaching and training, through simulation and guidance of interventions, to remote consultations and the provision of environments tailored to individuals' specific needs.

The problem of image segmentation and registration, although already satisfactorily solved for some methods of image acquisition, remain formidable obstacles to the creation of virtual and augmented realities from ultrasonic images (Sec. 12.3.1). Currently, the most attractive applications—and the ones that are most actively being researched—are in image-guided interventions. Much has already been achieved in skull base surgery and neurosurgery, where the skull acts as an effective constraint (Sec. 12.3.2). The perception and display of real-time images as they relate to unrestrained soft tissues, however, are immense challenges; but the rewards promise to be great and the solution of these problems will be fundamental to the move toward minimally invasive interventions.

13.2   Conclusions

For years, clinicians have been quite skeptical about the value of basic research in image perception.[13.19] Radiologists acknowledge that the studies are of scientific interest, but few can see their relevance to clinical imaging. Although small changes in performance can certainly be measured by ROC analysis, such changes seem to most people likely to be of questionable practical significance. Moreover, it is difficult to change existing patterns of human behavior and traditionalists argue that there is intrinsic redundancy in the diagnostic process. But these really are not good reasons for lack of optimism about the future. The advent of digital imaging technology and its adoption on a wide scale mean that these old-fashioned ideas have to change. There is too much to be gained to allow the great benefits now potentially within our grasp to be impeded by radiologists' natural reluctance to change their culture. We are beginning to understand enough about the perception of visual information to enable us to improve the interface between the morphology, pathology, and physiology of the patient and the mind of the clinician. But we still have much to learn.

13.3 References

[13.1]   Mahowald M.A., Mead C. The silicon retina. *Sci. Am.* 1991; 254(5): 40–46.

[13.2]   Dizhoor A.M., Ray S., Kumar S., Nierui G., Spencer M., Brolley D., Walsh K.A., Philipov P.P., Hurley J.B., Stryer L. Recoverin: a calcium sensitive activator of retinal rod guanylate cyclase. *Science* 1991; 251: 915–918.

[13.3]   Fesneko E.E., Kolesnikov S.S., Lyubarsky A.L. Induction of cyclic GMP of catatonic conductance in plasma membrane of retinal rod outer segment. *Nature* 1985; 313: 310–313.

[13.4]   Lester D.H., Ingleheam C.F., Bashir R., Ackford H., Esakowitz L., Jay M., Bird A.C., Wright A.F., Papiha S.S., Battacharya S.S. Linkage to D3S47(C17) in one large dominant retinitis family and exclusion in another: confirmation of genetic heterogeneity. *Am. J. Hum. Genet.* 1990; 47: 536–541.

[13.5]   Julesz B. A brief outline of the texton theory of human vision. *Trend. Neurosci.* 1986; 7: 41–45.

[13.6]   Kundel H.L., Nodine C.F., Carmody D. Visual scanning, pattern recognition and decision-making in pulmonary nodule detection. *Invest. Radiol.* 1978; 13: 175–181.

[13.7]   Shepherd M., Findlay J.M., Hockey R.J. The relationship between eye movements and spatial attention. *Quart. J. Exp. Psychol.* 1986; 38A: 475–491.

[13.8]   Gale A.G., Murray D., Millar K., Worthington B.S. Circadian variations in radiology. In: Gale AG, Johnson F, eds. *Theoretical and Applied Aspects of Eye Movement Research.* Amsterdam: North Holland; 1984: 312–321.

[13.9]   Kosslyn S.M., Thompson, W.L., Kim I.J., Alpert, N.M. Topographical representations of mental images in the primary visual cortex. *Nature* 1995; 378: 496–498.

[13.10] Berry M.V., Wilson A.N. Black-and-white fringes and the colors of caustics. *Appl. Opt.* 1994; 33: 4714–4718 & 4962.

[13.11] Pizer S.M., Zimmerman J.B., Staab E.V. Adaptive grey level assignment in CT scan display. *J. Comp. Asst. Tomog.* 1984; 8: 300–305.

[13.12] Haralock R.M. Statistical and structural approaches to texture. *Proc. IEEE* 1979; 67: 786–804.

[13.13] Halliwell M., Key H., Jenkins D., Jackson P.C., Wells P.N.T. New scans from old: digital reformatting of ultrasonic images. *Br. J. Radiol.* 1989; 62: 824–829.

[13.14] MacMahon H., Doi K., Sanada S., Montner S.M., Giger M.L., Metz C.E., Nakamori N., Yin F.-F., Xu X.-W., Yonekawa H., Takeuchi H. Data compression: effect on diagnostic accuracy in digital chest radiography. *Radiology* 1991; 178: 175–179.

[13.15] Wells P.N.T. Doppler ultrasound in medical diagnosis. *Br. J. Radiol.* 1989; 62: 399–420.

[13.16] Ohara K., Doi K., Metz C.E., Giger M.L. Investigation of basic imaging properties in digital radiography: 13. Effect of simple structured noise on the detectability of simulated stenotic lesions. *Med. Phys.* 1989; 16: 14–21.

[13.17] Wagner R.F., Brown D.G. Unified SNR analysis of medical imaging systems. *Phys. Med. Biol.* 1985; 30: 489–518.

[13.18] I.C.R.U. *Medical Imaging—the Assessment of Image Quality.* Report 54. Bethesda: International Commission on Radiation Units; 1996.

[13.19] Kundel H.L., Hendee W.R. The perception of radiologic image information. *Invest. Radiol.* 1985; 20: 874–877.

Index